BADLANDS

A Geography Of Metaphors

BADLANDS
A Geography Of Metaphors

KEN DALGARNO

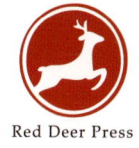

Red Deer Press

Published in Canada by Red Deer Press, 195 Allstate Parkway, Markham, Ontario L3R 4T8

Published in the United States by Red Deer Press, 311 Washington Street, Brighton, Massachusetts 02135

www.reddeerpress.com

10 9 8 7 6 5 4 3 2 1

Library and Archives Canada Cataloguing in Publication

ISBN 9780889955202

Cataloguing data available from Library and Archives Canada

Publisher Cataloging-in-Publication Data (U.S.)

ISBN 9780889955202

Data available on file

Red Deer Press acknowledges with thanks the Canada Council for the Arts, and the Ontario Arts Council for their support of our publishing program. We acknowledge the financial support of the Government of Canada through the Canada Book Fund (CBF) for our publishing activities.

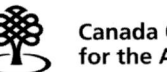 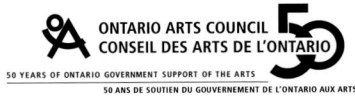

Design by Kerry Designs

Printed in China by Sheck Wah Tong Printing Press Ltd.

To Lucia, Jasmine & Justine

CONTENTS

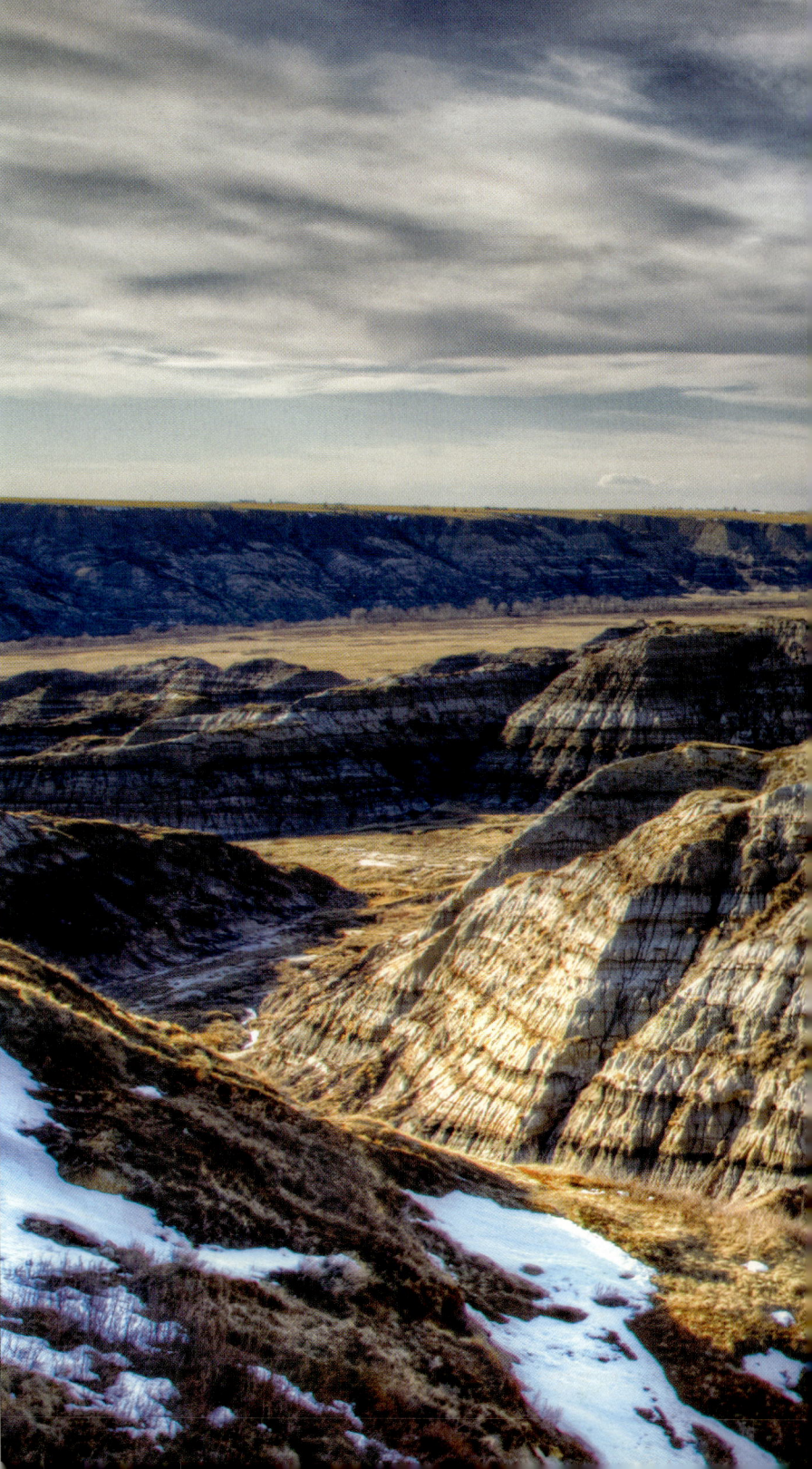

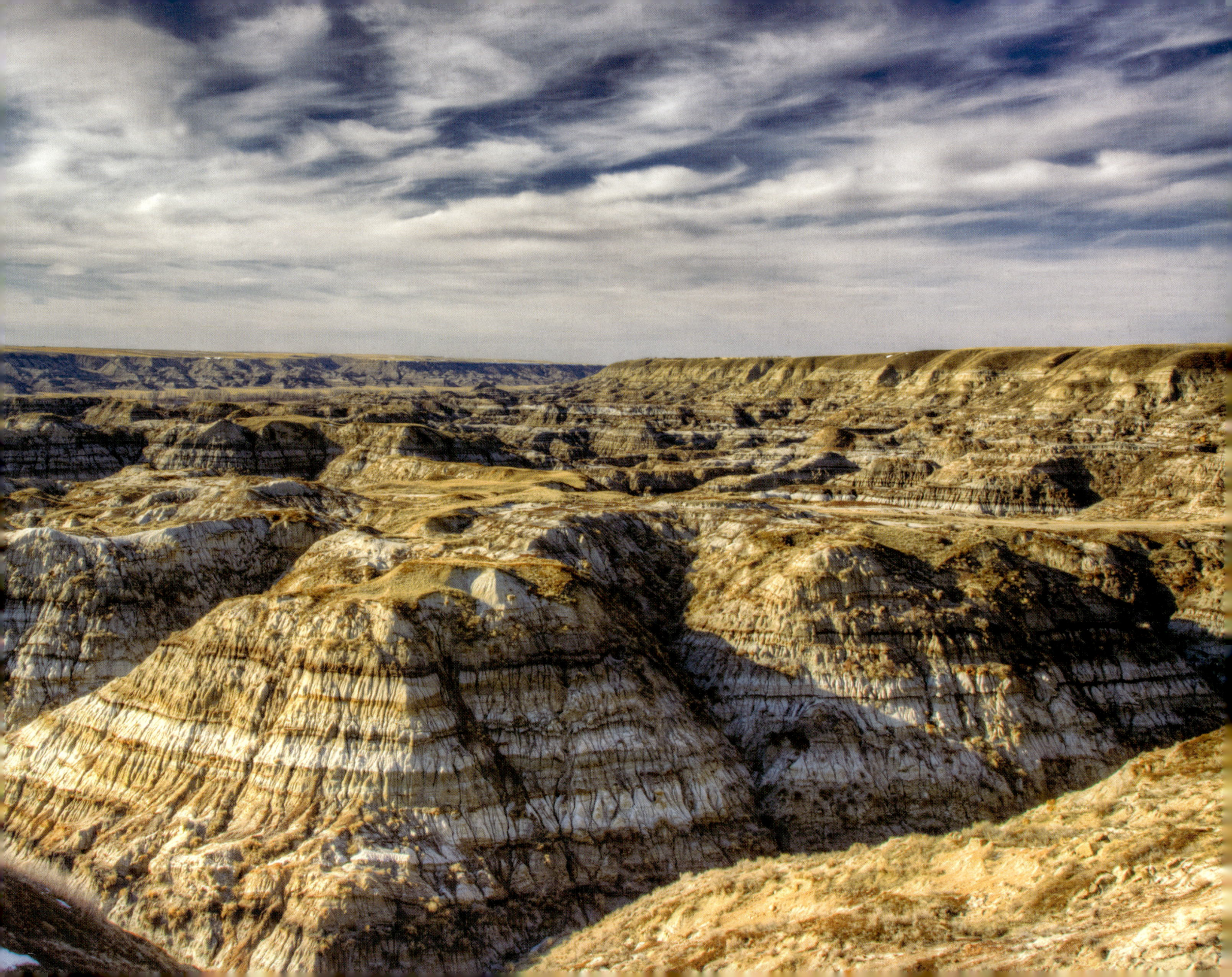

FOREWORD

Ross King

The badlands of southern Saskatchewan, Alberta, and the Northern Great Plains emphatically defy the clichés of the prairie landscape as unrelievedly flat, featureless, and everywhere sown with wheat. For six years I rode the school bus along Highway 39, past some of those endless vistas of wheat and sky. But a dozen miles southeast of Estevan, the highway dipped into the Souris Valley and the landscape dramatically changed. Not only did we pass through the man-made spill piles—the leftovers from the 1930s and 1940s, when electric shovels strip-mined coal and left these little sugarloafs behind—but a few miles to the west, and even more remarkable, were the hoodoos of Roche Percee. Slabs of wizened rock, spindly ochre archways, wonky pulpits of stone—they looked like the ruins of a lost civilization, or the efforts of a whimsical giant run amok with modelling clay. It was, and still is, a truly magical landscape.

With his sensitivity to the haunting and sometimes uncanny qualities of the prairies, Ken Dalgarno is the perfect photographer for these badlands. He captures what might be called "prairie baroque"—the ornate and sometimes fantastic geological and arboreal forms found in tucked-away corners of the plains. I first came to know him through his paintings and photographs of the Crooked Trees of Alticane, those looping trunks and writhing branches that seem to flout all botanical logic, and that he caught beautifully in their arrestingly improbable poses. His badlands series is the perfect continuation of these pictorial meditations on the eerie grace of this landscape.

Though relatively new to photography, Dalgarno uses sophisticated techniques to great effect. As a painter, he is a colourist with a vivid palette and boldly charged brush. Here, in his photographs, he prints on an aluminum substrate and blends several exposures to achieve a greater spectrum of colour. The result is an enhanced luminosity: the soaring blue skies, tawny flanks of cliff, and fiery sunsets leap out of the frames. The impact of the landscape—and of Dalgarno's photographic capture of it—is undeniable. His photograph "The Stone Angel," showing a boulder in the Big Muddy magnificently poised as if for take-off, recently won the Viewers' Choice Award at the Great Plains Art Museum in Lincoln, Nebraska.

We owe the name "badlands" to the French fur traders who supposedly called these entrancing moonscapes "mauvaises terres à traverser" (bad lands to cross). Regarded by the early settlers as economically unfeasible, some of them even became a "Devil's Playground" (as Butch Cassidy called the Big Muddy) for fugitive outlaws. But in a fitting turnabout, this other-worldly topography is now, as Dalgarno shows, a beautiful and important part of the visual heritage of the prairies.

INTRODUCTION

There can be few places in the world where the visual impact of the landscape is as hauntingly captivating as the badlands of the Northern Great Plains. Unique to the prairies, and encompassing several isolated areas in Alberta, Saskatchewan, North Dakota, and Montana, these amazing badlands contain some of the most surreal and magical terrain you can imagine. It's truly a place where reality seems to unravel as the badlands reveal themselves. I first visited Dinosaur Provincial Park in Alberta a few years ago, mainly out of curiosity. I was instantly struck by the serrated cliffs, plunging gorges, pregnant buttes, bizarre spires, and eerie sinkholes. The undulating fissures or crevices that scar the broken hills like fault lines are almost everywhere you look. But perhaps the most spectacular feature is the mystical hoodoos. Carved by erosion, these stone angels rise up from the ground like sculpted monuments. Perched on top of the hoodoo is a caprock of hardened sandstone, balancing precariously on a pedestal like a jest of God.

For me, the badlands are a geography of metaphors. As well as the astonishing pictorial elements in the landscape, there are also the encoded symbolic ones. Entering the badlands is like walking into an archive where stories have been exiled. This becomes instantly apparent when travelling along the vast open prairie and, without fanfare, the landscape literally collapses into a valley of secrets. It's as if you are swallowed whole by the land and transported into its layers of history—and that history goes back millions and millions of years. The badlands are essentially a place where the past and the present have collided. It's for these reasons I've chosen to name the images in this collection after books by contemporary Canadian and American authors.

Scientific evidence suggests that the earth has undergone numerous cataclysmic events that have drastically altered the landscape. Entire continents have moved, mountains have risen, seas have drained and, because of this, life on earth has been radically transformed. In fact, the badlands of the Northern Great Plains have been covered, at one time or another, by such widely diverse ecosystems as a sea, a tropical forest and, on four separate occasions, a gigantic ice field. Each of these events, spanning several million years, are recorded as layers in the earth's strata, like chapters in a book. Over time, the debris from these layers consolidates to form sedimentary rock like shale, sandstone, ironstone, or bentonite clay. Each layer, called a formation, will have unique characteristics and, important to badlands topography, varying degrees of hardness. Typically, formations are named after the area where they were first discovered. For instance, the Edmonton formation located in Alberta is composed of sediments that were deposited primarily in a water environment and include a number of dinosaurs such as Triceratops.

These buried formations were ripped open and exposed during the break-up of the last ice age, approximately 18,000 years ago. Large glaciers and, later, the resulting huge amounts of meltwater gouged the landscape, thus forming the canyons, valleys, and rivers we see today. The creation of the badlands is a result of this violent landforming. But in addition, it is the result of a few chance variables present in the geology of the area that combine with the environment to allow this unique ecosystem to evolve into what it is today—an outdoor art gallery of geological wonders.

The key component necessary for a badlands region to form is a semi-arid climate. In this type of dry environment, the bedrock is poorly cemented so it is more susceptible to the effects of erosion. This climate is also prone to infrequent but heavy amounts of precipitation, which usually occur in the form of storms or cloudbursts. This type of precipitation is more erosive than nourishing. The soil, typically bentonite clay, is also a major contributing factor in the bizarre landscapes created in the badlands. There are different varieties of bentonite, but essentially this soil is volcanic ash. When bentonite clay is dry, it is loose, crumbly, and unremarkable, but when it gets wet, it swells in size and becomes extremely slippery and treacherous to walk on. However, when bentonite becomes saturated, it actually repels water like an umbrella. This is how the intricately etched fissures in the hills are formed. All of these variables make it difficult for vegetation to become established. Thus, without roots to help stabilize the soil, erosion and weathering have free reign in the semi-arid badlands.

Consequently, the artistry of erosion creates monument-like formations. Of course, wind and water take much of the credit but, more specifically, ice wedging and frost break up the rocks and create the most dramatic effects. The climate in the Northern Great Plains typically goes through a number of freeze and thaw cycles throughout the winter months. In this scenario, when water gets into a crack, it expands when it freezes and contracts when it thaws. This repetition turns cracks into gaps and gaps into crevices, much like a pothole on a paved road. Eventually a sandstone column will emerge and separate from a rock wall. If the conditions are right—and the column has a hard enough capstone—with luck, erosion might carve away the softer lower layers of sandstone and turn it into a beautifully sculpted hoodoo.

There is so much more to the badlands than remarkable geology. History is well-documented here. The Badlands contain a treasure trove of Cretaceous-period dinosaur fossils. The bones at Dinosaur Provincial Park are so numerous it has been called a dinosaur fossil warehouse. First Nations history is also well-documented in the Badlands. Rock art, dream beds, stone effigies, and pictographs, some dating back 3,500 years, can be found within the jagged terrain. Writing-on-Stone Provincial Park contains perhaps the best examples of this. The Badlands from Saskatchewan to Mexico were also stopping points along the Outlaw trail set up by notorious bank robber Butch Cassidy. The Big Muddy in Southern Saskatchewan was the Canadian point where fugitives like Sam Kelly, Dutch Henry, and Butch Cassidy and the Wild Bunch "cooled their heels" in the caves near Peaked Butte when they needed to be invisible.

It is easy to hide in the midst of the surreal geomorphology of the Badlands. It's a magical place where nature is the storyteller, a shaman masquerading as both creator and destroyer, exhaling geographic metaphors, shaping echoes into form. It's a perfect theatre to embody the dance of life itself.

AVONLEA BADLANDS

With no mention on a map, no sign to provide directions or even indicate their existence, the Avonlea badlands are a well-kept secret buried amidst the wide-open Saskatchewan prairie. Several of my Avonlea badlands photographs were published in the Saskatchewan magazine, *Prairies North* (Fall 2012), and a number of people responded by saying they had never heard of these badlands—especially remarkable since they are located so close to Regina and Moose Jaw. They don't reveal themselves, like most badlands do, with a grand descent into a sprawling and breathtaking valley. Instead, these badlands are curiously etched into a small parcel of the prairie about eight kilometres east of the town of Avonlea and alongside the Avonlea Creek. The rugged semi-arid Avonlea badlands are a bit of a geological oddity since they are situated only a kilometre northeast of the lush green Long Creek Golf Course.

The town of Avonlea is located in south central Saskatchewan, approximately 60 kilometres equidistant from the capital city, Regina, and neighbouring Moose Jaw. The best way to explore the Avonlea badlands is by guided tour, since the land is on private property and there is often livestock roaming in the area. Tours can be arranged during the spring and summer months by contacting the Avonlea Heritage Museum in the town. There is no trail but it's a moderately difficult hike directly north off Highway 334, 700 metres one way.

While the anonymity of the Avonlea badlands is a bit disheartening—especially since there is a wealth of geological and cultural history found both here and in the nearby Dirt Hills region—the museum does a good job outlining this history. Ultimately, however, you will need to visit this stunning badlands region to appreciate its ravaged beauty. I believe the Avonlea badlands should be designated a Provincial Historic Site (much like the pierced rocks at Roche Percee,

Saskatchewan) so it can be better experienced and be included in Saskatchewan's rich history.

Academics have been aware of this area for quite some time. An archaeological excavation near the Avonlea Creek Valley, undertaken in 1956 by Bruce MacCorquodale and A. E. Swanston of the Saskatchewan Museum of Natural History, discovered that this area was utilized by First Nations peoples as early as 400 A.D. As evidence, they unearthed a uniquely notched projectile point called the Avonlea Arrowhead. Apparently the arrowhead originated in this area and was distributed from here over a 700-kilometre area both east and west. A follow-up symposium on this excavation was published in 1988, called "Avonlea Yesterday and Today: Archaeology and Prehistory," edited by Leslie B. Davis and published by the Saskatchewan Archaeological Society, which further affirms the significance of the area. The 1956 MacCorquodale and Swanston dig also discovered a buffalo jump site here. The area tested was found to contain numerous bone fragments and decomposed bone. Even without corroborating data, when you see the sharp drop-off from the flat prairie into the rugged badlands below, it's easy to believe that this area would be an ideal place to capture and kill buffalo.

Jim Ryder, from the Department of Indian Studies at the University of Regina, also helped interpret a number of interesting Aboriginal sites. In the nearby Dirt Hills, south of Avonlea, there are numerous stone circles, but Ryder located a uniquely large encampment, elliptical in shape and measuring 43 metres by 34 metres. Most notably, however, Ryder designated a large grouping of piled stones as a battle graveyard. He hypothesized, with no time to bury their dead, families returned to cover the bodies with buffalo skins and secure them with rocks.

Avonlea was also situated along the main thoroughfare known as the Wood Mountain Trail. This trail linked Fort Qu'Appelle, east of Regina, to Wood Mountain, located just 30 kilometres north of the U.S. border. Wood Mountain was a trading post for the Hudson's Bay Company and, later, a detachment for the Northwest Mounted Police (NWMP). Both places were also ideal wintering spots for First Nations people. In Canada, Sitting Bull spent much of his time travelling to and from the Wood Mountain area and would have used this trail on numerous occasions. From the 1850s onward, the trail was utilized by everyone from explorers, hunters, traders, the NWMP, missionaries, and, later, pioneers. In fact, according to *Arrowheads to Wheatfields*, the local history book for the area, up to 700 carts per day would traverse the 250-mile trail between Fort Qu'Appelle and Wood Mountain.

What makes the Avonlea badlands a perfect destination is that it contains all of the classic elements found in what French fur traders termed *mauvaises terres à traverser* or "bad lands to cross." There are numerous hoodoos rising up from the semi-arid strata, including the haunting "Dispossessed" (pages 7 & 8), which looks like a shackled spirit encased in sandstone. To my mind, the solitary "Dispossessed" is one of the finest examples of a spire (a hoodoo without the caprock) found in any badlands. The artistry of erosion has etched fantastic trench-like fissures into the hills—so much so that some of these forms separate from the hillside and become sculptural. An example of this is seen in the photograph, "A Gathering of Saints" (page 10), where abstract forms appear to be congregating at a place for worship. The Avonlea badlands are a must-see when exploring the surprisingly remarkable Saskatchewan landscape.

RIGHT: CASTLE ROCK
The morning light transformed this wall of rock a bright golden yellow. The contrasting blue sky with rippled clouds accentuates the fissured channels.

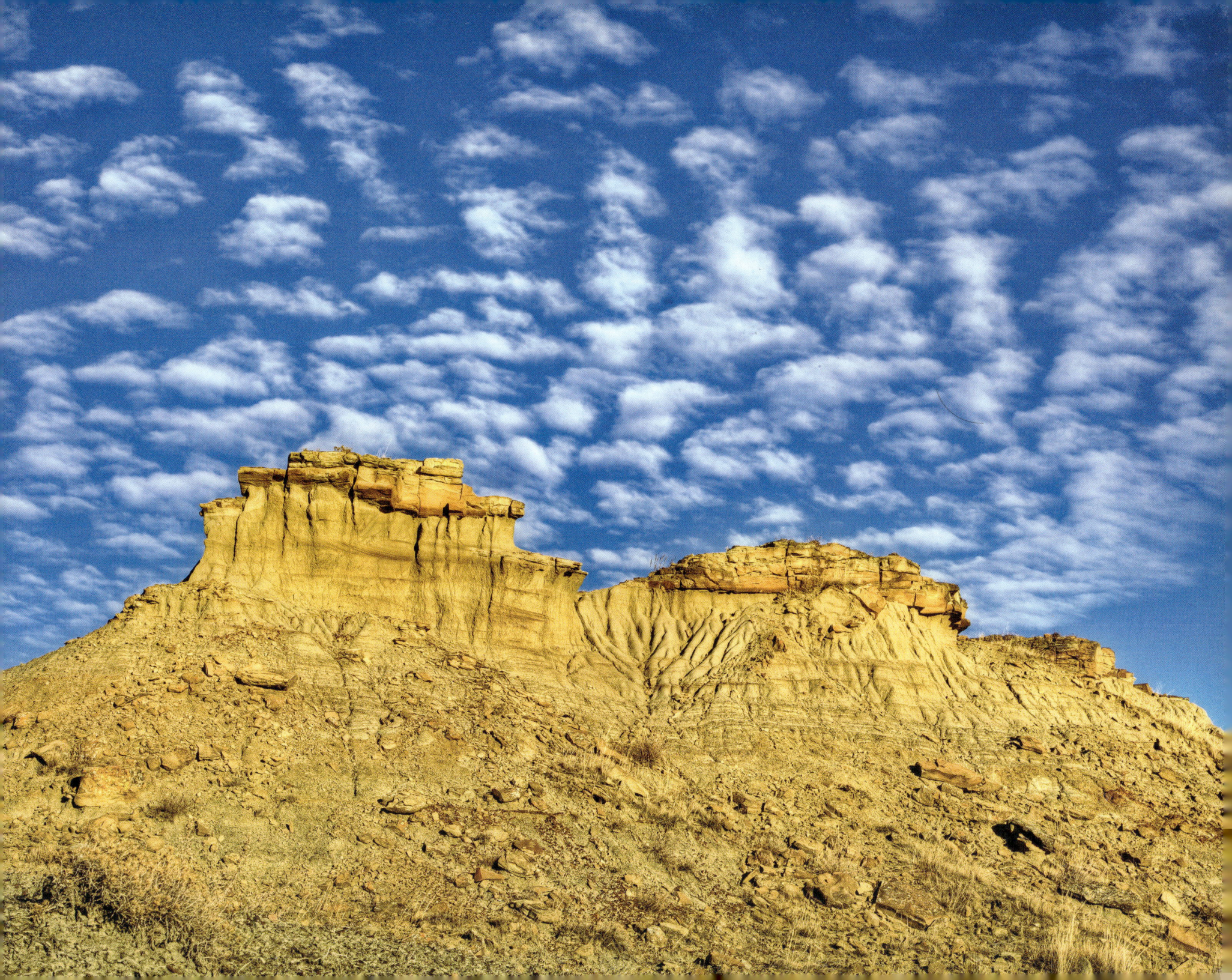

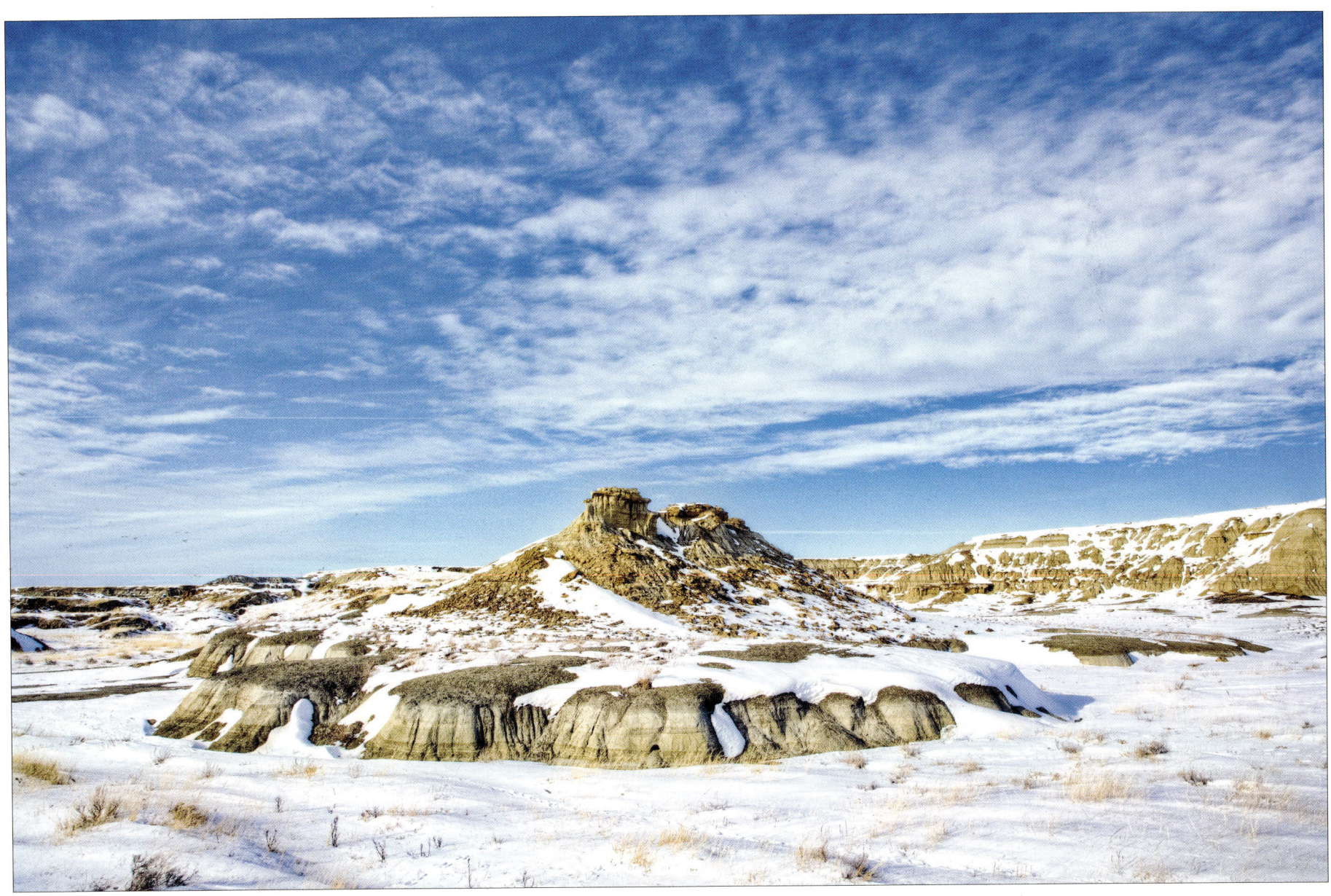

CASTLE ROCK (IN WINTER)

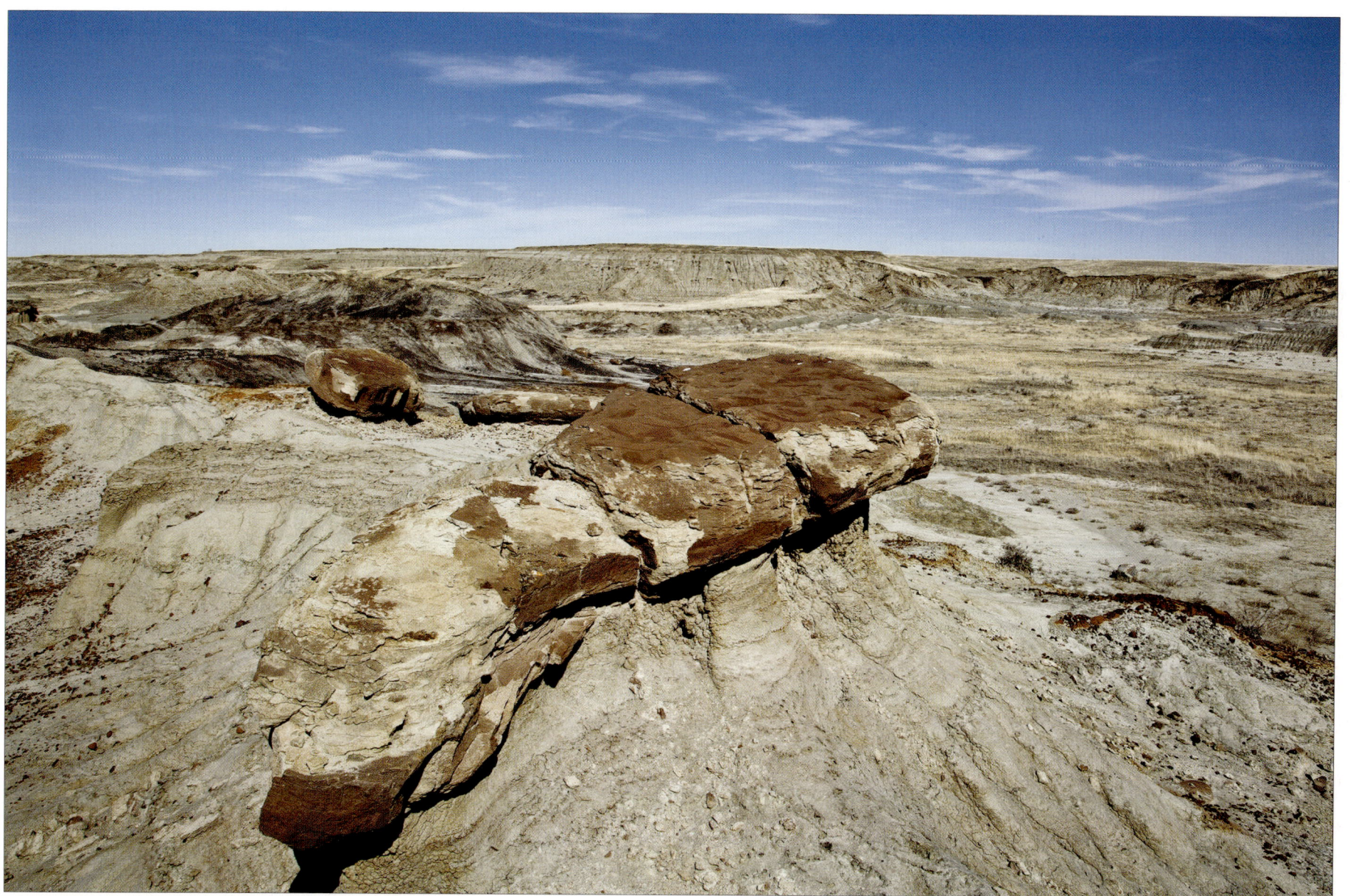

DIVISADERO

The Avonlea Badlands are small in geographical terms, yet they contain all the classic badlands features such as hoodoos, eroded rock outcroppings, fissured hills and bentonite clay.

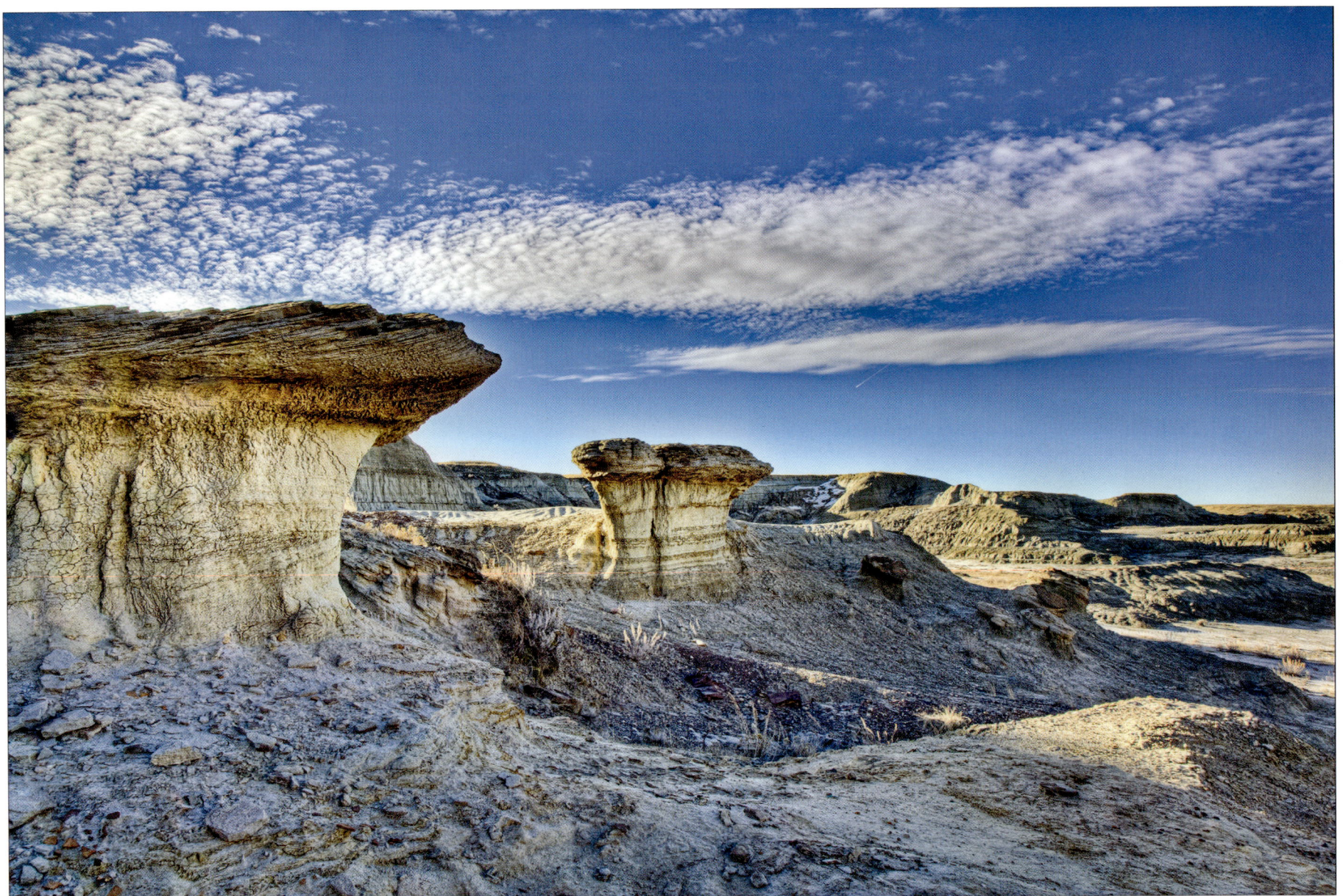

FATAL PASSAGE
The solitary silence is always a bit unnerving when exploring the badlands.
While I was waiting for the morning light to showcase these hoodoos, I noticed two coyotes only 75 metres away.

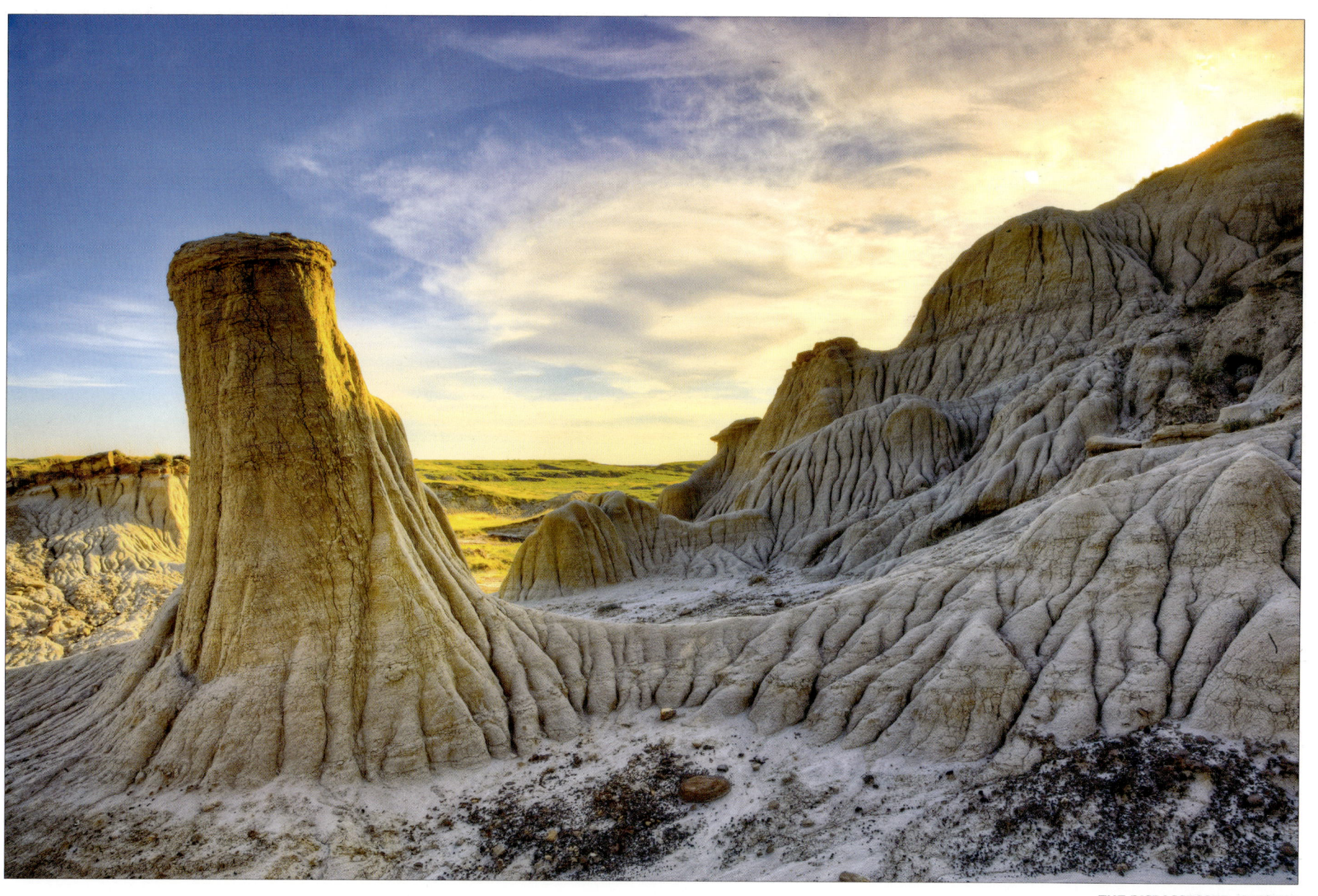

THE DISPOSSESSED (SUNSET)
For me, this large column is the jewel of the Avonlea Badlands. It seems
to mutate as your angle of view changes.

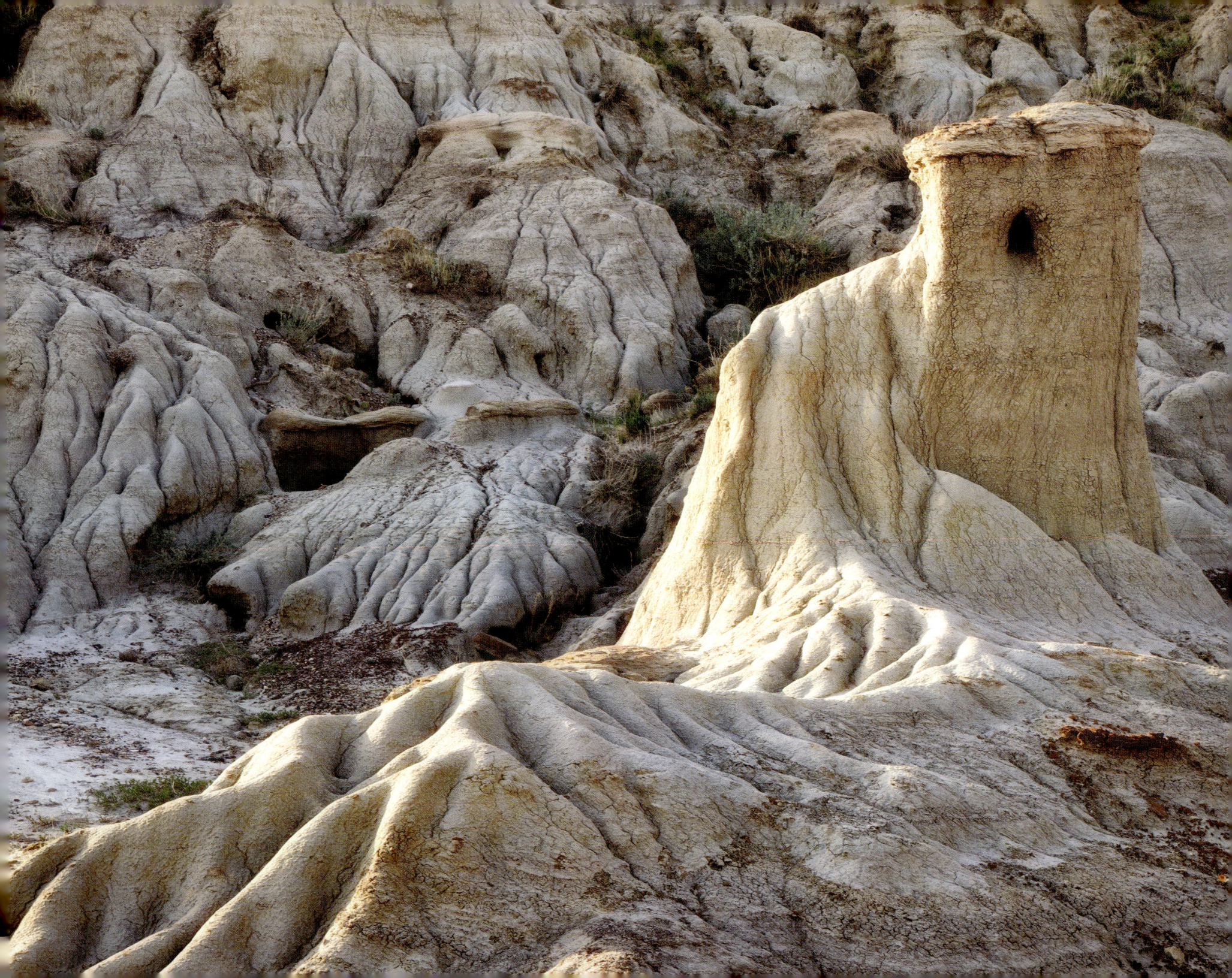

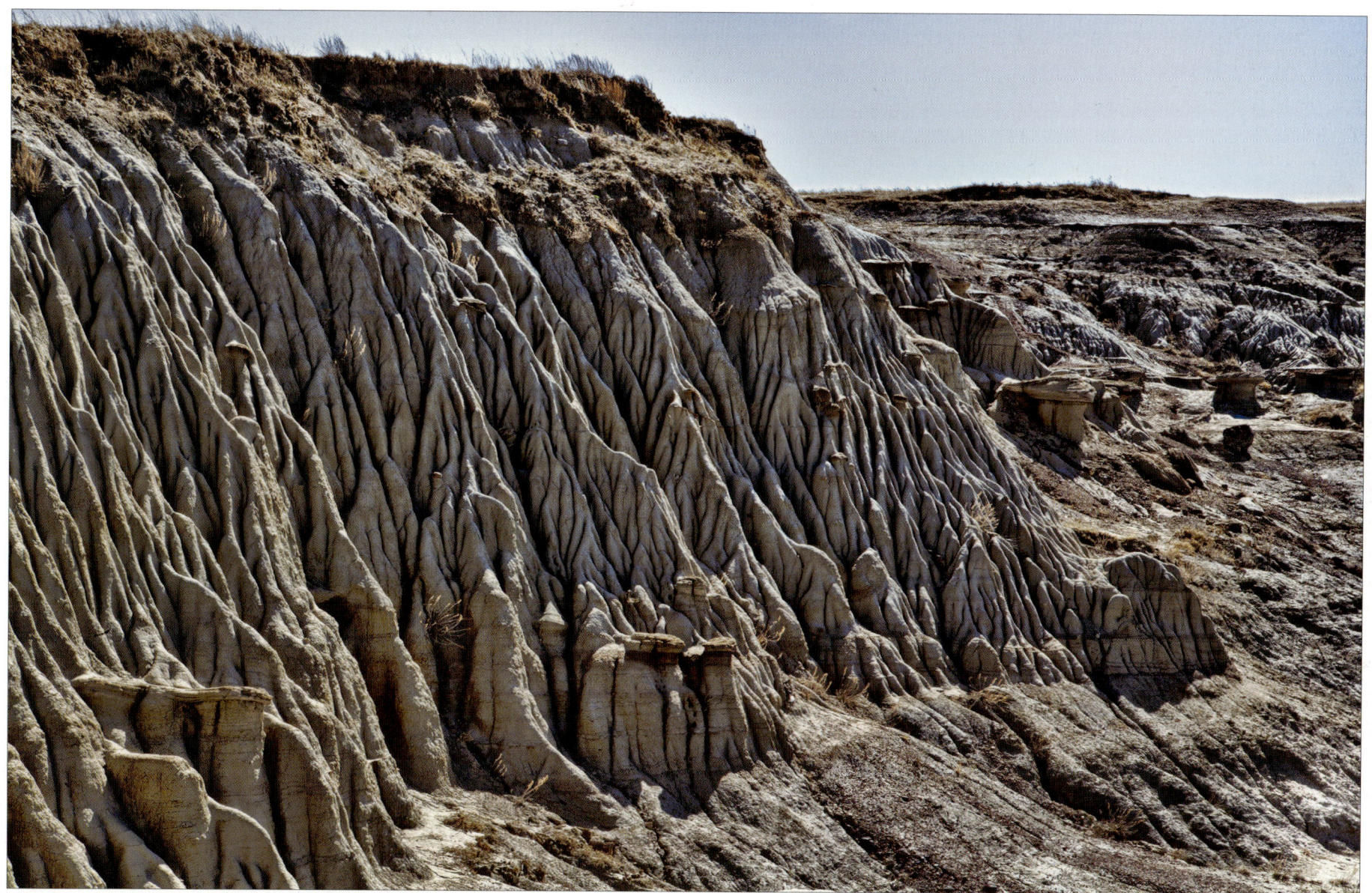

LEFT: THE DISPOSSESSED (FRONT VIEW, LOOKING FROM THE SOUTH)
From this angle this structure transforms into a spirit encased in stone.

ABOVE: THE COLONY OF UNREQUITED DREAMS
While climbing to photograph the hoodoos (on far right), my jaw dropped when
I came across this fantastic wall of etched sandstone.

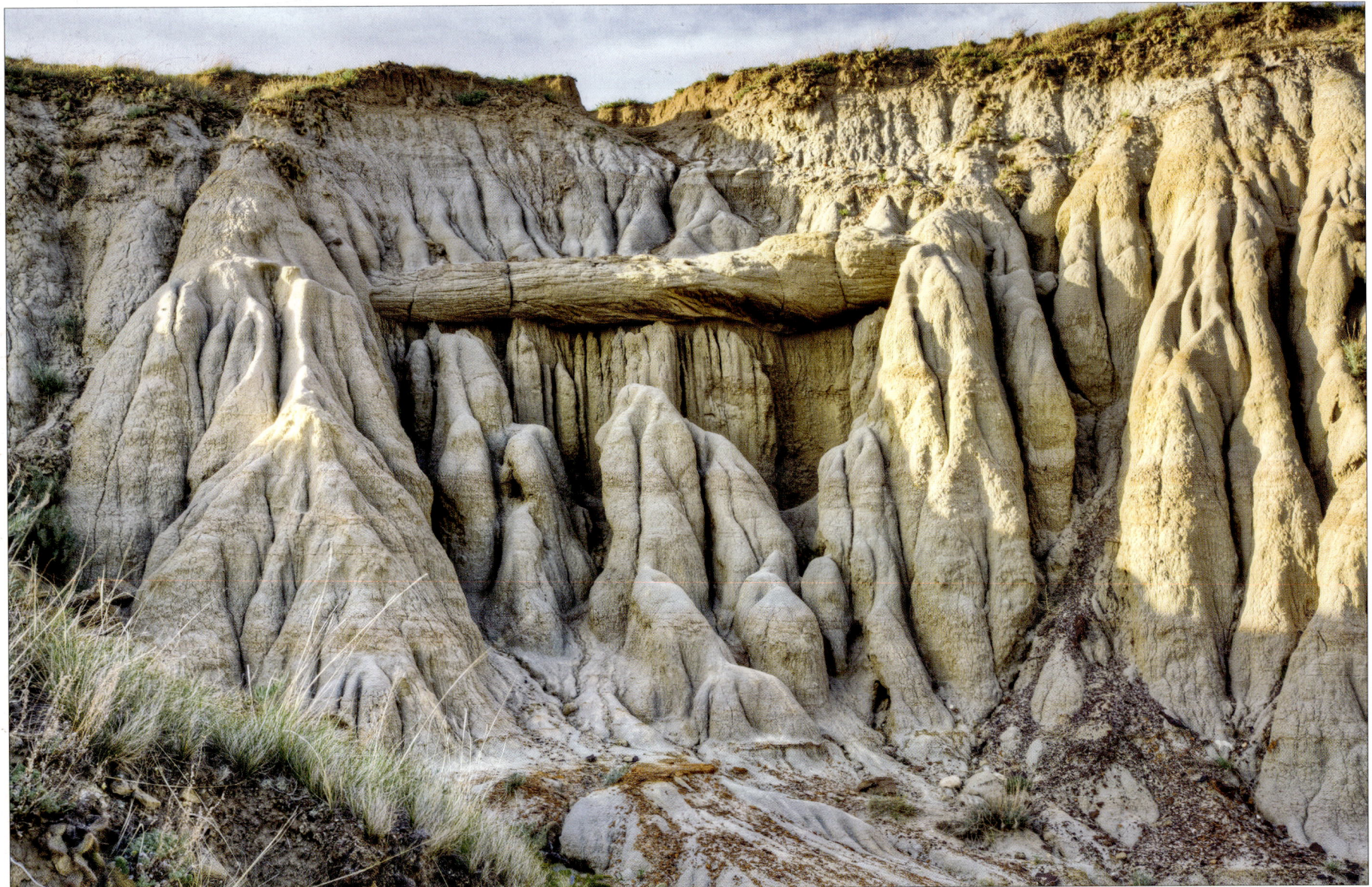

ABOVE: A GATHERING OF SAINTS
This image has the sombre feel of a mourning procession about to enter a place of worship.

RIGHT: VOICES IN TIME
Reminiscent of Hugh MacLennan's novel, this view provides a glimpse back in history toward remnants of a past world. The sun is a foil to the slow but constantly changing badlands landscape.

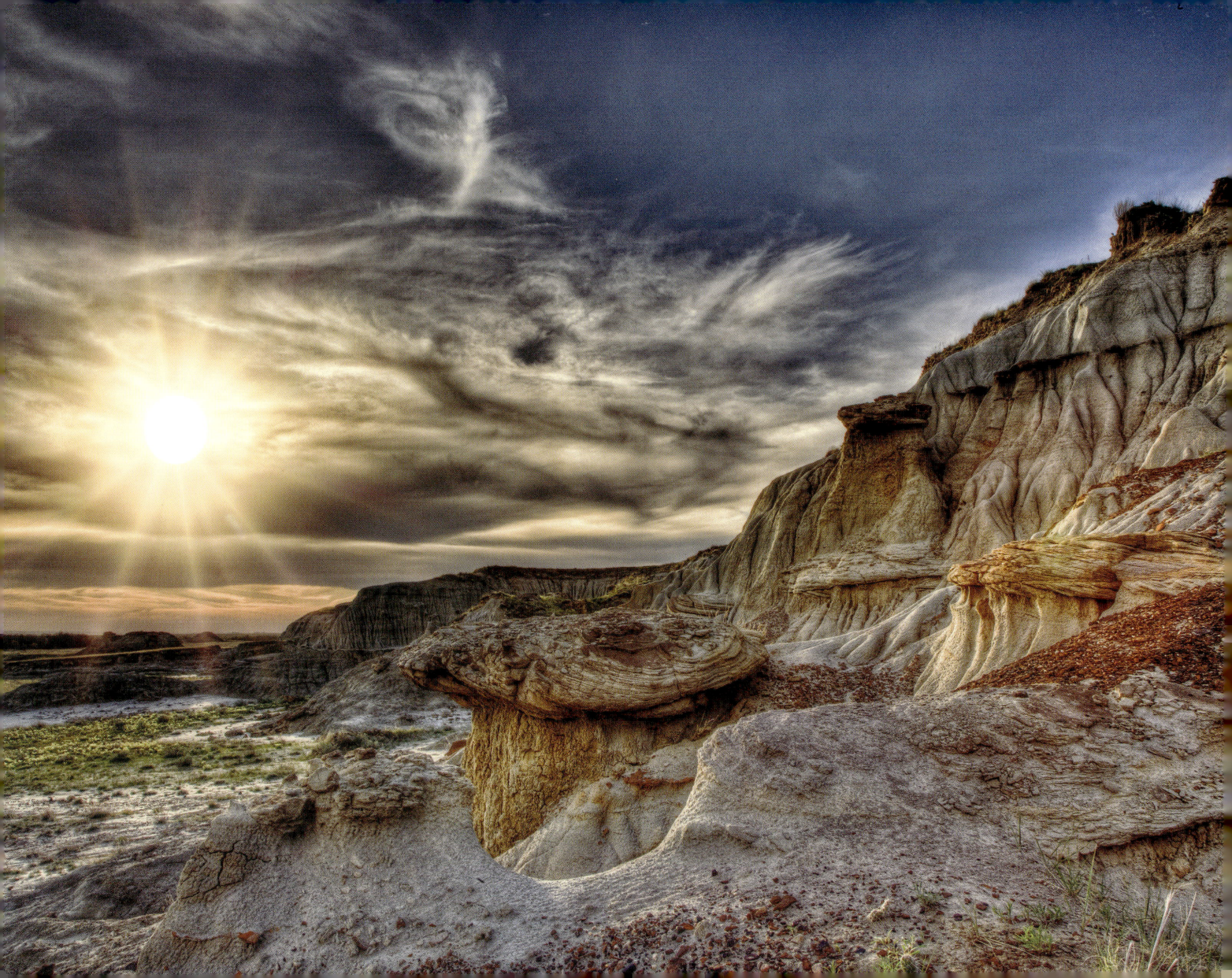

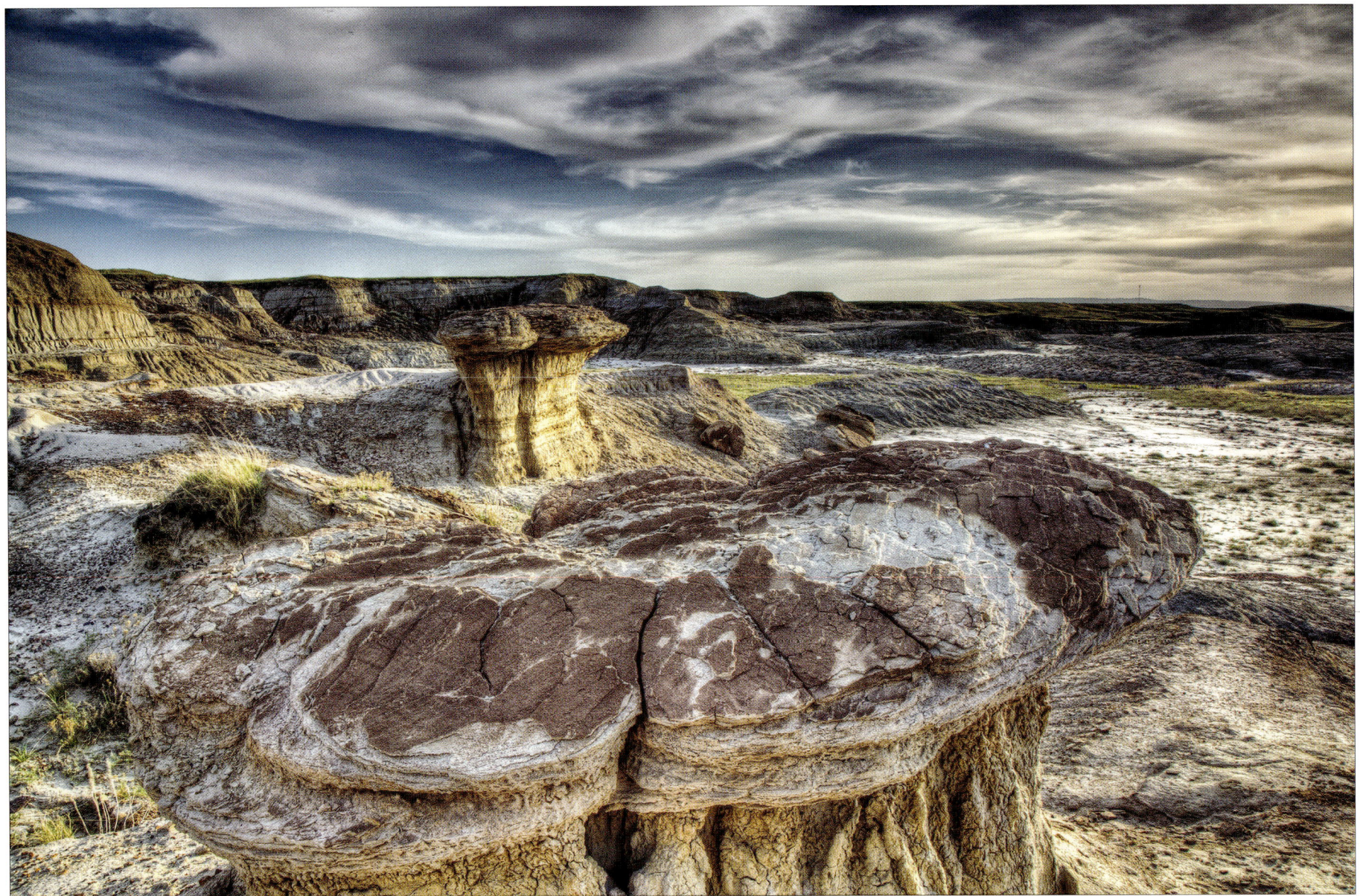

WORLD OF WONDERS

Some of the photographs collected in this book are high dynamic range (HDR) images. This involves taking three or more images and blending them together in photo editing software, thus creating a greater degree of detail. Because the images are slightly delayed, cloud movement is accentuated. The result is a swirling sky that balances well with the broken landscape.

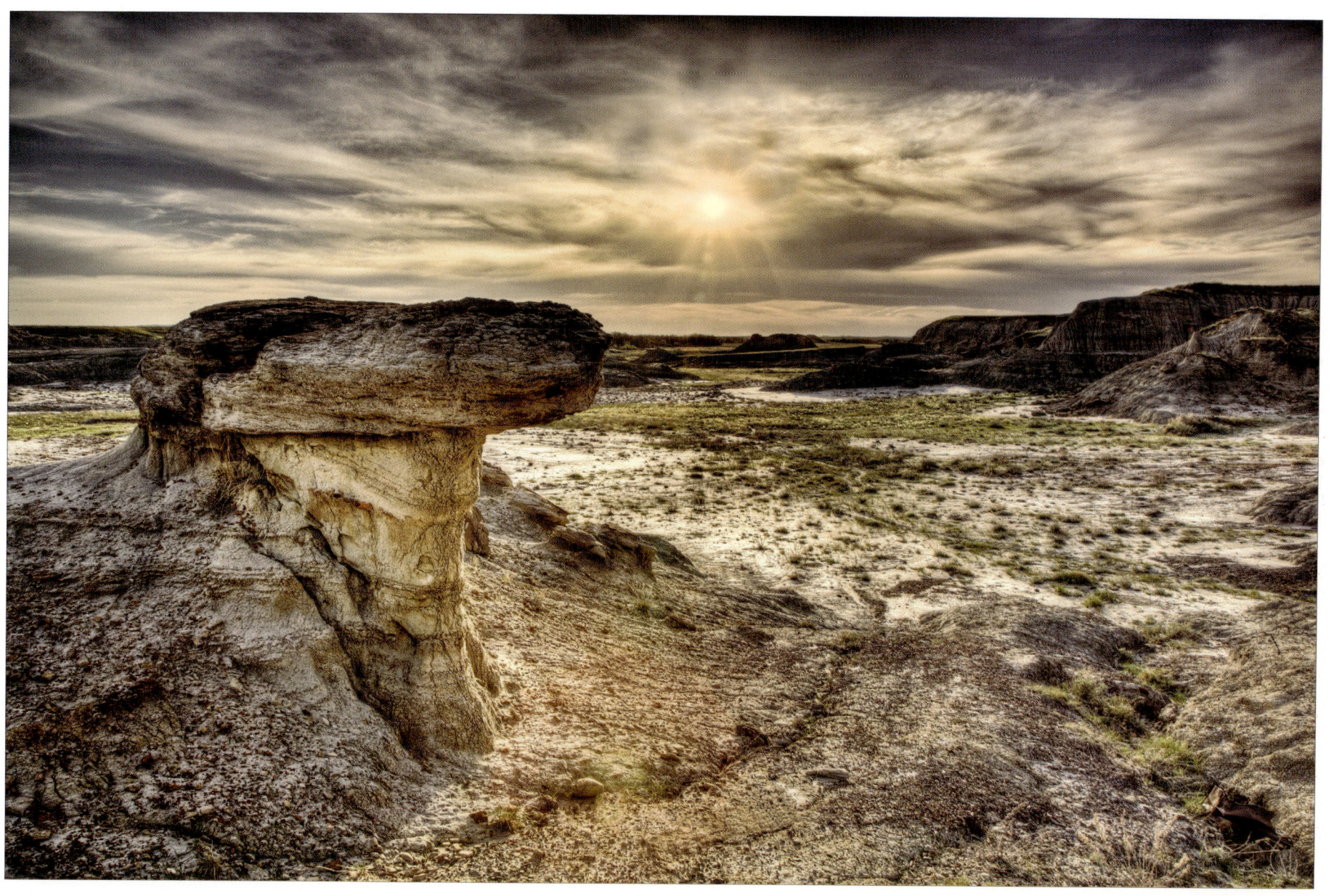

THE REBEL ANGEL

"Vogue la galere. Let your ship sail free." From *The Rebel Angels* by Robertson Davies

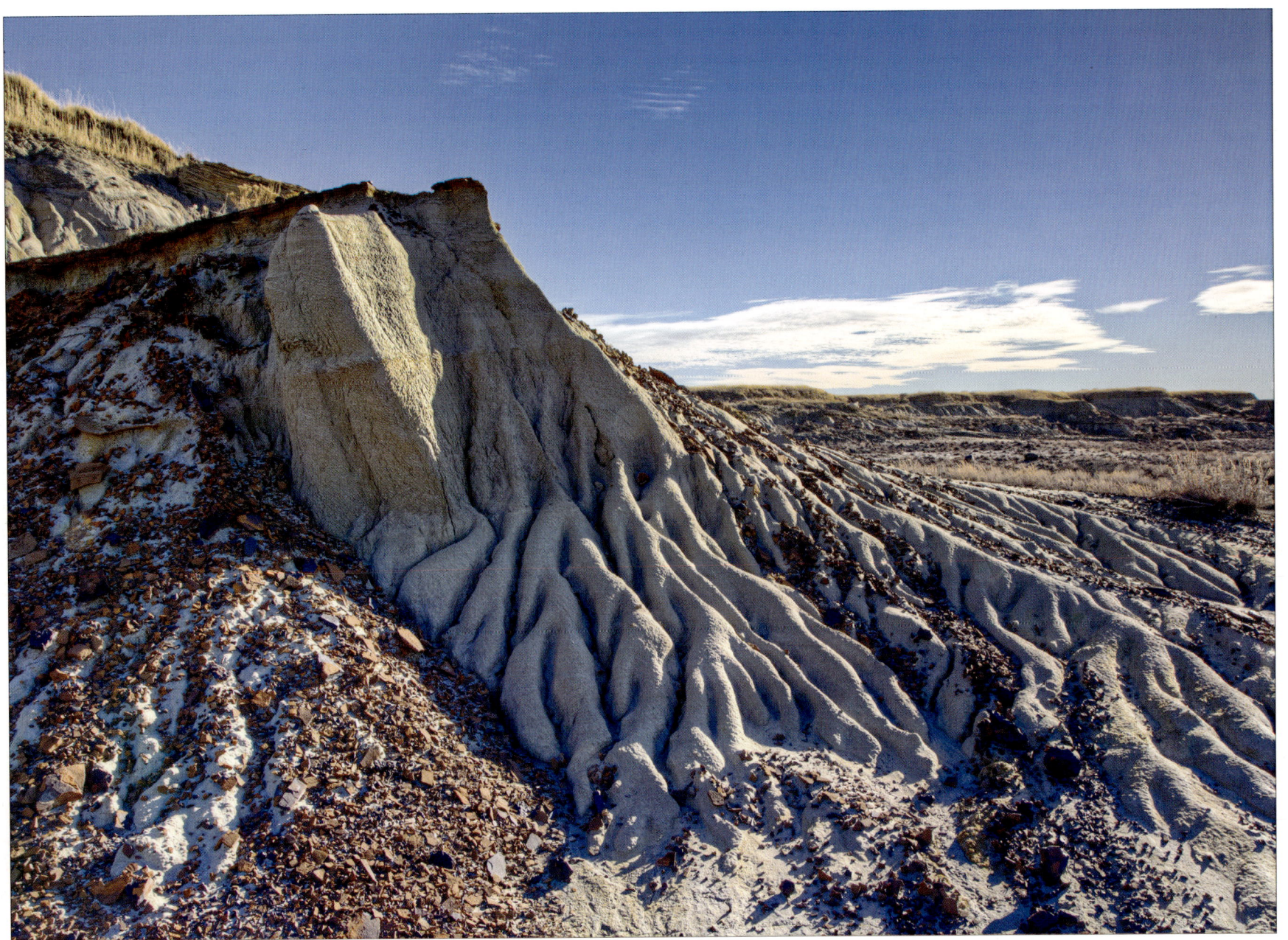

SANCTUARY LINES

The earth seems to be in creation mode as a new landform appears to emerge from below its surface.

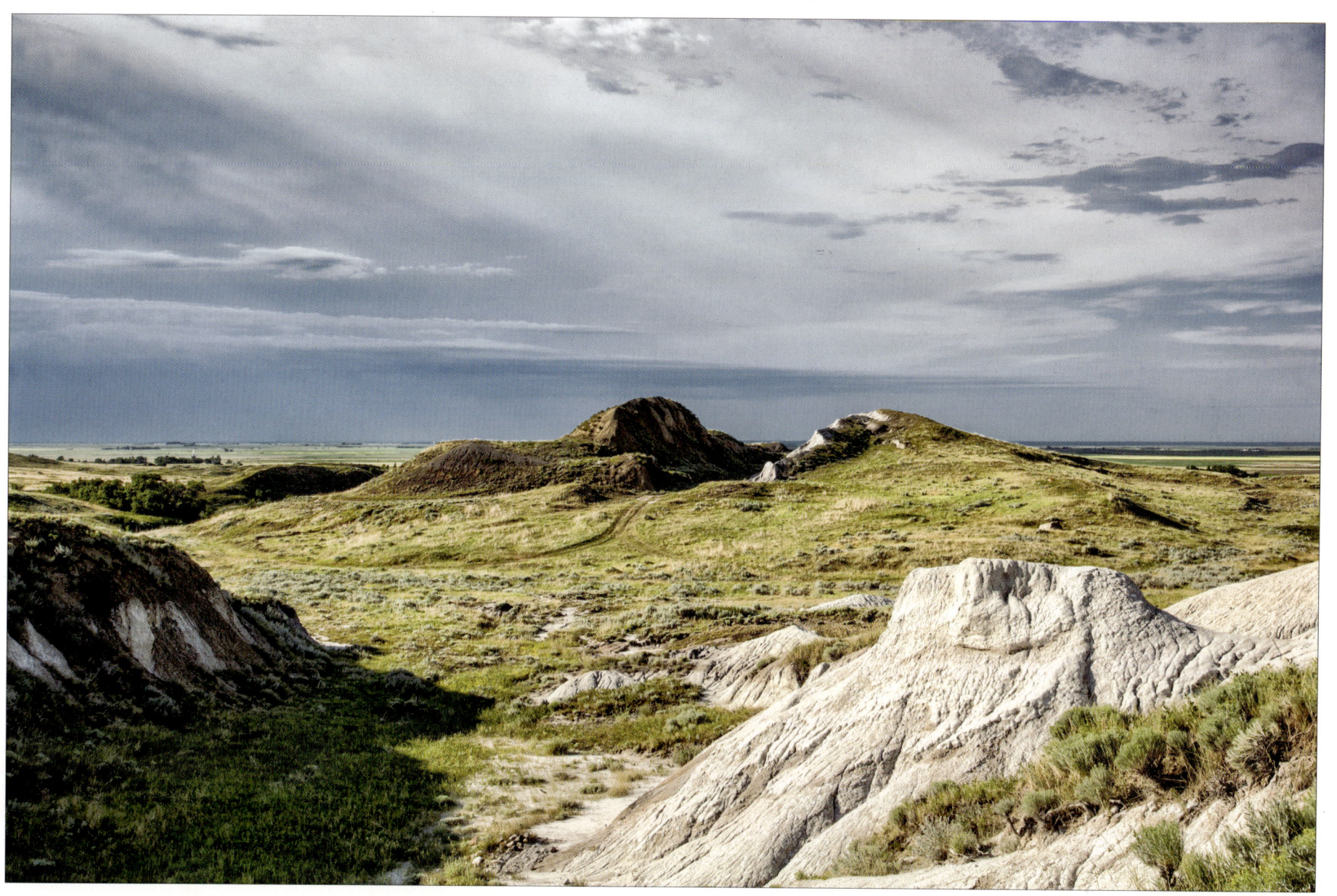

FOR THOSE WHO HUNT THE WOUNDED DOWN (Massold Clayfields, near Avonlea)

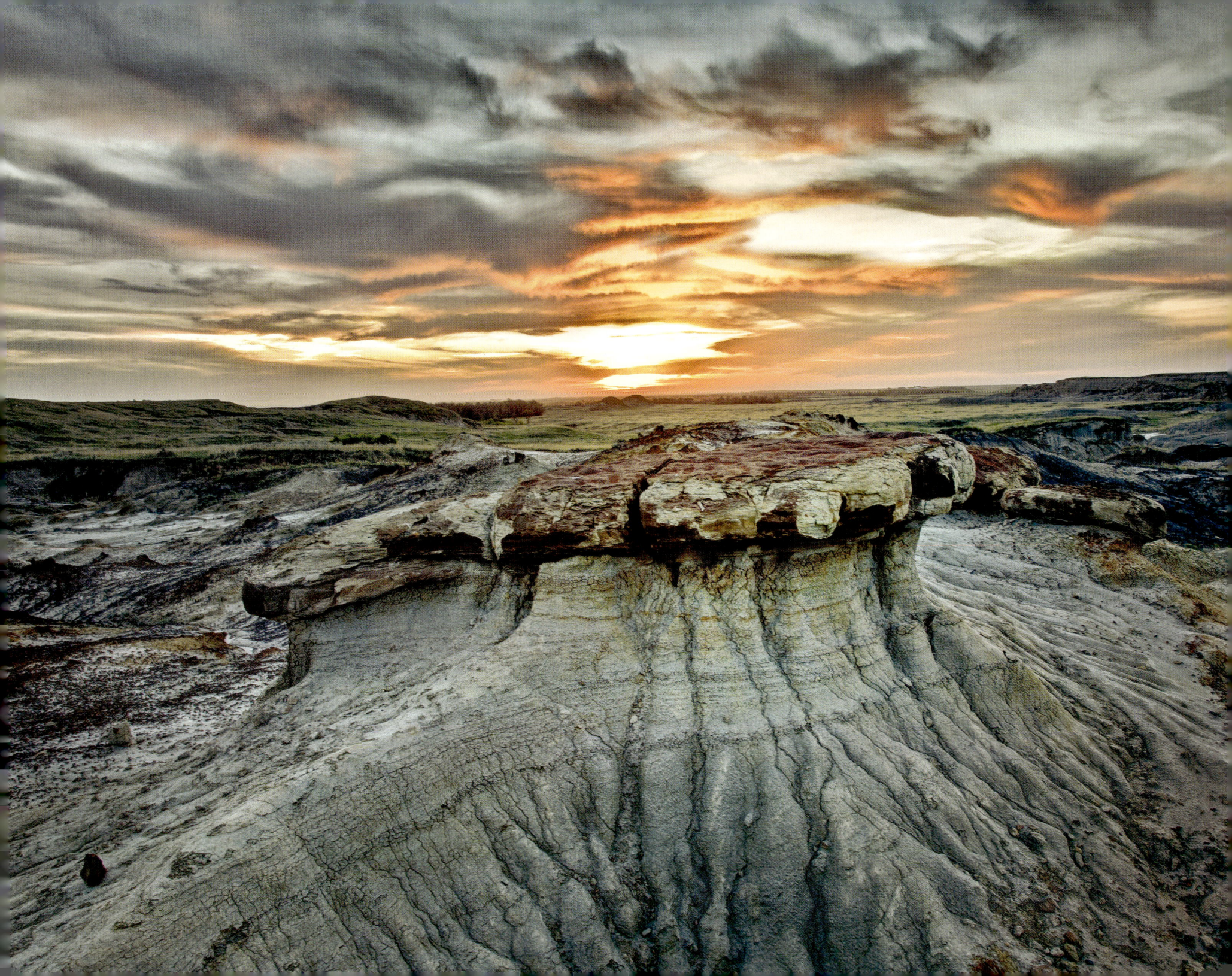

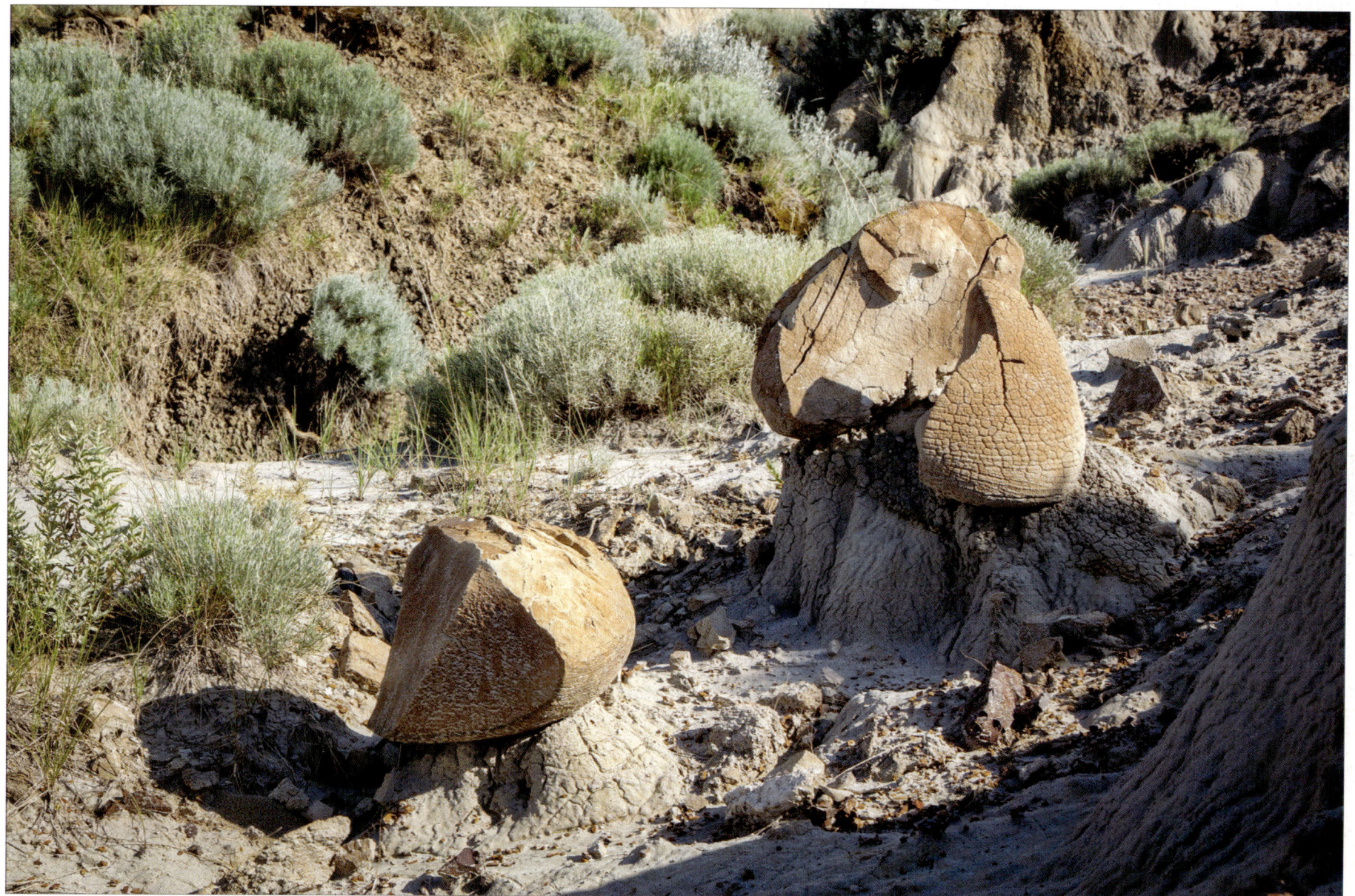

LEFT: LIFE BEFORE MAN
A fantastic Avonlea Badlands hoodoo; a dramatic Saskatchewan sunset; the sprawling prairies visible in the background… the only thing missing from this HDR photograph is a dinosaur wandering through this prehistoric scene.

ABOVE: THE SACRIFICE (Massold Clayfields, near Avonlea)
As if displayed in an art gallery, both rocks appear perfectly placed so the smaller stone looks as if it has been cut from the larger.

BIG MUDDY BADLANDS

The sparsely populated Big Muddy region cuts a deep swath through south central Saskatchewan, beginning near Willow Bunch and extending south of Bengough and into Plentywood, Montana. The prairie topography here is largely made up of rolling hills, deep ravines, and hidden gullies, so there is no question this rugged country can hide secrets. The Big Muddy Valley was originally a meltwater channel that carried vast quantities of water eastward. It is about sixty kilometres long and, in places, three kilometres wide. Classic badland features like hoodoos, rugged rock outcropping, and layered strata are scattered throughout the valley. The main badlands, consisting of extensive barren hills, fissured bentonite, and rock concretions, can be found adjacent to Big Muddy Lake, but are only accessible via back roads.

While this area has sometimes been referred to as the loneliest place in Saskatchewan, it has some of the most spectacular landscape vistas found in the province. The iconic Castle Butte is a wonderful example. Probably a landmark for centuries, this sentinel rises up from the valley floor, towering 70 metres over the prairie. The best time to view Castle Butte is just before dusk when the setting sunlight ignites the rock with colour. To add to the allure of this scene, just a short walk into the hills is the riveting "Stone Angel" rock formation. This, in my opinion, is one of the best examples of a hoodoo anywhere. (The photograph "Stone Angel" (page 19) won the Viewers' Choice Award at the group exhibition "Contemporary Indigeneity" at the Great Plains Art Museum, University of Nebraska, and is now part of their permanent collection.)

While the Big Muddy boasts some impressive landscapes, the region is probably best known for its infamous tales of notorious outlaws, horse thieves, and cattle rustlers, who wreaked havoc in this area in the 1890s. The Big Muddy was Station No. 1 on the Outlaw Trail that traversed south through Miles City, Montana, past Deadwood, South Dakota, and all the way to Old Mexico. This was a fairly well-defined escape route and hideaway set up by legendary outlaws Butch Cassidy and the Sundance Kid. Way stations were located every 24 kilometres or so, where fresh horses and supplies could be obtained. Often these stations were simply ranches where the owner was either sympathetic to the outlaws or had little choice but to acquiesce. Whatever the case, the Outlaw Trail began to be used by a number of criminals like Dutch Henry and the Wild Bunch, and Sam Kelly of the Nelson–Jones Gang. The Sam Kelly caves, located only 400 metres north of the U.S. border, are still showcased as part of the Big Muddy badlands tours, originating in the town of Coronach, Saskatchewan.

RIGHT: THE STONE ANGEL
This photograph was included in a group exhibition at the Great Plains Art Museum at the University of Nebraska. It won the viewer's choice award and now is part of the permanent collection.

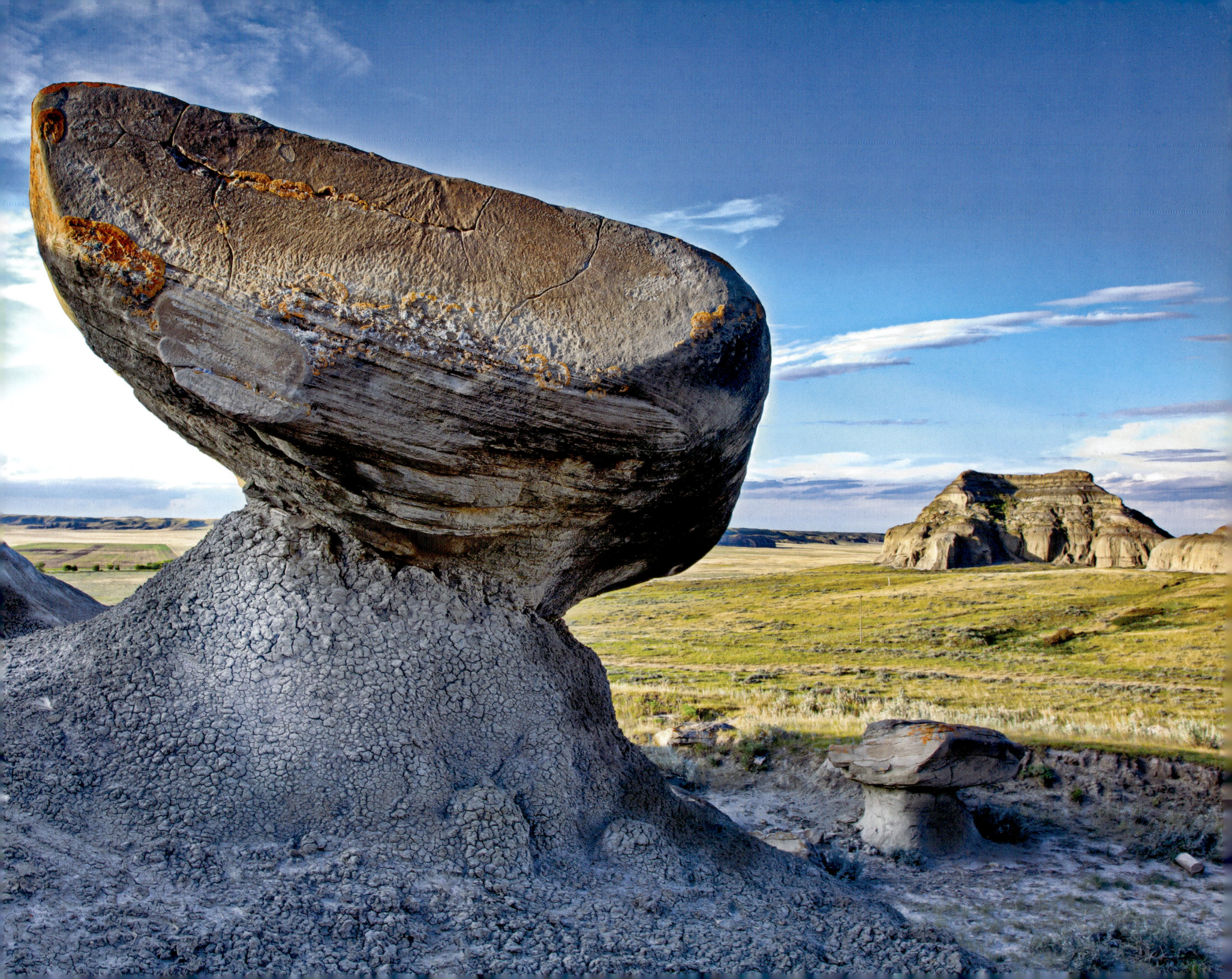

While American outlaws were gaining notoriety from high-profile bank and train robberies in the United States, it is believed they sometimes crossed into Saskatchewan just to "cool their heels" and seek refuge in the concealing terrain of the Big Muddy. That may have been their intention, but with the closest Northwest Mounted Police (NWMP) detachment in Wood Mountain over 100 kilometres away, temptation outweighed idleness. In Canada, their main activity was horse rustling. Often the gangs would crisscross the border, stealing horses in one country, rebranding them and selling them in the other country. This came to a head when Inspector Jas. O. Wilson of the NWMP informed his superiors in Regina (sometime prior to 1902) that, "Valley County, Montana, is the crookedest county in the Union and that Big Muddy is the worst part of it, and nearly every man in that section is on the rustle."

The death of Frank Carlyle, which is mentioned in the local history book *Big Muddy Badlands: Just North of the 49th*, is a story that seems to corroborate these comments. Carlyle, from Toronto, joined the NWMP as an officer at the age of 22. He somehow became acquainted with Dutch Henry and shortly after resigned from the force. The story goes that Carlyle was supposed to blow up a bridge in Montana so that Dutch Henry could rob the approaching train before the Nelson–Jones gang—who had also planned the same heist further on up the tracks. Carlyle got drunk and failed to complete his task. Not long afterward, while in Canada, Carlyle met up with two riders and was escorted to a coulee where he was subsequently shot and killed. His body was left to rot in an isolated spot now called Carlyle Coulee. It was because of this and other crimes that the NWMP finally set up a detachment in the Big Muddy in 1902. The new NWMP detachment effectively put an end to outlaw history in the Big Muddy.

While outlaws wreaked havoc in the area around the turn of the 20th century, prior to this, the most famous resident of the Big Muddy region was Chief Sitting Bull. The legend of Sitting Bull began to grow sometime around 1864, when his Sioux tribe started to spar with the white men who were increasing settlement into traditional Indian Territory and who were beginning to over-hunt the buffalo. When gold was discovered in South Dakota in 1874, the tension escalated. The United States Army arrived with the intention of moving Indians onto reserves. This tension finally culminated in the Battle of the Little Bighorn in 1876, where Sitting Bull and his Sioux wiped out the 7th Cavalry Regiment of the United States Army under the command of General George Custer. Custer and approximately 265 of his U.S. soldiers were slaughtered. Sitting Bull and 2000 of his Sioux tribe fled Montana and found temporary sanctuary in Canada. They lived fairly peacefully in southern Saskatchewan from 1876 to 1881—primarily in the Wood Mountain and Willow Bunch area, but also venturing as far north as Moose Jaw and east to Fort Qu'Appelle in an attempt to lobby for a reserve.

There were numerous political pressures urging Sitting Bull to surrender to the United States government but, ultimately, the scarcity of buffalo was probably the overriding factor. On July 10, 1881, Sitting Bull and a caravan of between 200 and 400 emaciated Sioux (actual numbers seem to vary, but it was a small fraction of his original tribe) left Willow Bunch for the last time. They travelled over 250 kilometres southeast through the Big Muddy region, crossed the border at Plentywood, Montana, and arrived at Fort Buford (near present-day Williston, North Dakota) where they officially surrendered to the United States government.

The town of Coronach, Saskatchewan, offers three Big Muddy Badlands tours. These can be pre-booked at the Tourism Centre, beginning the long weekend in May, and ending September 30.

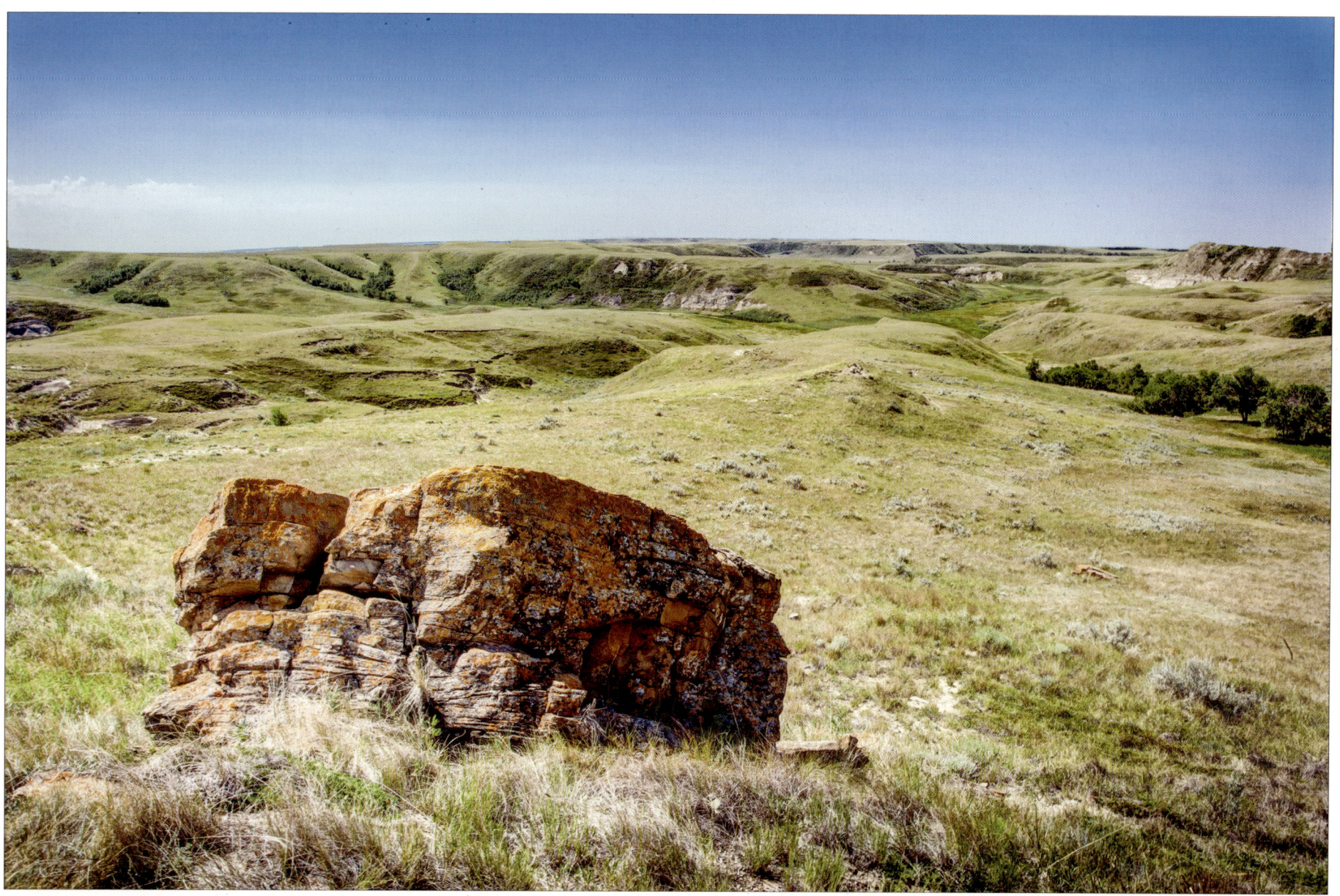

THEY SHALL INHERIT THE EARTH
This solitary rock sentinel overlooks the Big Muddy Valley. It's a perfect metaphor for what some have called the loneliest place in Saskatchewan.

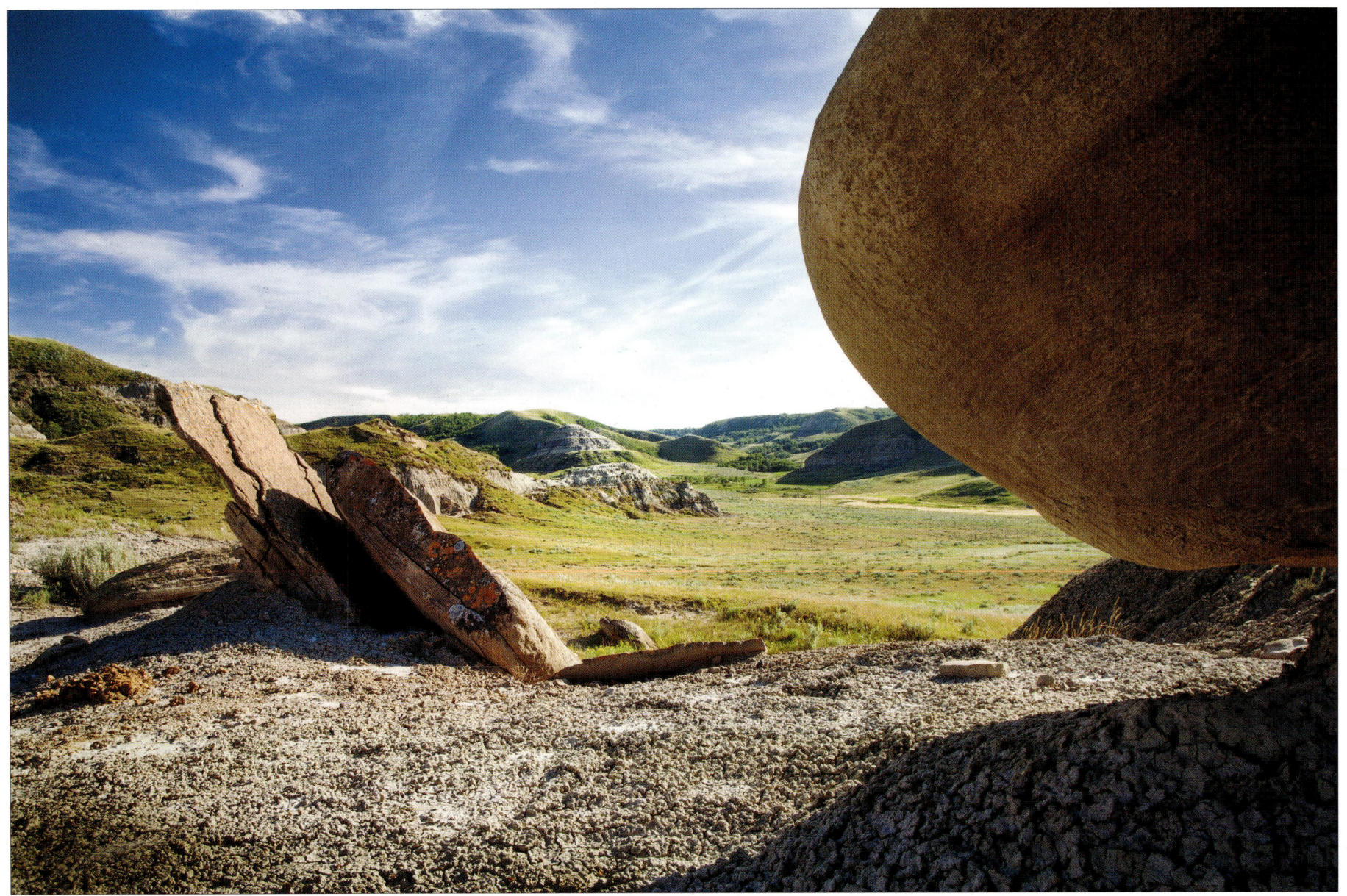

THE VALLEY (BAY) OF LOVE AND SORROWS
I love the rich and varied textures captured from this angle. The only thing missing? A long and ominous rattlesnake skin that hung from a barbed-wire fence just a few metres outside the frame of view.

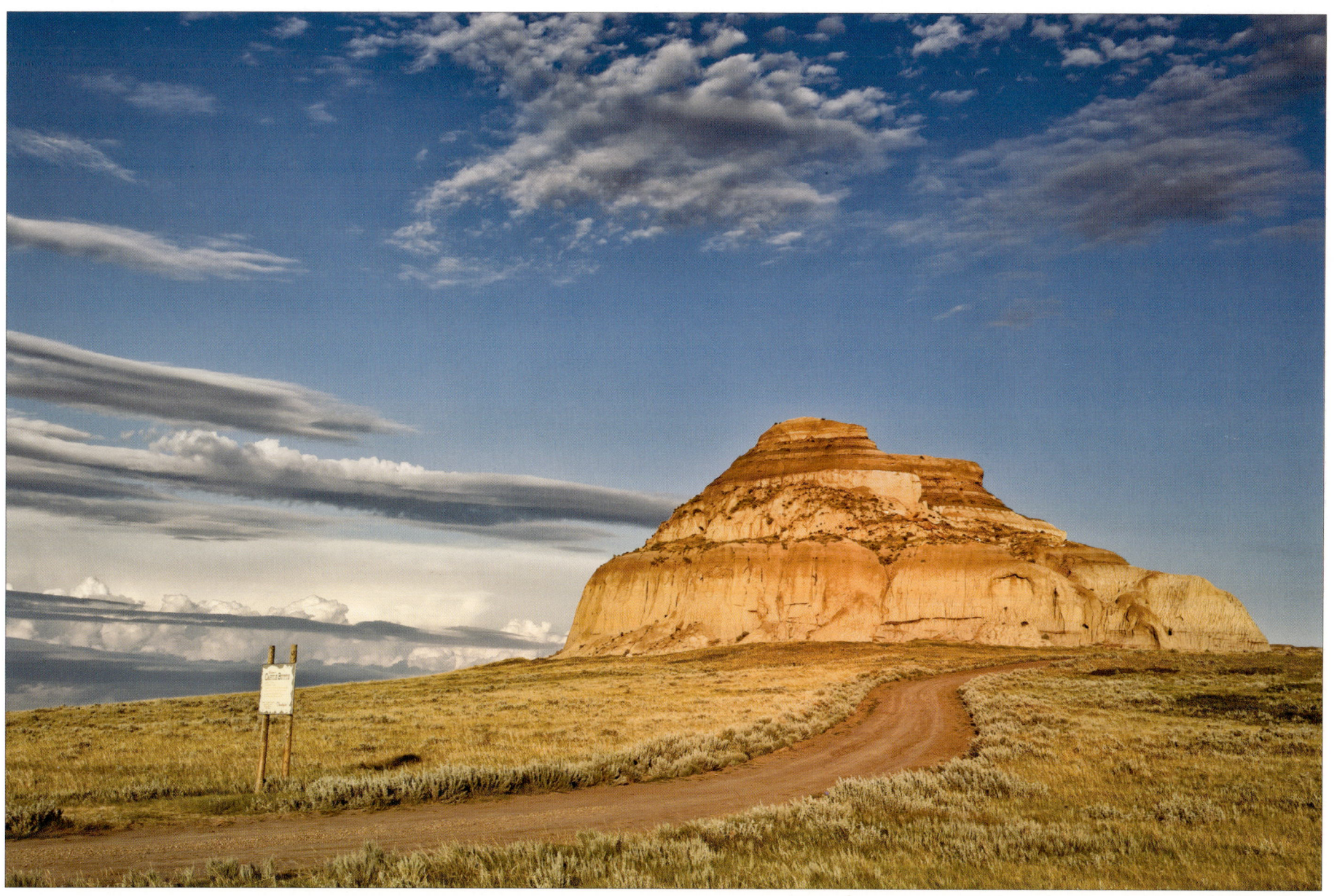

CASTLE BUTTE

This landmark towers 70 metres above the surrounding prairies. During the day the butte retains its natural earth-tone colour but when its walls are struck by the setting sun, the yellows and oranges explode with vibrancy.

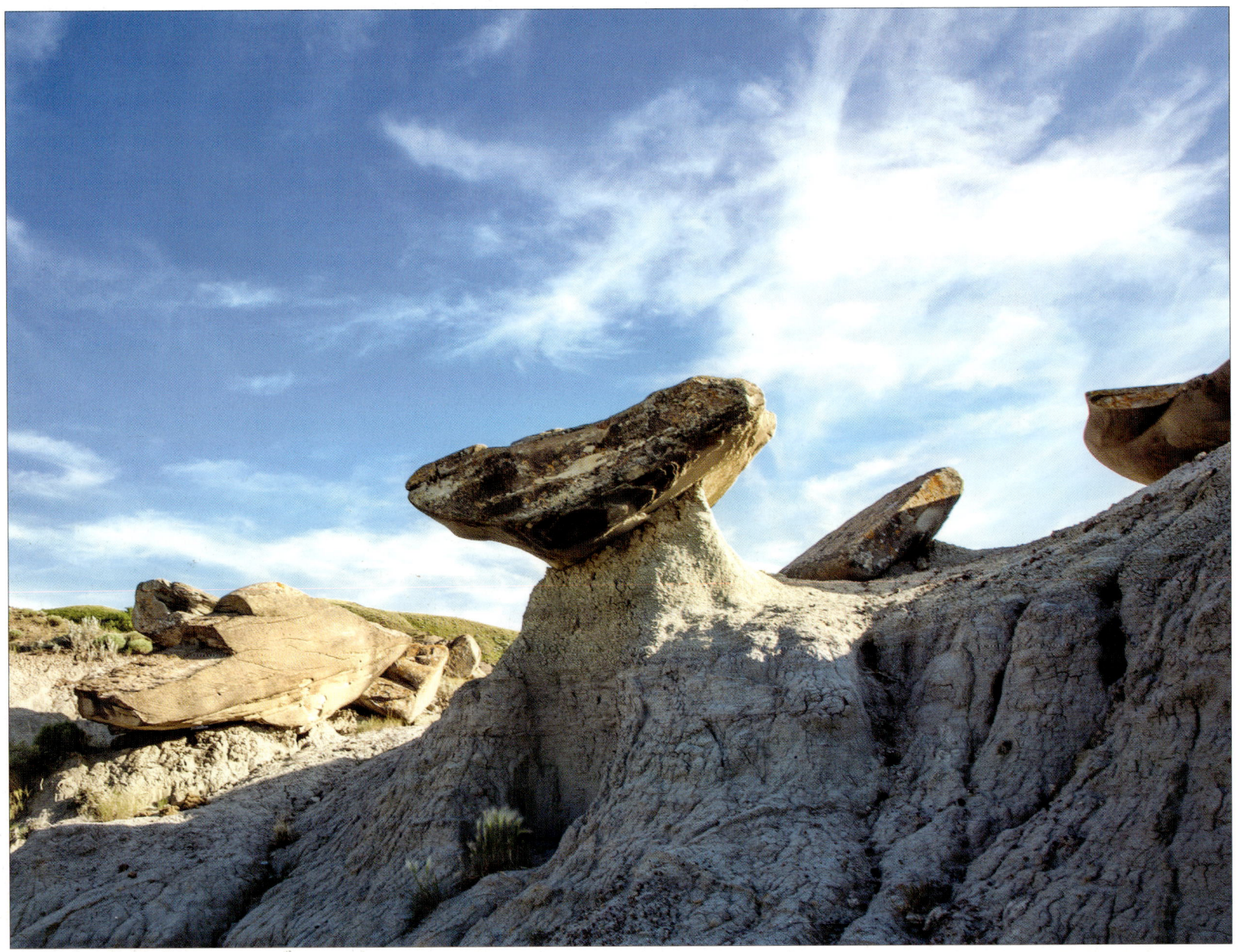

ABOVE: A FINE BALANCE

RIGHT: A FATAL GRACE

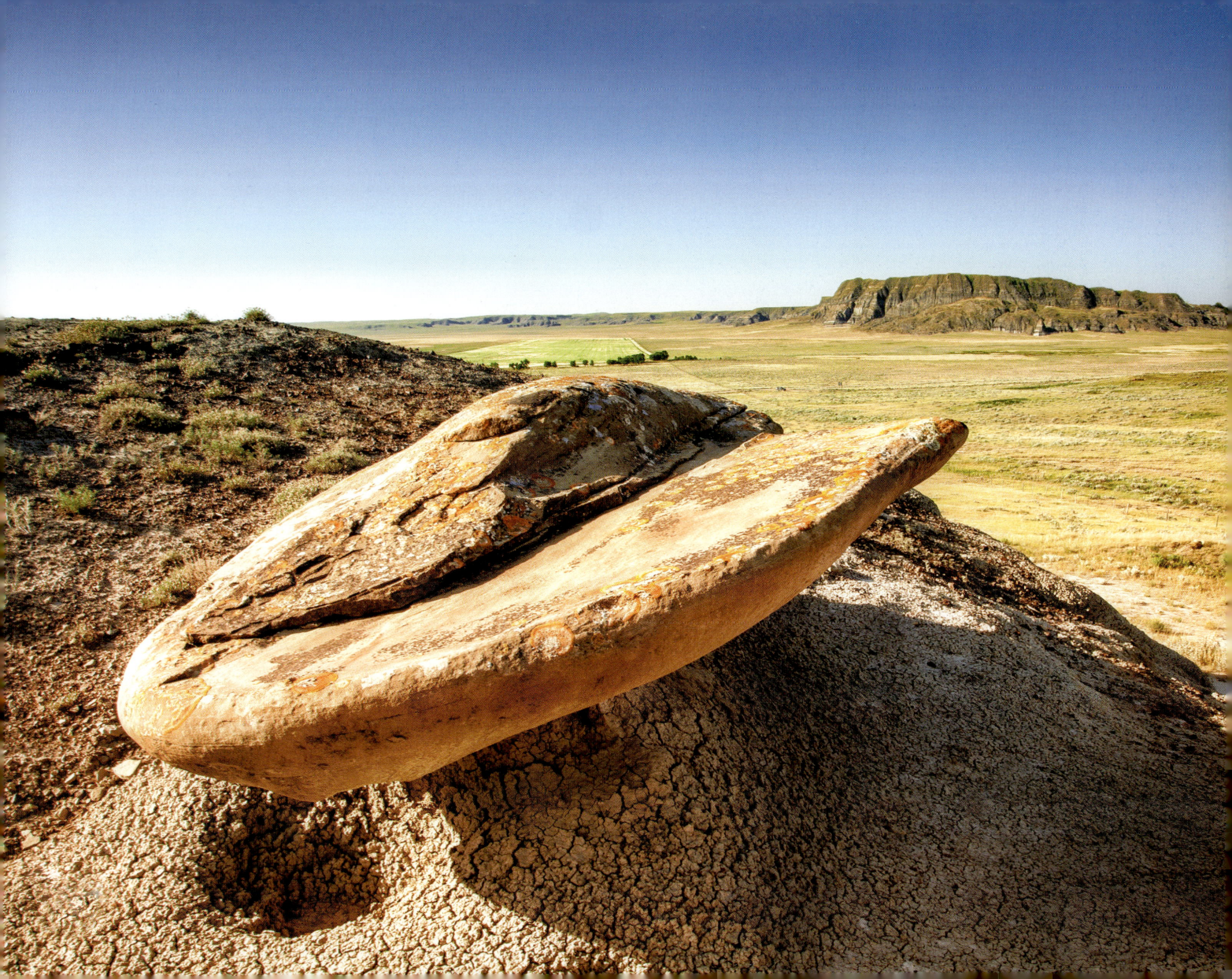

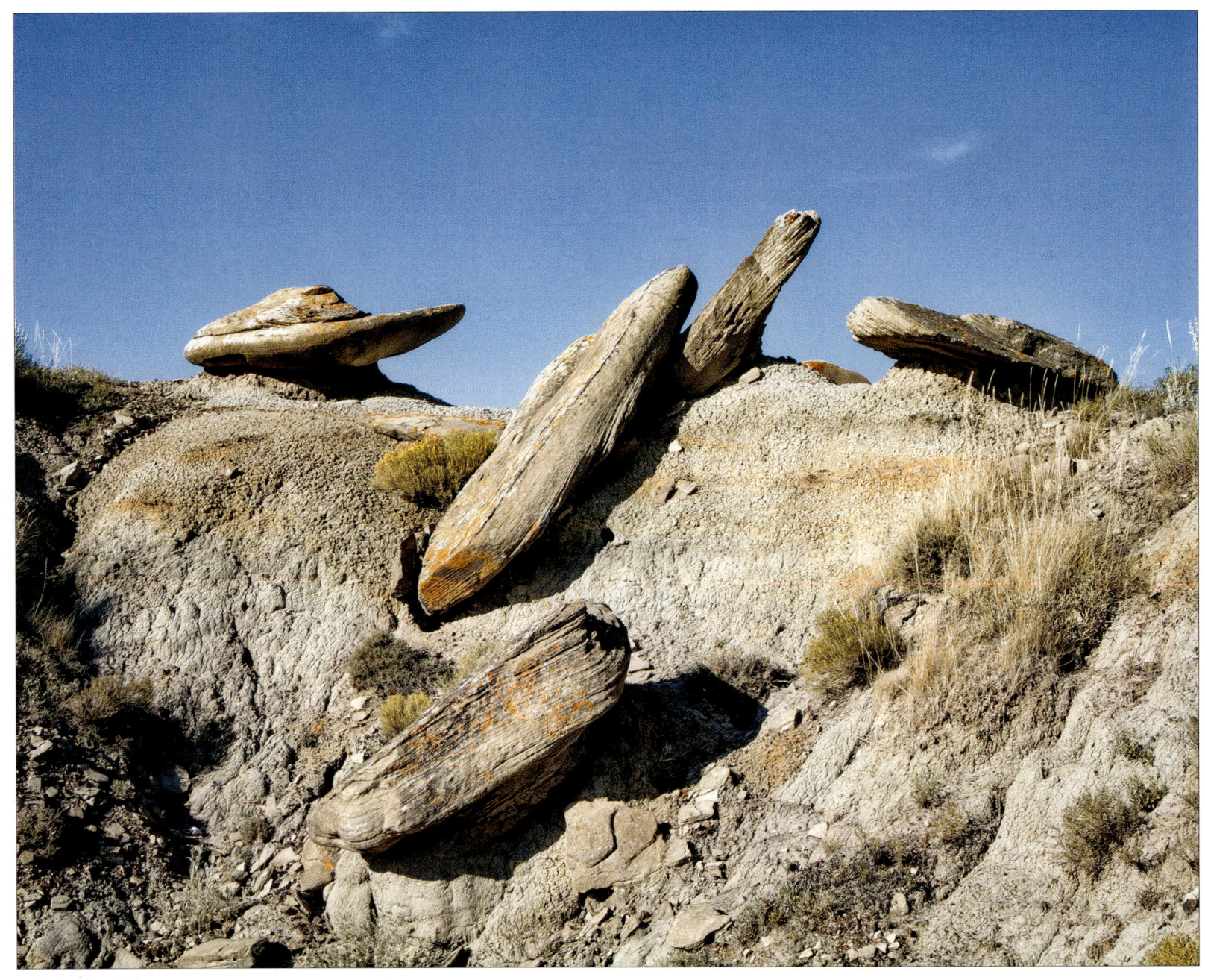

LYRICS OF EARTH

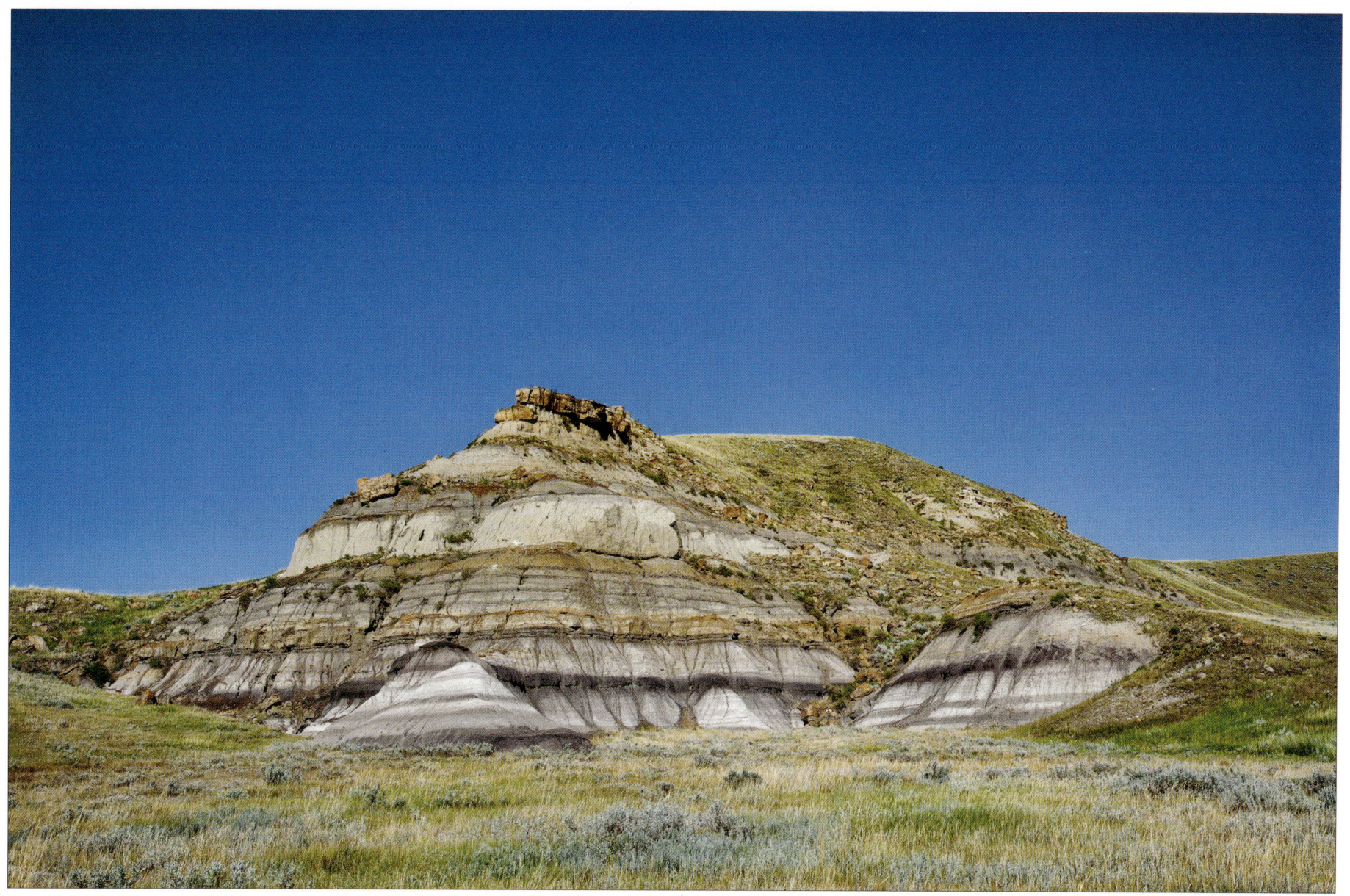

BLACKFOOT RIDGE
When photographing the badlands I often look for dramatic skies to enhance the harsh character of the land. However, a clear blue sky also works well, especially to intensify the ridge's dynamic multicoloured striations.

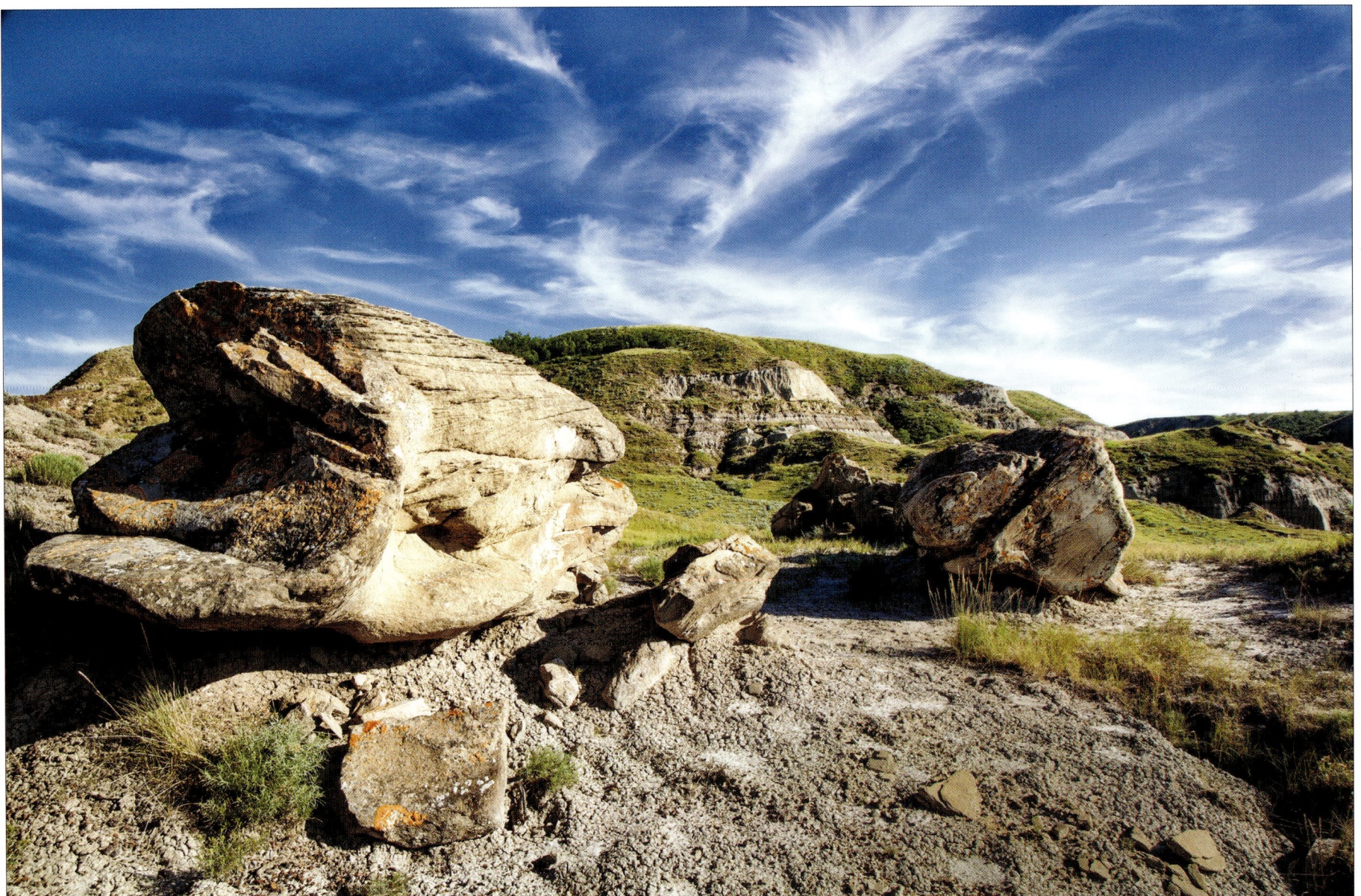

THE DIVINERS

These two large boulders bookend the storied hills of the Big Muddy Badlands. Just beyond these hills is Montana. I'm reminded of the rich history of outlaws and cattle rustlers who crossed back and forth along the 49th parallel—and of course Sitting Bull and his band of Sioux who lived in this area from 1876 to 1881.

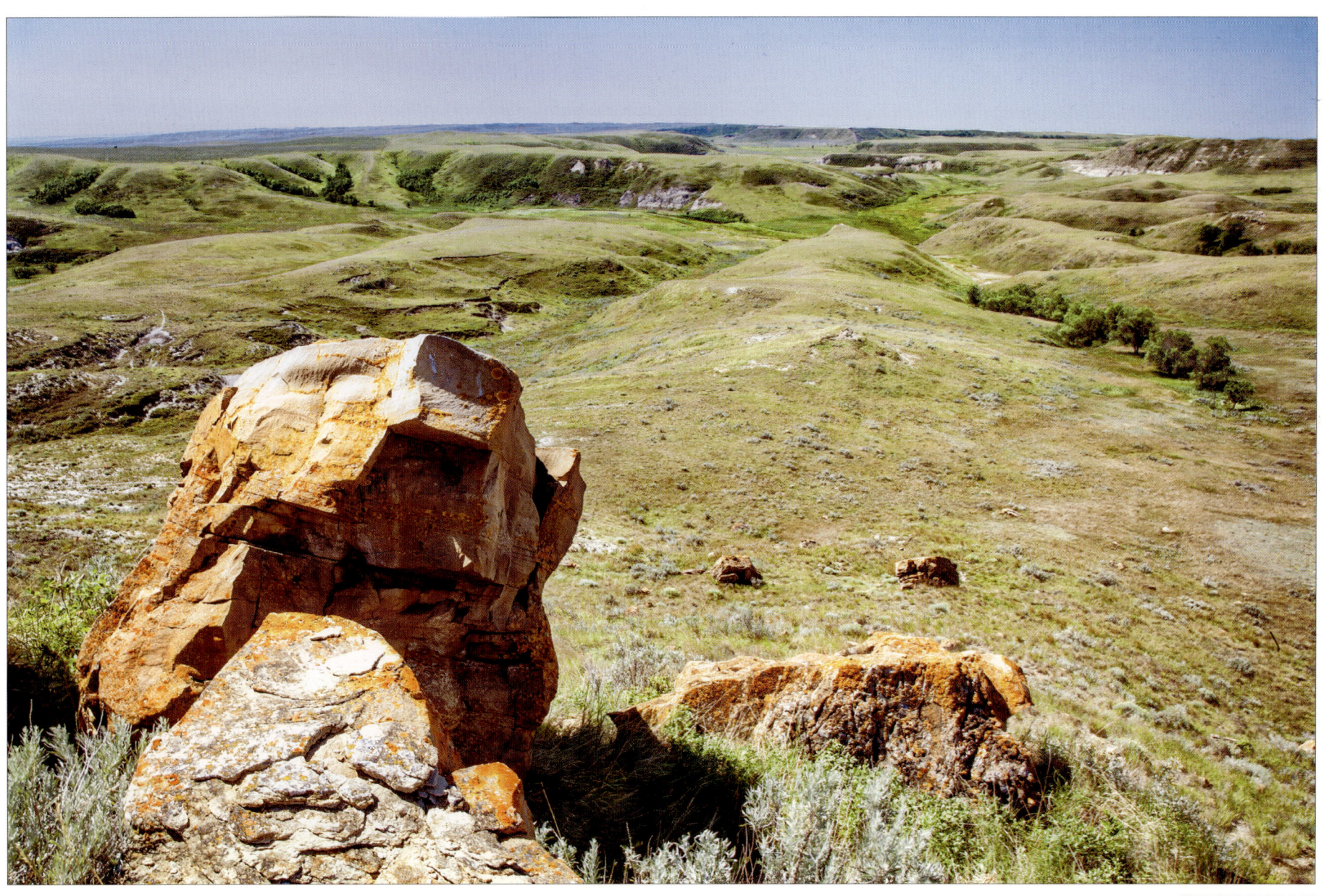

EMANCIPATION DAY

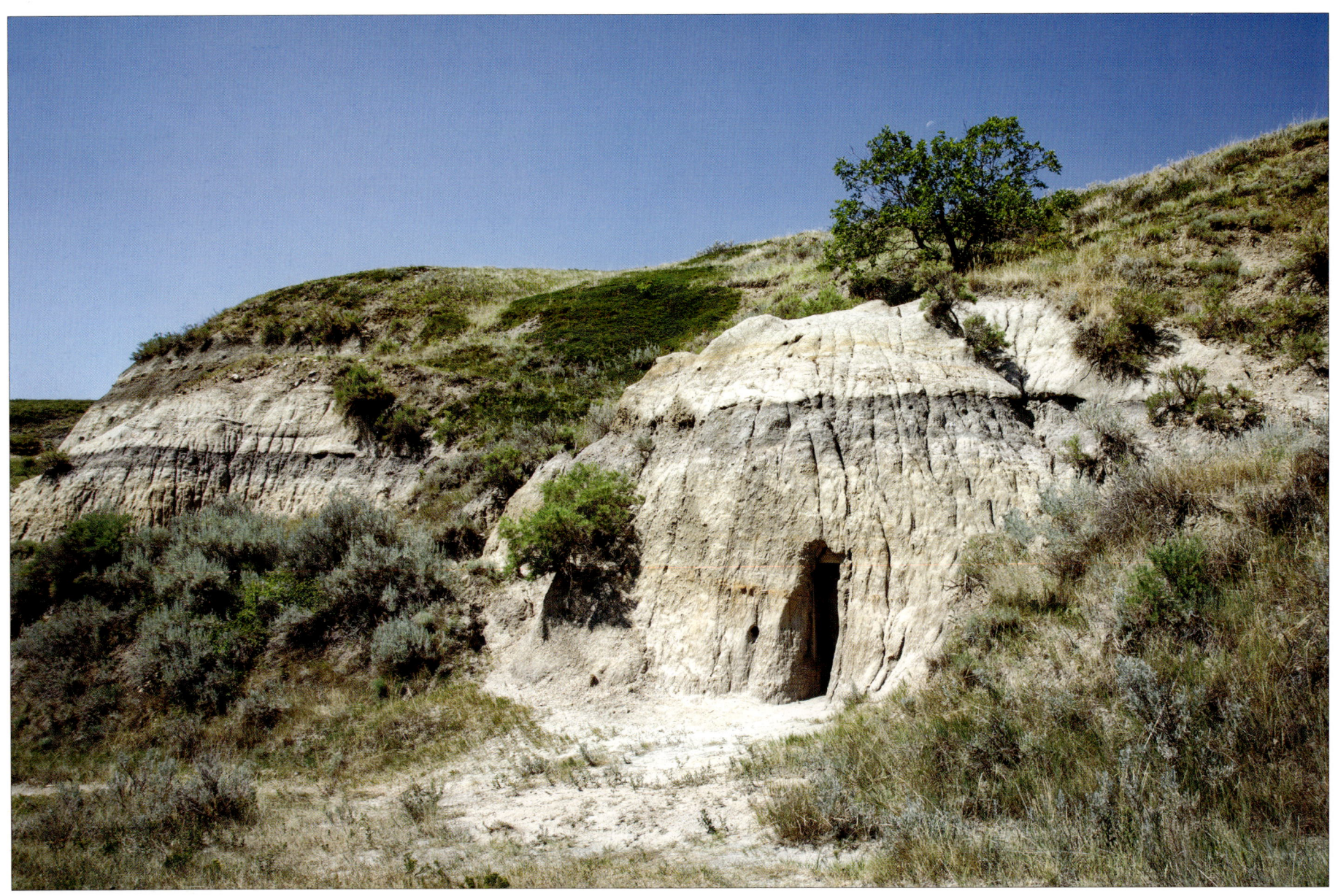

SAM KELLY OUTLAW CAVES

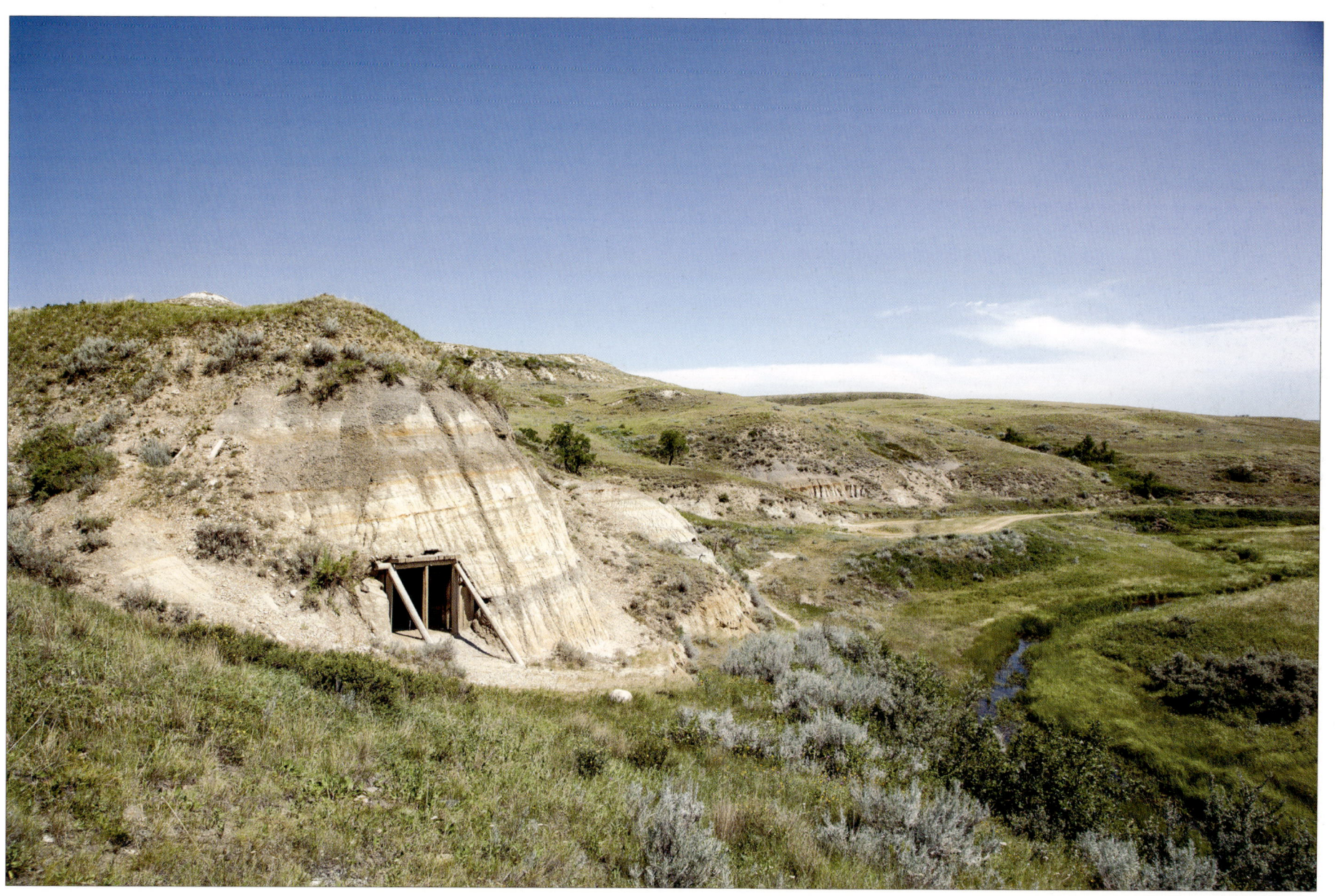

SAM KELLY OUTLAW CAVES II

31

ROCHE PERCEE

It isn't exactly clear when, but the naming of this collection of mysterious rock outcroppings, located along the lush Souris River Valley, just north of the Saskatchewan–North Dakota border, is attributed to the Métis sometime in the 1800s. La Roche Percee translates from French to mean simply, "the pierced rock" (the adjacent village is also called Roche Percee). "The Rocks," as locals call them, have been a prairie landmark for as long as anyone can remember. It's also a spiritual place for First Nations people.

According to *The Encyclopedia of Saskatchewan*, Assiniboine Chief Dan Kennedy said that his people camped at Roche Percee in the late 1700s and returned often to worship until the 1950s. Explorer James Hector, from the Palliser Expedition, visited the area in 1857, and the Northwest Mounted Police (NWMP) camped at The Rocks during their westward trek across the country in 1874. The village of Roche Percee was established soon after when the first coal mine opened in 1891. The banks of the Souris Valley near the village are dotted with interesting outcroppings, especially the north bank. Other sandstone formations can be found further east along the valley as well, but the feature rocks at Roche Percee are made up of three groupings located just southeast of the village.

As their name indicates, most of these rocks have piercings of various sizes due to erosion. The uneven cementing of the sandstone allows erosion to create pits, which eventually erode into a hole or piercing. The first grouping situated adjacent to the road is a good example of this. There are several piercings and broken fragments that make this outcropping resemble a shattered fortress wall. While it looks impressive today, it was much more so in 1917. A postcard from the University of Alberta Library's Peel's Prairie Provinces collection (#11319) show this wall of rock located on the crest of a hill, substantially larger and with a gaping hole in the sandstone approximately two metres in diameter. Posing in front of the enormous rock is a congregation of 38 people dressed in formal attire—five of whom are standing in the huge eroded cavity and three who have actually climbed on top. They look as if they have just come from a church service. Unfortunately, this outcropping with its large piercing was shattered by lightning a few years later. Today it's a mere fraction of its original stature.

The second grouping is located approximately 15 metres to the east in a depression or hollow. Some of these rocks are concealed by fairly extensive tree growth so they are difficult to photograph. Nevertheless, there are some outstanding formations

RIGHT: FAMOUS LAST WORDS
Erosion at work: in a University of Alberta photograph, circa 1920, this sandstone outcropping was substantially larger and looked much like a fortress wall.

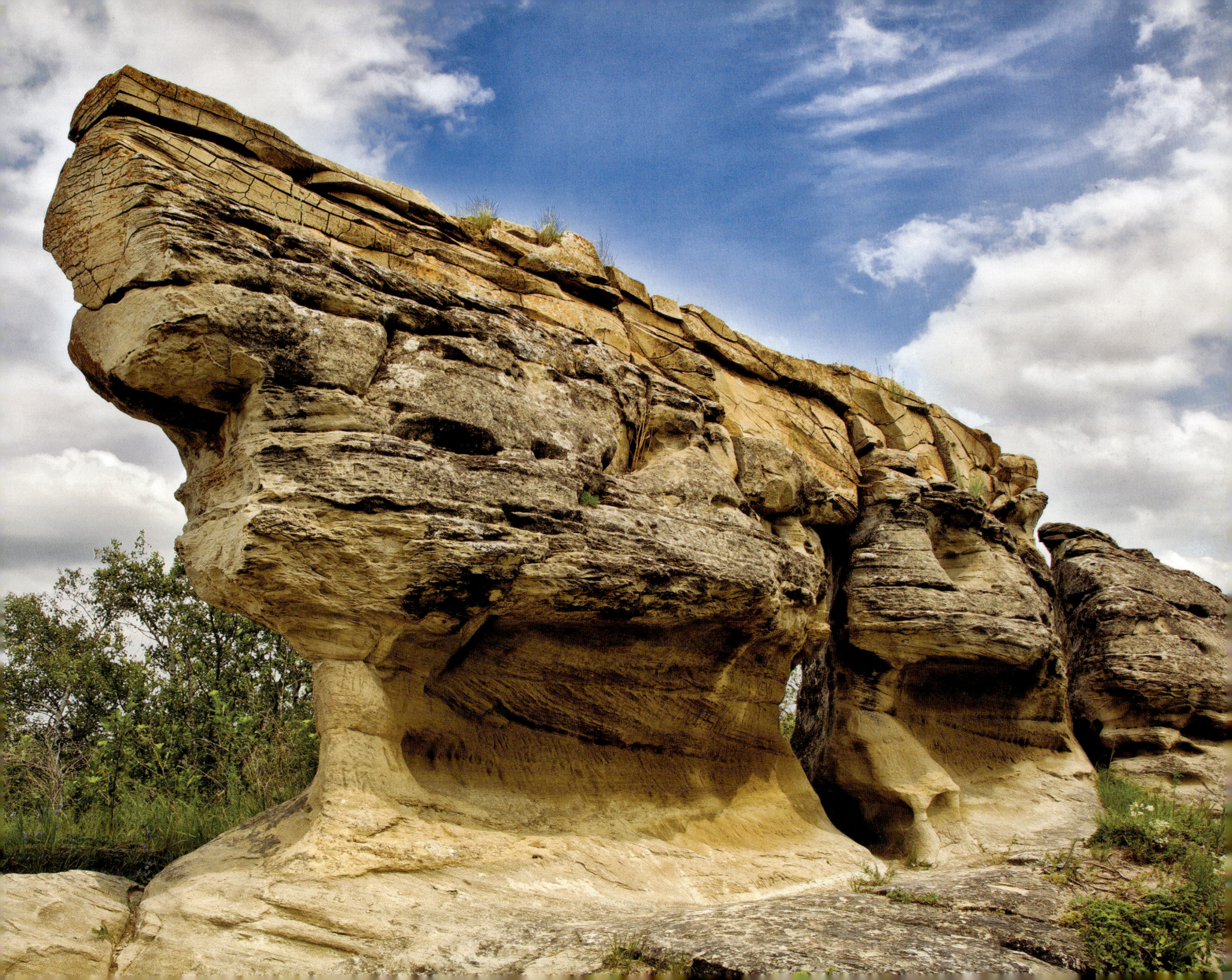

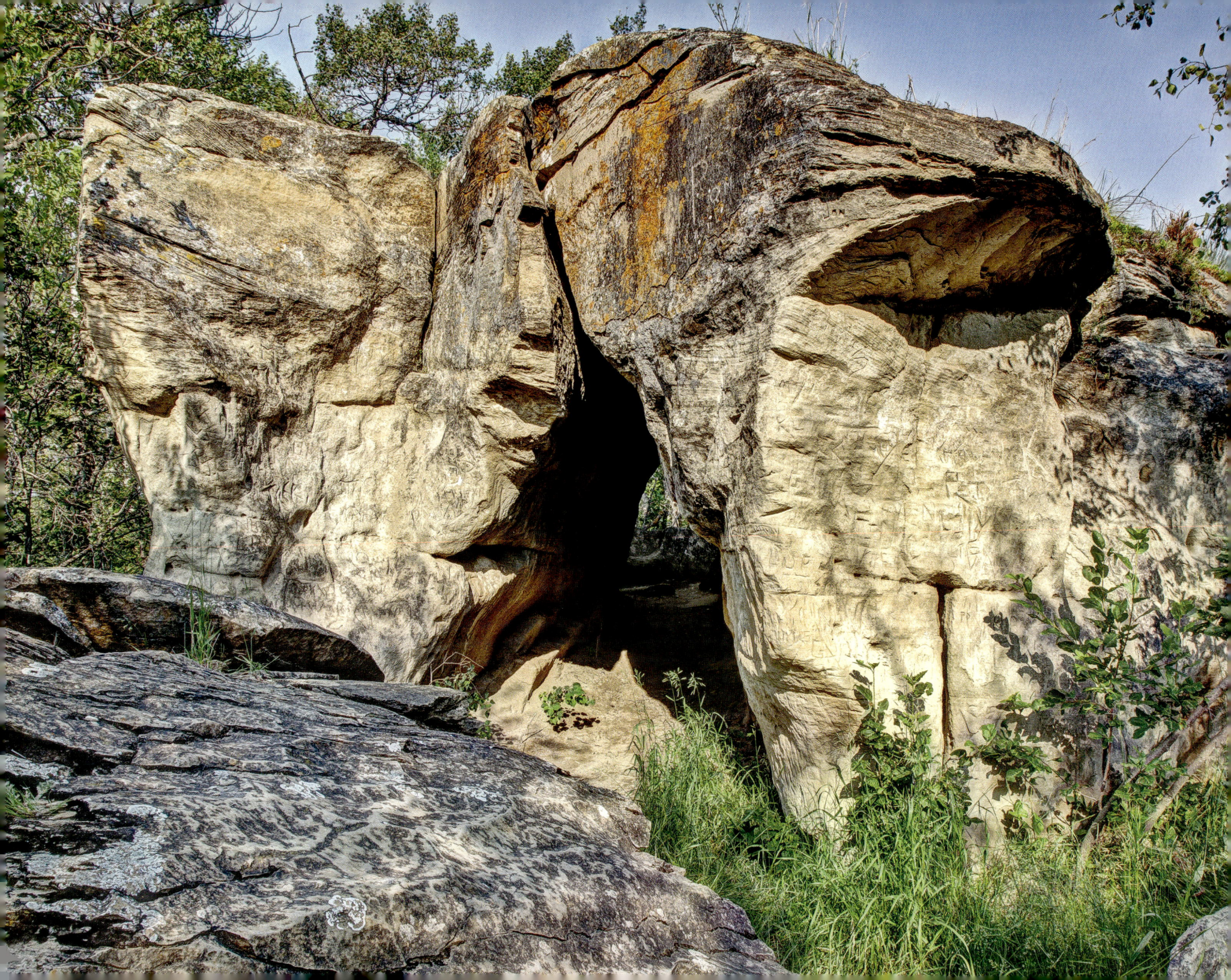

found here, including some with shallow caves (see photograph "The Archive" page 34).

The third grouping is located approximately 50 metres still further to the east. The photograph "A Mixture of Frailties" (page 36), named after the novel by Canadian author Robertson Davies, perfectly showcases this grouping. It is arguably the most picturesque, since the large and dramatic "Deadhouse Gate" archway is located here (see page 38). It looks like a gateway into another world, and in some ways it is. The rock is covered with names and marks from the past. In fact, according to the diary of Henri Julien, who travelled here with the NWMP in 1874, the rocks at Roche Percee had numerous symbolic Indian petroglyphs etched into them— similar to those in Writing-on-Stone. The NWMP followed suit and left their marks. Early settlers did the same and, unfortunately, visitors today do so as well. So much so that some of the rocks read like a guest book written in stone.

Although the rocks at Roche Percee are sandstone formations, which are one of the main geological characteristics of any badlands, they can't be designated as a badlands region because they are so thinly distributed throughout the valley. Nevertheless, these strange pierced rock outcroppings and their alluring caves attract attention. They defy logic. They look as if they have somehow been dropped and placed incongruously along the verdant banks of the Souris Valley. Even the devastating effects of erosion have produced two oddly contrasting results. While some of these formations have demurely wilted into flattened mounds, other rocks are jagged and pierced, courageously battling the elements.

This battle is especially poignant when you travel through the village of Roche Percee today.

While the wind and the rain slowly eroded The Rocks over centuries, nature decimated this village of close to 170 people in a matter of days during a flood in 2011. The village was, for the most part, submerged. Over half of the homes in the community were destroyed. The village looked like a deserted ghost town when I passed through it. The flood is reminiscent of the sudden lightning strike that fractured The Rocks early last century, and is a very harsh reminder of the power of nature and how quickly the landscape can change.

Notwithstanding, the rocks at Roche Percee are instilled with mystic chronicles. They are an archive where stories have been exiled. And when you stop and listen to the wind swirling and enveloping these remarkable transfigurations, you can almost hear the enchantment and the sorrow.

LEFT: THE ARCHIVE
Much like Writing-on-Stone in Alberta, the rocks at Roche Percee are considered a sacred place by First Nations people.
According to Assiniboine Chief Dan Kennedy, his tribe never passed these unique pierced rocks without leaving an offering.

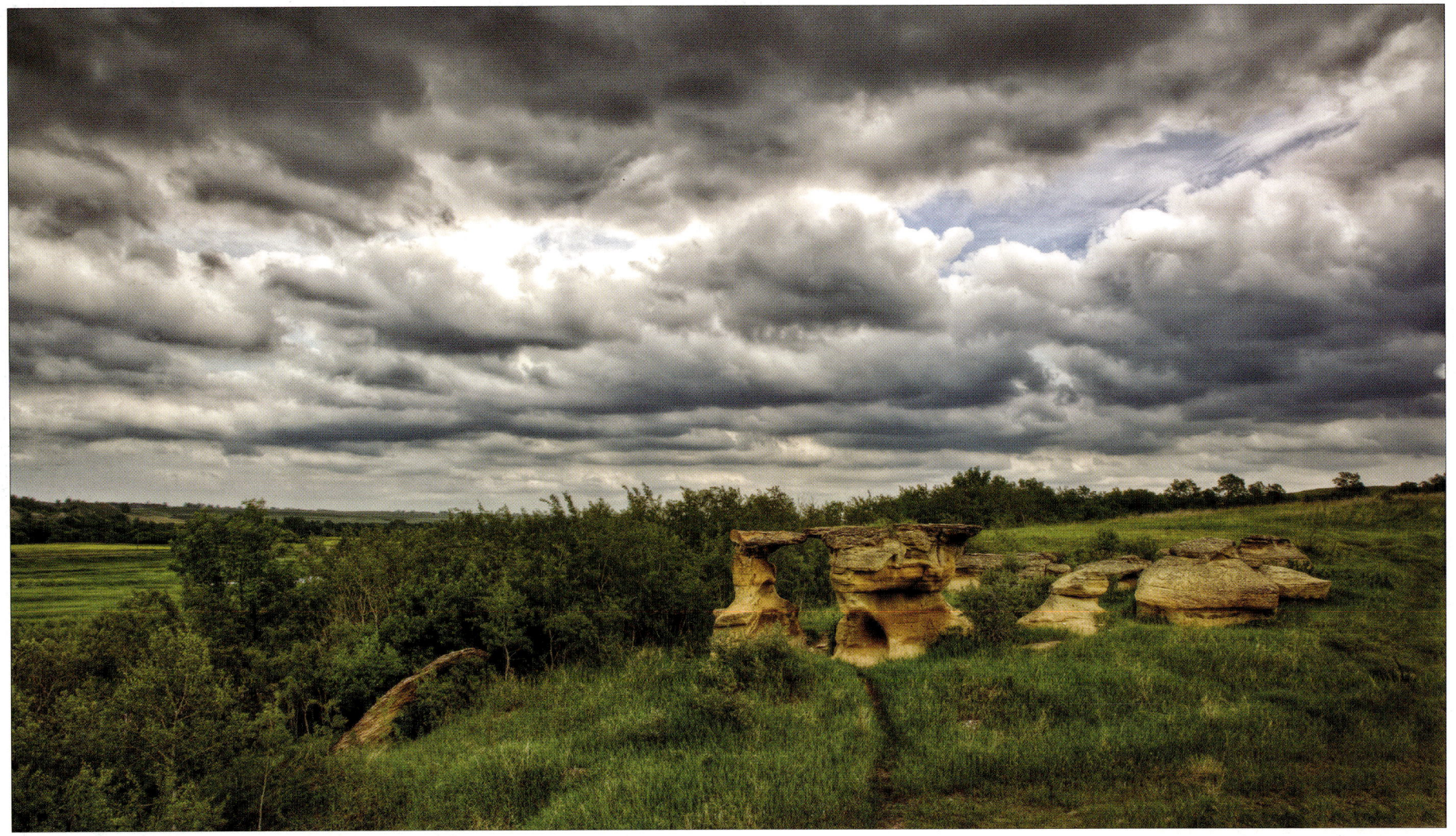

ABOVE: A MIXTURE OF FRAILTIES
These outcroppings overlook the Souris Valley. Remarkably, erosion has produced some rocks that are jagged and battle-scarred while others are smooth and flattened.

RIGHT: THE LOST AND FOUND STORIES
Graffiti found carved in soft sandstone.

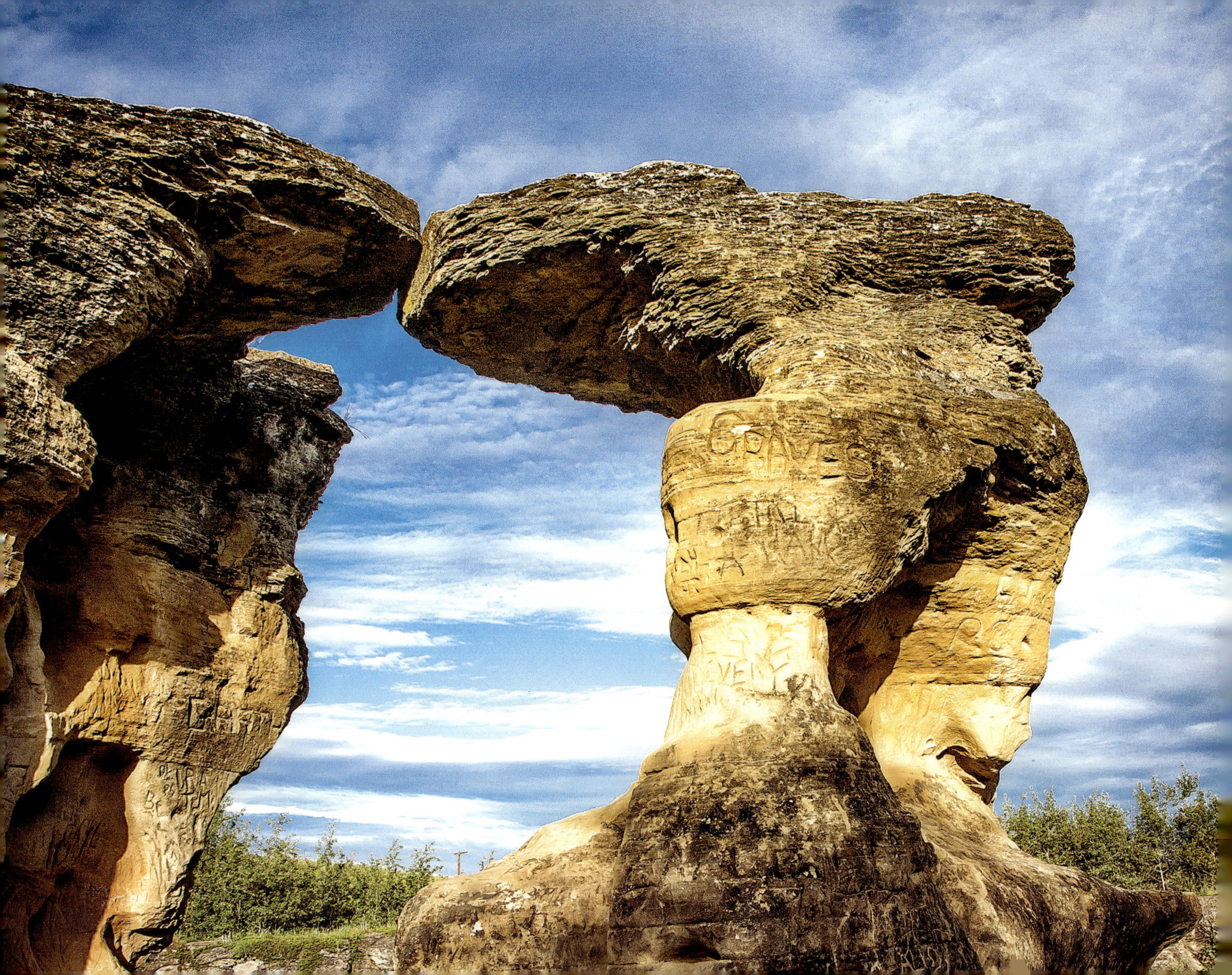

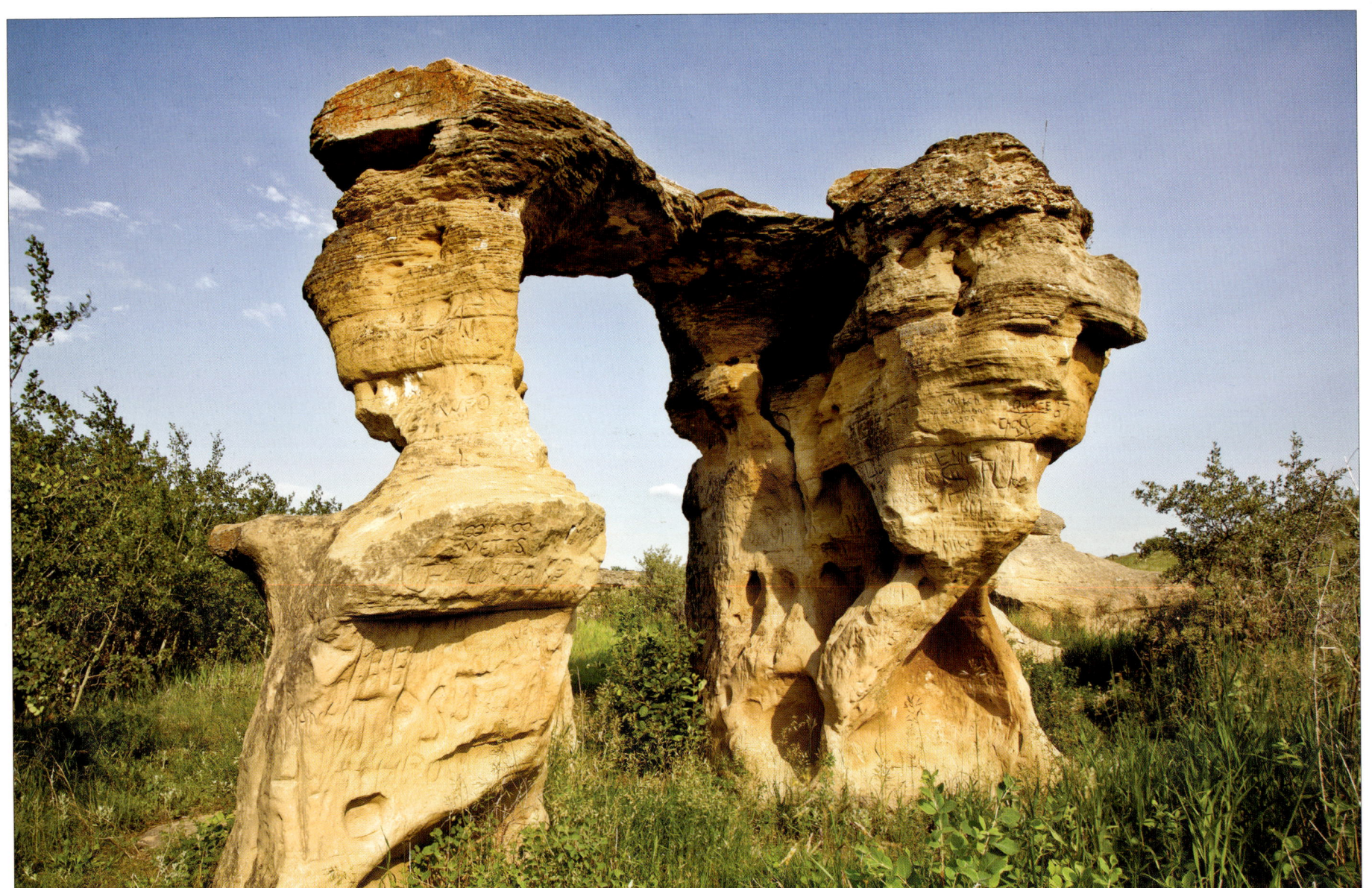

DEADHOUSE GATE
Reminiscent of a triumphal archway from ancient Rome, this abstract stone structure, with its strange twists and crevices, acts like a memorial, commemorating some unknown conquest. It is actually two separate formations that touch at the top.

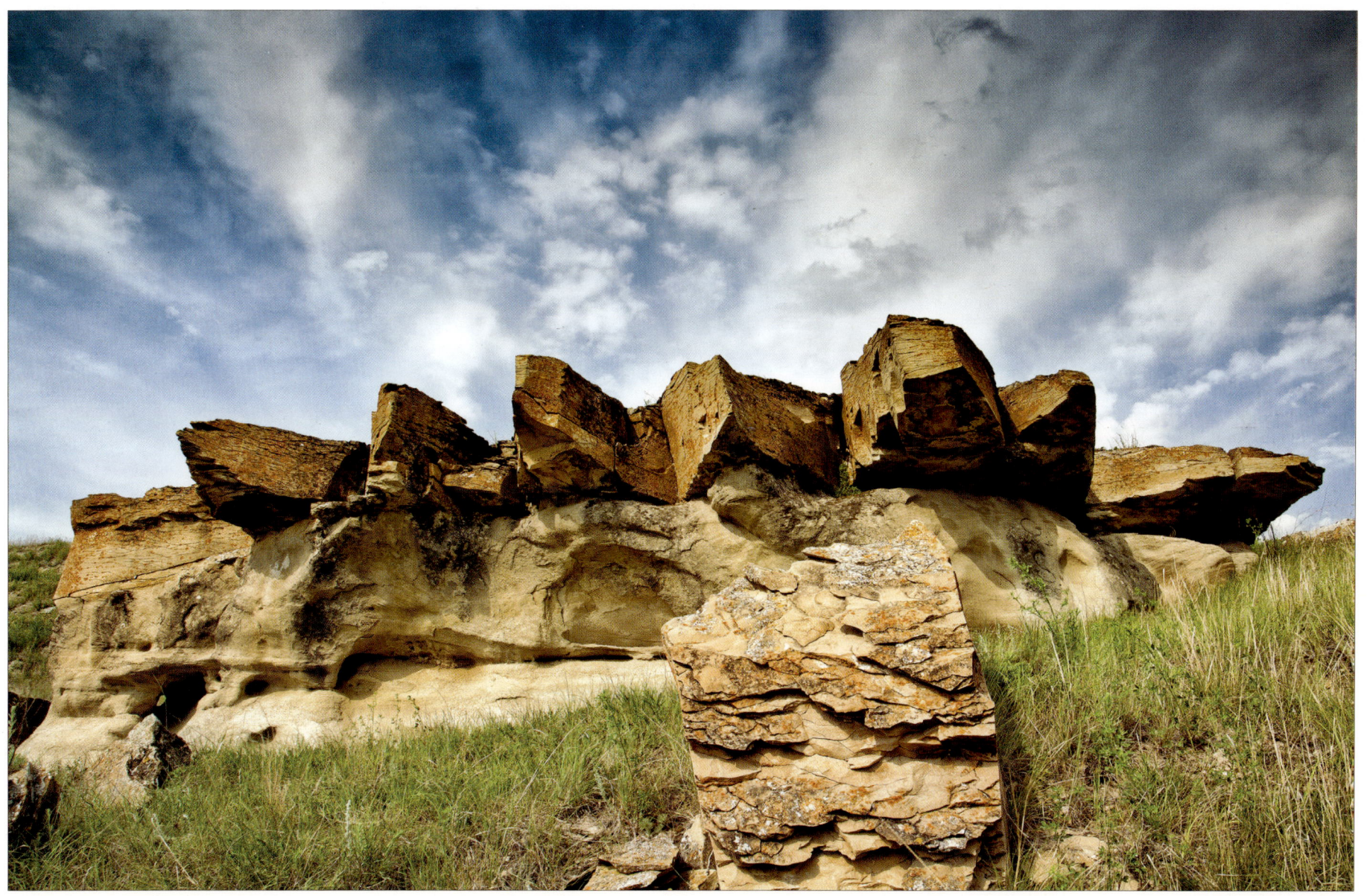

ANCESTRAL CROWN
The radiating clouds enhance this already regal-looking rock. If you look closely at the bottom left of the rock you can see a small hole, or piercing, where the light shows through.

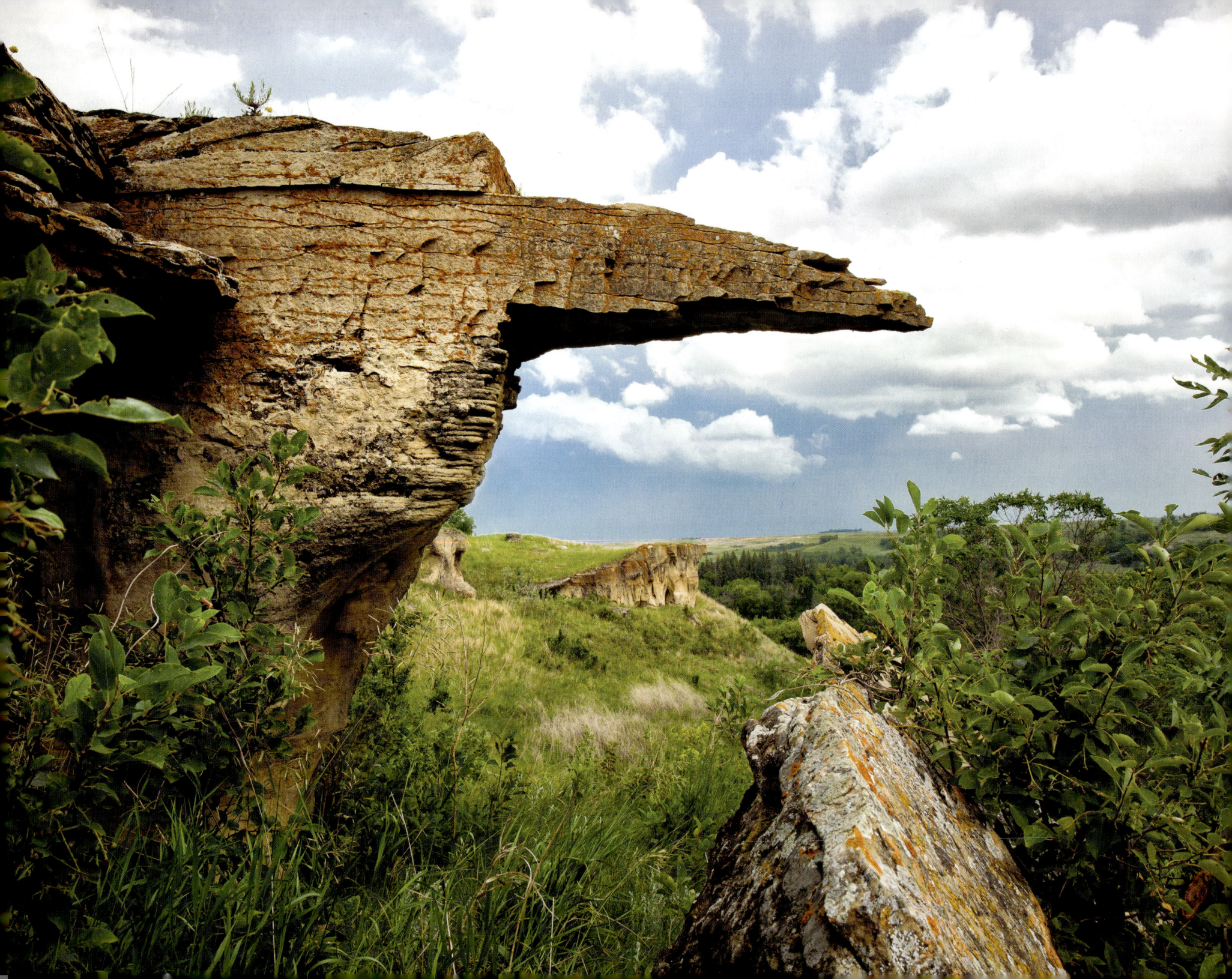

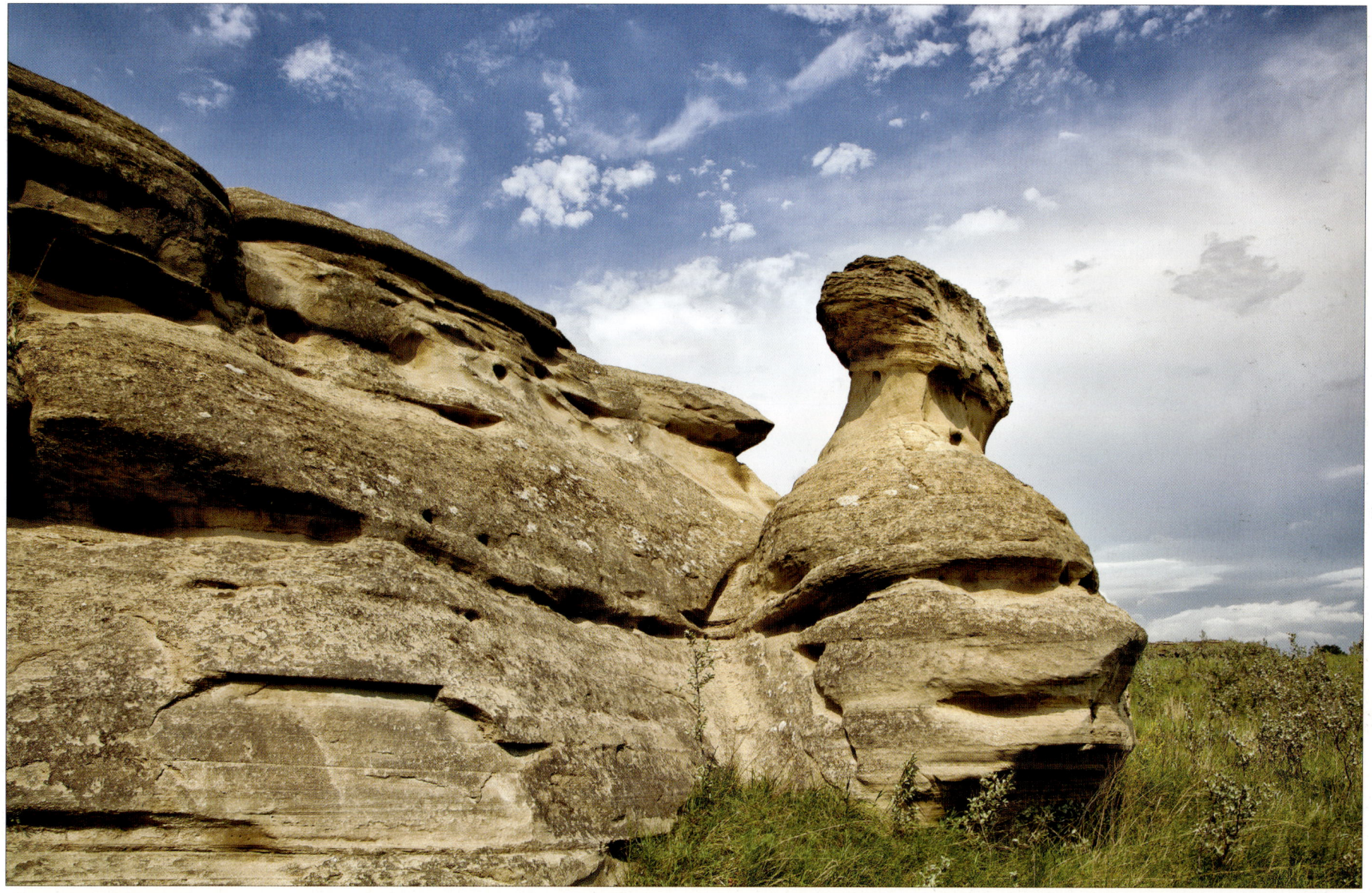

LEFT: THE TELLING OF LIES
Located on the north side of the Souris Valley, this pointed rock and what seems a former section of it, frame the outcropping in the background.

ABOVE: THE STONE PRINCE
This outcropping with its haunting face-like features is reminiscent of those found in Writing-on-Stone in Alberta. Both regions are situated just north of the 49th parallel.

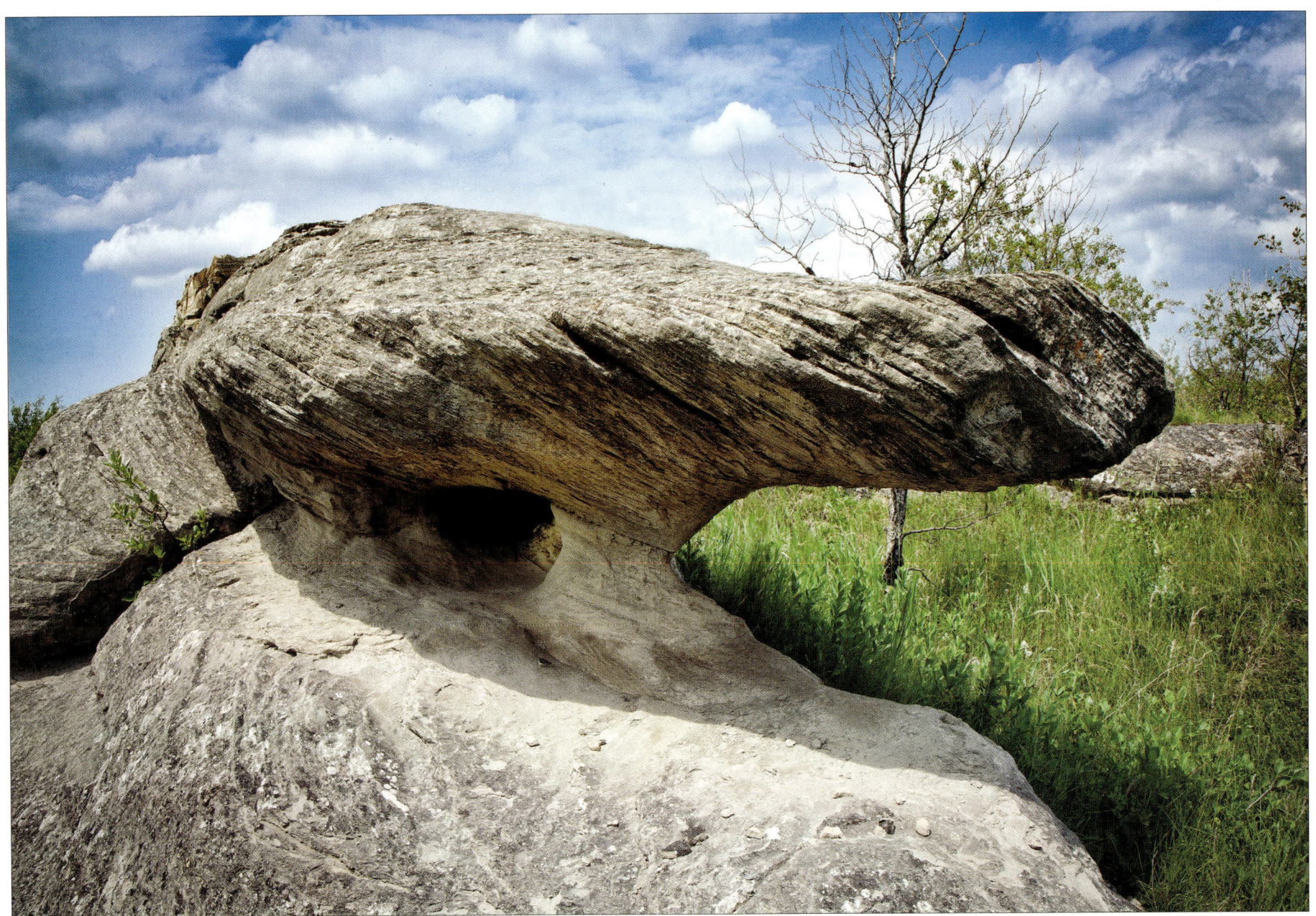

ENCHANTMENT AND SORROW
This photograph is a fantastic example of the many pierced rocks that populate the area and why the Métis named it Roche Percee (pierced rock).

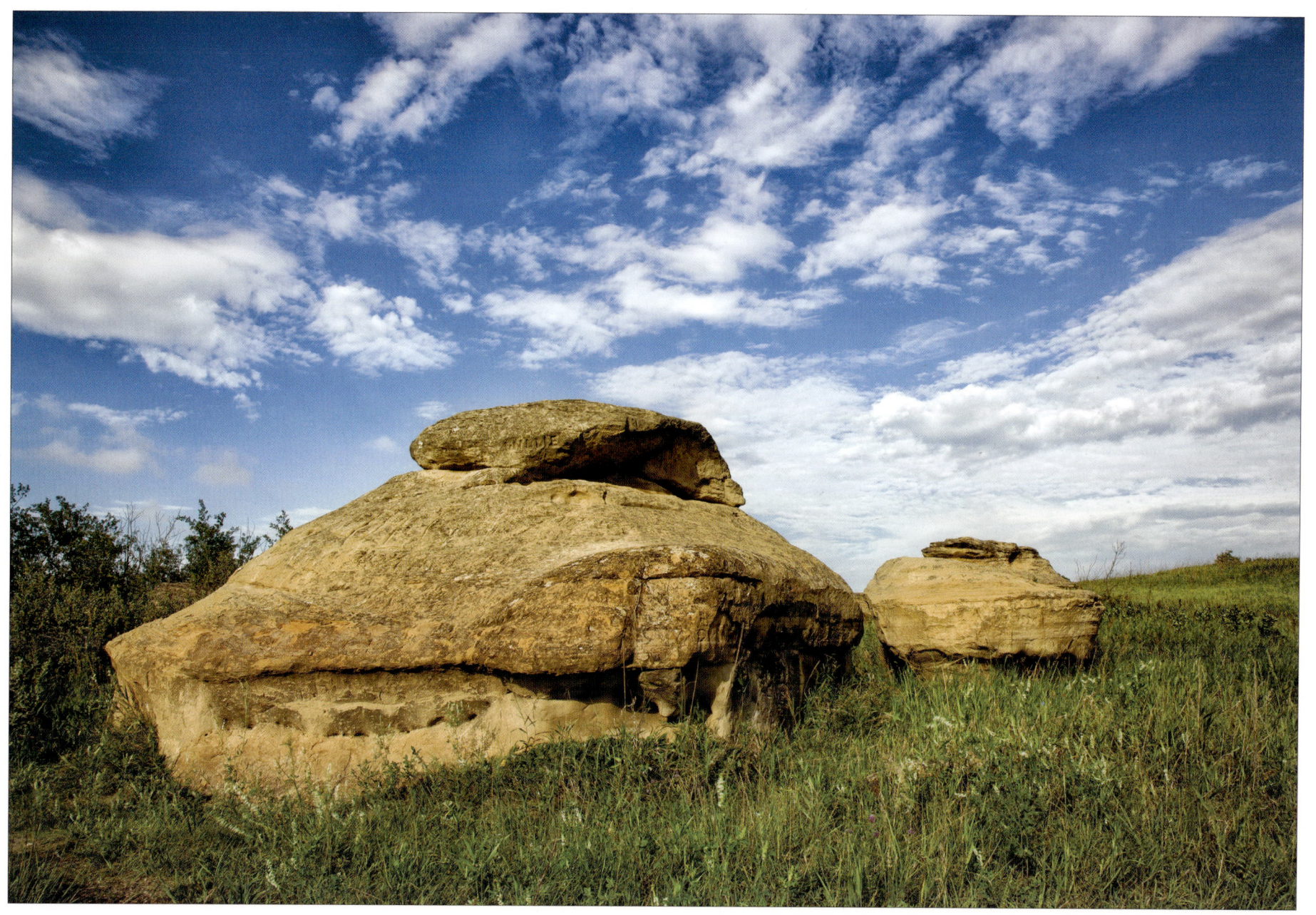

ANCIENT LINEAGE
This smooth and flattened mound is part of Roche Percee's third grouping of outcroppings.

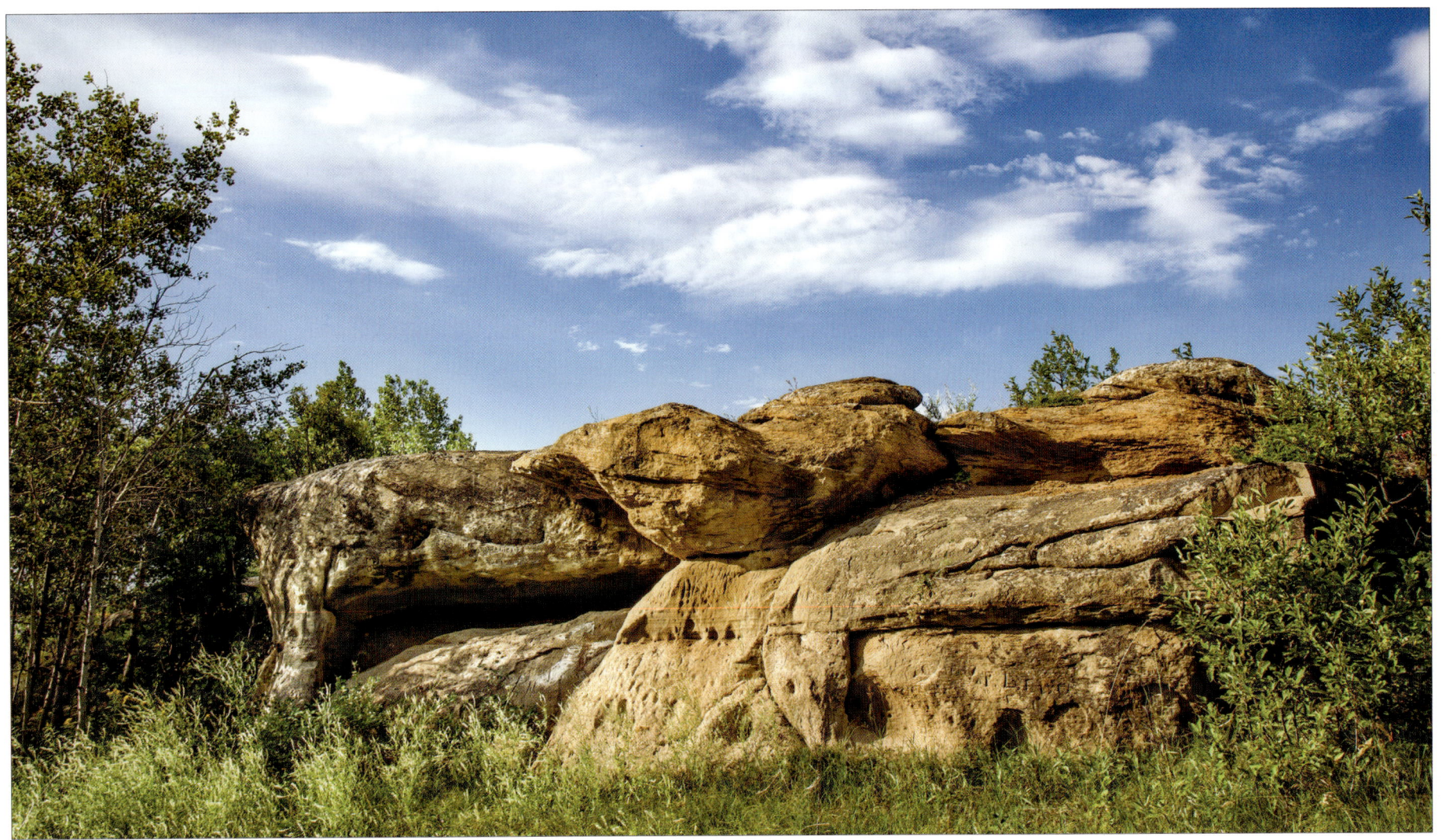

THINGS AS THEY ARE
This folded mass of sandstone is part of the second grouping of outcroppings. The crevice on the left runs quite deep and, once again, there is a small opening, or piercing, barely visible in the upper right portion of the rock.

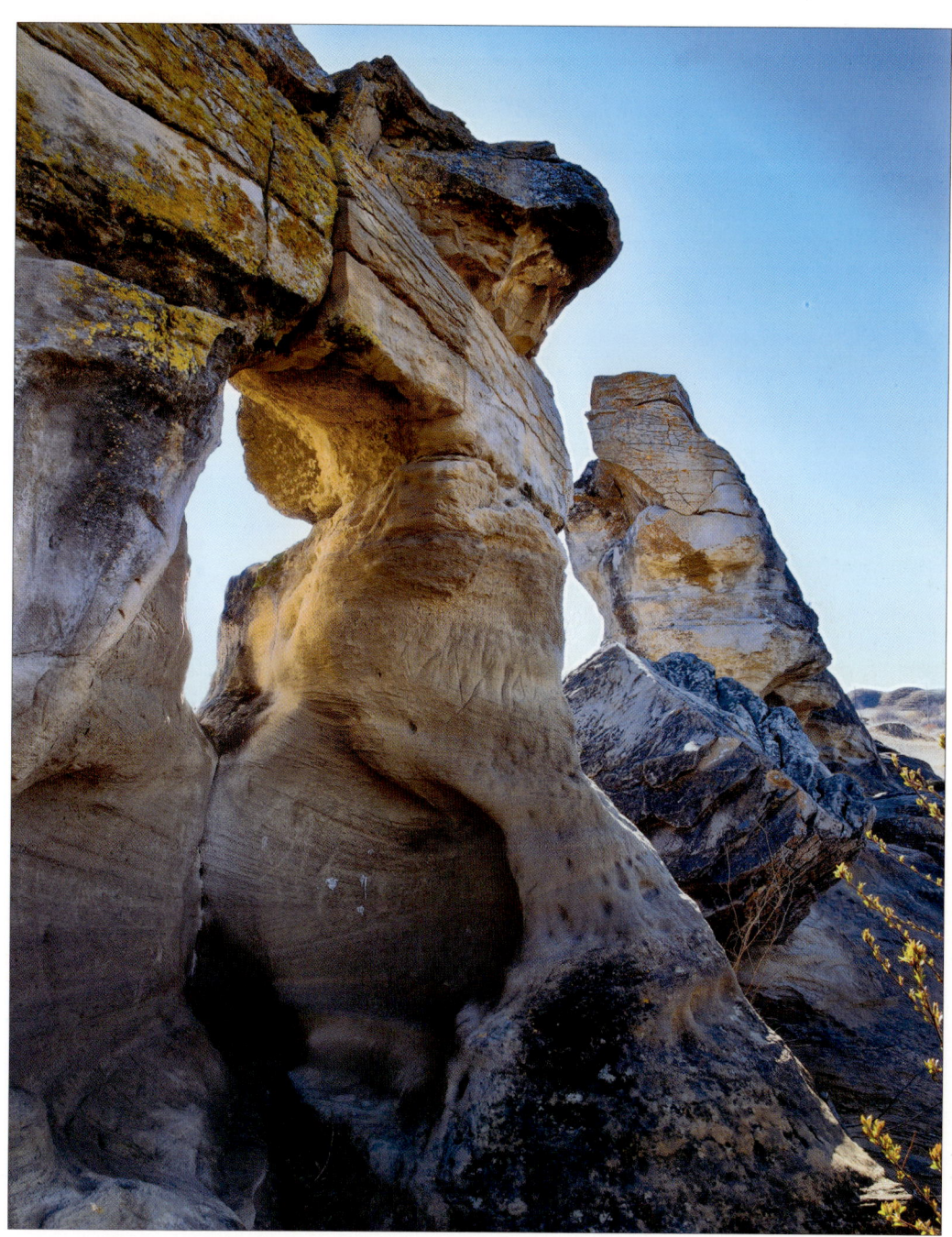

TRANSFIGURATIONS
From the first grouping of rocks. Shape, texture and light define this abstract torso, sculpted by erosion.

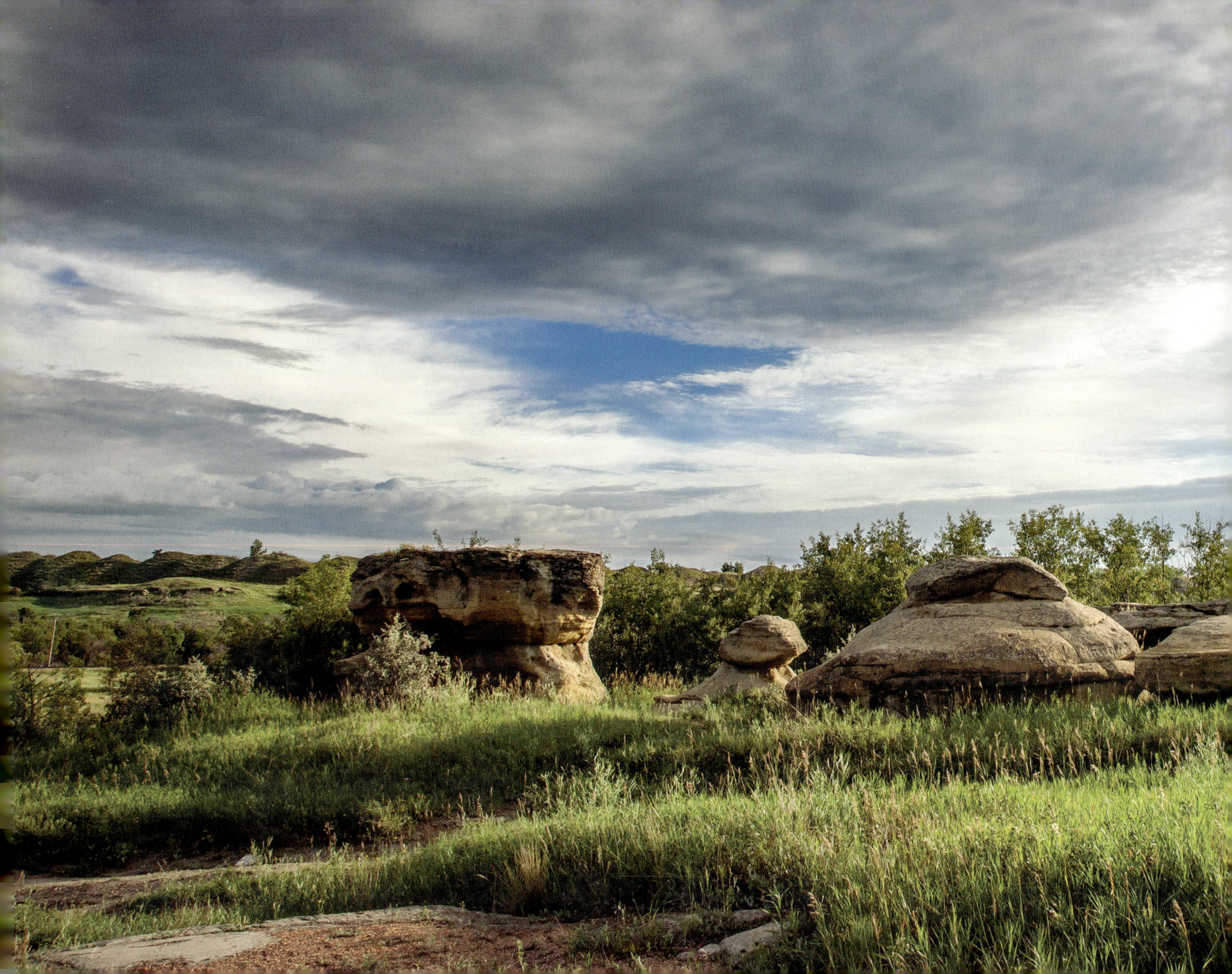

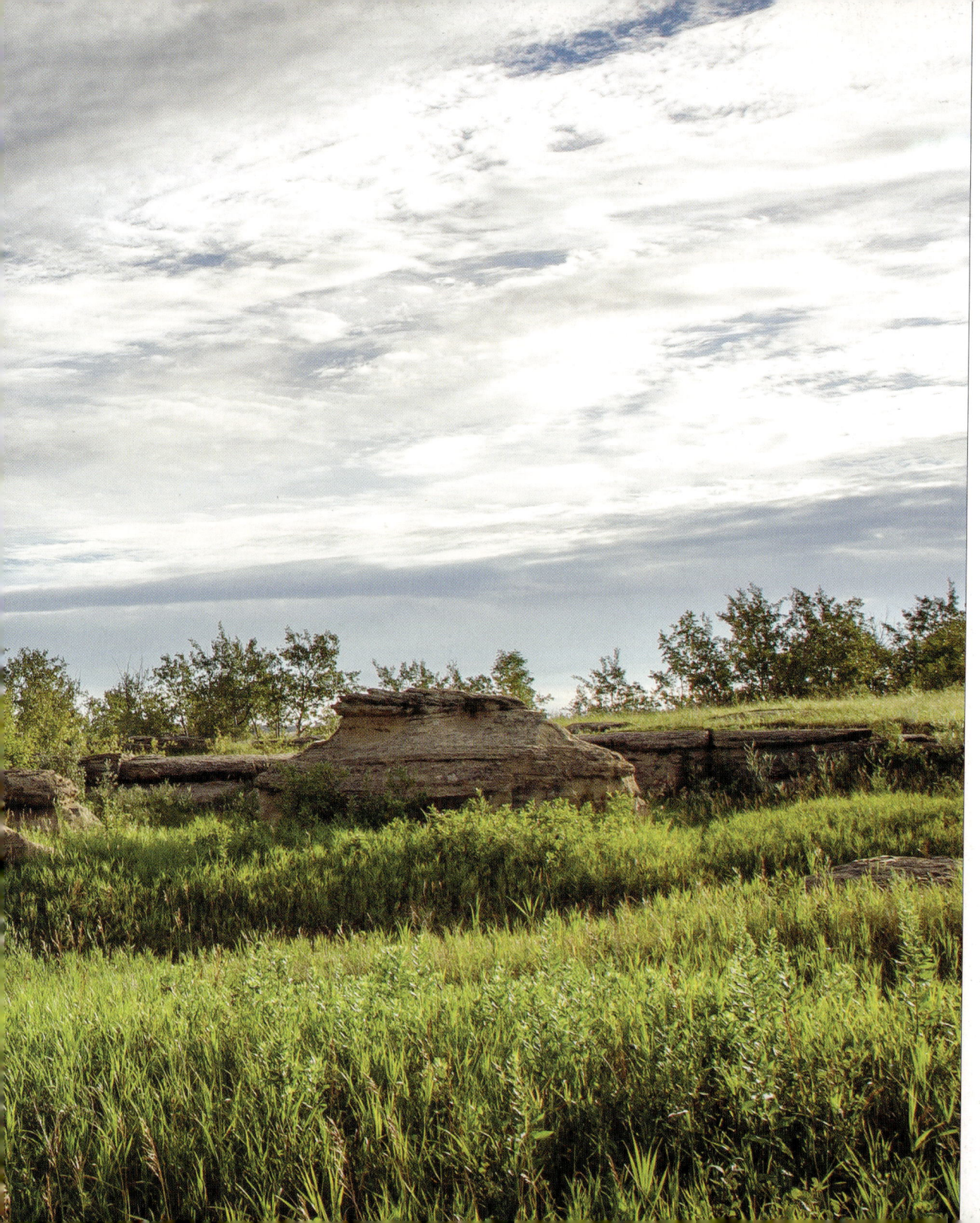

CREATION

These flattened mounds are completely out of character here. They look like they have been left behind and are waiting patiently for the artistry of erosion to turn them into something grand.

KILLDEER BADLANDS

Over the span of millions of years, this landscape has undergone a complete metamorphosis several times over. It was once an ancient ocean. Later, it was a tropical jungle inhabited by large carnivorous dinosaurs, like the well-known Tyrannosaurus rex and Edmontosaurus. In more recent times, sculpted by glaciers from the last ice age, this terrain has been shaped into an intriguing but desolate badlands landscape. Located along the southern fringe of Saskatchewan—in fact, the fringe of civilization—it is the most rugged badlands in Saskatchewan. Here, carved into the East Block of Grasslands National Park, and a coyote cry from the Montana border, free-standing rogue buttes randomly punctuate the terrain. The locals call them "dobies," which is an abbreviated version of adobe, or clay. Some of the buttes are grey and sterile; others somehow sustain a small cross-section of prairie vegetation. Intermingled between the buttes are exposed layers of strata and semi-arid terrain containing dry creeks and alkali bogs. While the surrounding Grasslands National Park is home to around 70 species of grass and over 50 species of wildflowers (Theodore Roosevelt National Park has over 400 plant species), the Killdeer badlands appear especially desolate.

There are no signs to direct you to the Killdeer badlands. The best way to find them is by driving south on Highway 2 for almost as far as you can travel before leaving Canada. Going west on a gravel road about 300 metres before the U.S. border, you will eventually see a sign that says, "Poverty Ridge Park Station." Further on is Dawson's Lookout, which provides a commanding view of the Killdeer badlands. If the weather is wet, proceed with caution. The dirt road leading into the badlands can get extremely muddy.

From the elevated vantage point at Dawson's Lookout, the eroded buttes of the fabulous Killdeer badlands stretch into the horizon like ancient pyramids on the plains. With few rugged stone outcroppings and even fewer cliffs, there is a natural fluidity to this landscape that evokes a subtle sense of calm. You might even get the impression that these badlands aren't really that bad. This assessment would be a mistake.

Once you venture forth and immerse yourself in these badlands, you very quickly acquire a solid understanding of just how difficult and deceiving they can be. First of all, there are no posted trails here. The only pathways are those made by deer,

RIGHT: SCAR TISSUE

On this particular overcast day I explored the Killdeer terrain but the light was uncooperative. Tired after climbing this butte, I set up my camera, composed the shot and waited. Remarkably, a narrow shaft of sunlight broke through the clouds. Finally, the light hit the peak and I knew I had the shot.

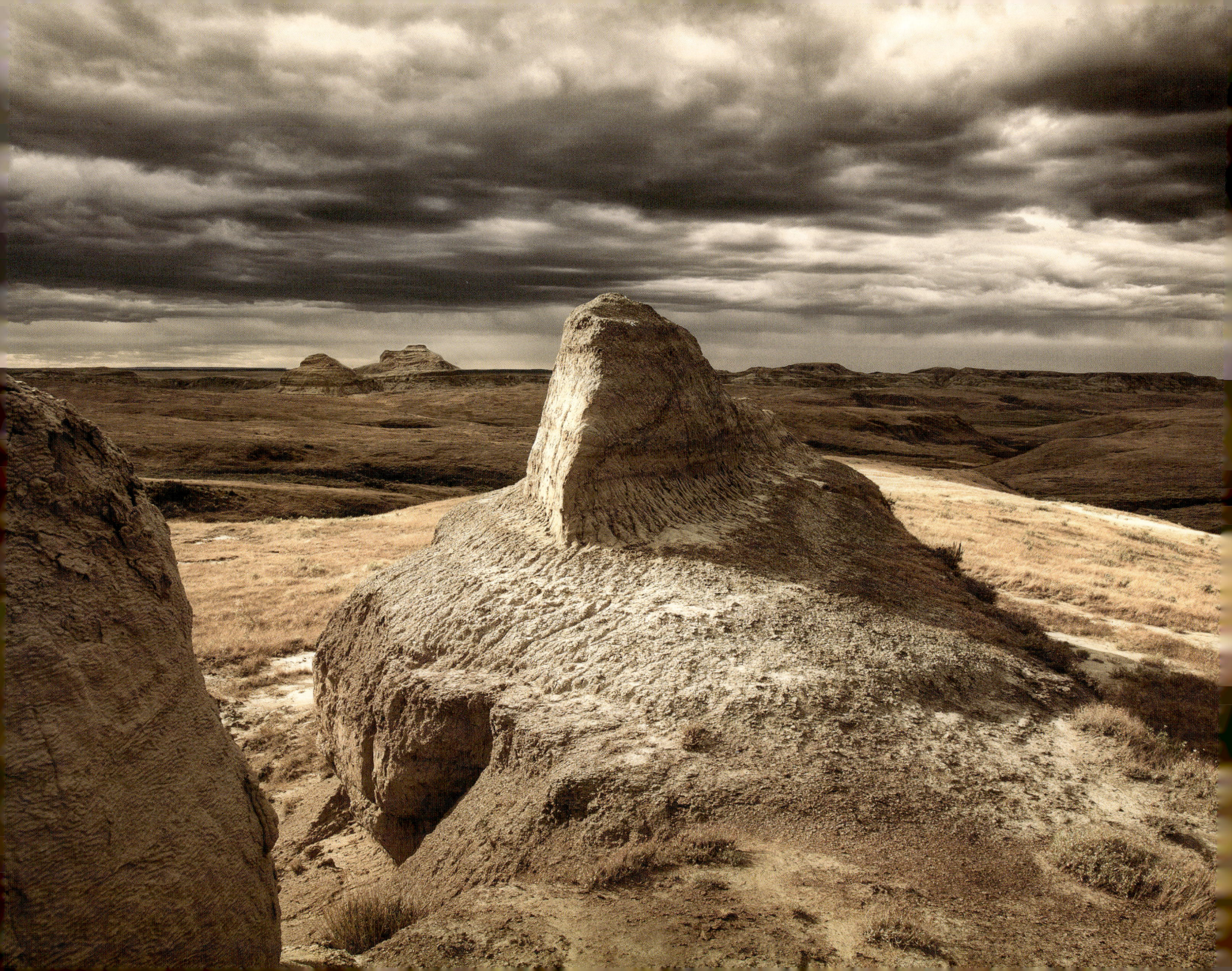

and these invariably end in thick brush. Although the terrain looks fairly pedestrian from the prairie rim, it is very much furrowed with meandering creeks or runnels etched into the land like veins. These are frequently camouflaged by bushes or tall grass. Then again, what often looks like a creek may actually be a dry riverbed. It's tempting to cross, but these dry riverbeds sometimes contain quicksand. (Thankfully, the park has placed signs to identify this hazard.) You need to take stock of the land and plan your route accordingly before blindly pushing forward.

As difficult as the Killdeer badlands are to traverse, you are more than rewarded with what is always a visually stimulating adventure. During a guided hike along the dry Rock Creek coulee, we were shown what were purportedly the exposed bones of a Triceratops lying partially buried on the crest of a butte. Our guide used the almost imperceptible variations in the nearby barbed-wire fence to lead us to the location. On a separate hike, he took us to a grasslands area just northwest of the Killdeer badlands. While we stopped to rest and look around,

our guide began to rummage underneath some large rocks. As we watched him, unsure of what he was doing, we were shocked to see him pull out fragments of bone that he announced were human! We huddled around him in amazement as he continued to pull out more fragments, which were passed around and inspected. He told us that he had found these human remains about 20 years ago while hiking. At the time, he contacted the University of Regina, who dated the bones to be from the late 1800s. The fragments were then consecrated beneath a large rock.

While I was surprised to be shown human bones, it's not unusual to find dinosaur bones at Killdeer. The East Block badlands have one of the richest deposits of dinosaur fossils in Canada. The first dinosaur bones in Canada were discovered at Killdeer in 1874 by George Mercer Dawson, a Canadian scientist and surveyor who, at the time, was working with the International Boundary Survey. Not all of the history here is millions of years old. Grasslands National Park has a large number of First Nations archaeological sites, including an astounding 12,000 tipi rings. Keep

an eye out for a number of rare animals as well: black-footed ferrets, prairie dogs, burrowing owls, greater short-horned lizards, and rattlesnakes all reside in Grasslands. Seventy-one Bison were re-introduced into Grasslands National Park in 2006. Eight years later, that number has risen to 400.

Lastly, these badlands can be eerily isolating. In response to a soundscape study commissioned by the Park, *Canadian Geographic* magazine called Grasslands "one of the world's last great quiet places." This may be the case, but I found that when the wind blows across these badlands, the sound can be deafening. Grasslands is also one of Canada's darkest places and has been designated a dark sky preserve. This means there is next to zero ambient light to contaminate the night sky. Consequently, Grasslands National Park is a fantastic place for star-gazers and astronomers to converge and view the constellations. So, whether your desire is to discover stunning scenery, paleontology, natural or First Nations history, the Killdeer badlands is a spectacular place and well worth the time and effort to explore.

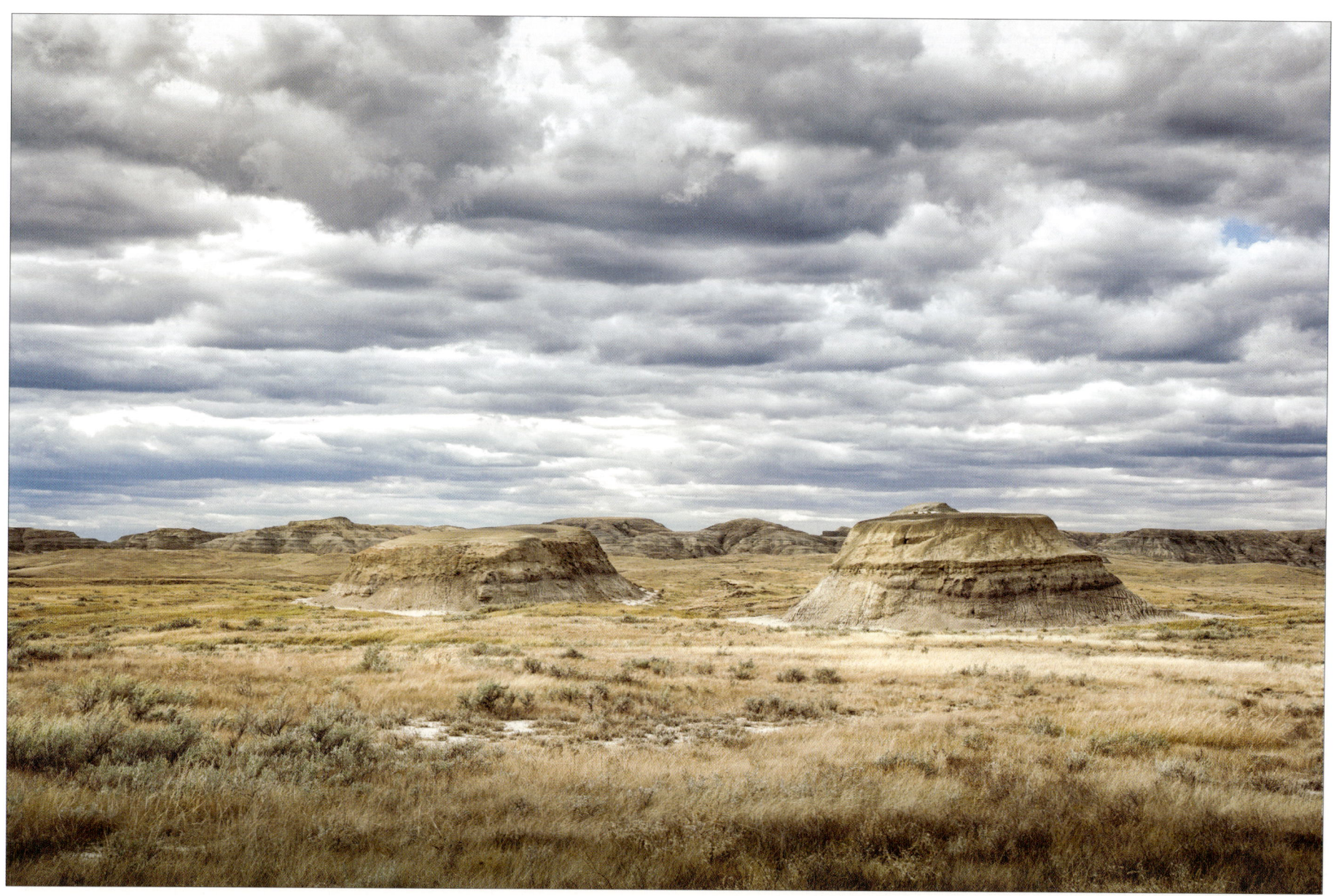

THE DISINHERITED
Big sky and open terrain. Two free-standing buttes look like weather-worn pyramids dotting the prairies.

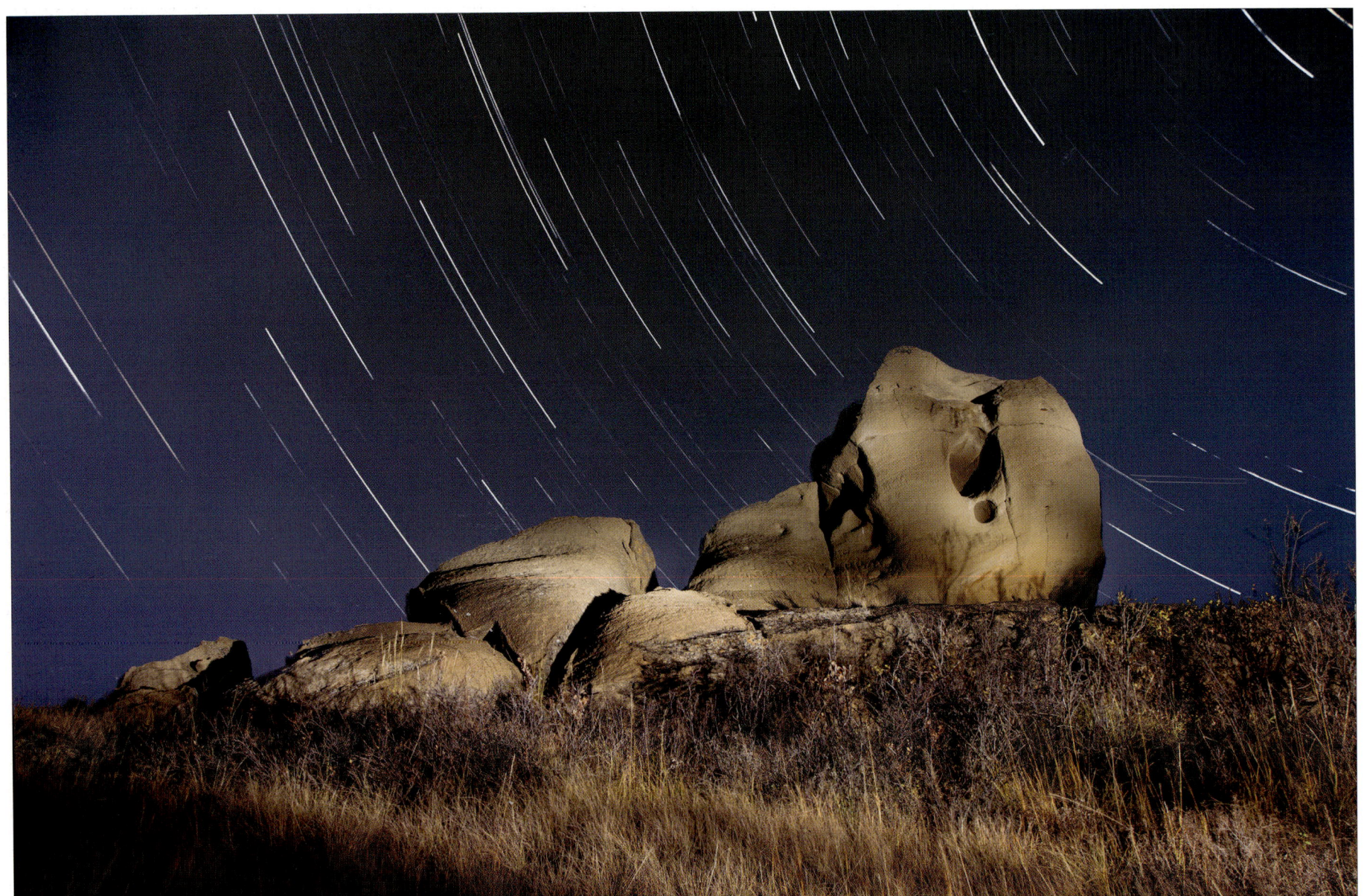

THE LIBRARY AT NIGHT

If you are in Toronto you might want to check out the Art Gallery of Ontario's collection of sculptures by the modern master Henry Moore. If you are in Saskatchewan, you might want to check out the Killdeer Badlands. It has an extensive collection of sculpted rock formations created by nature.

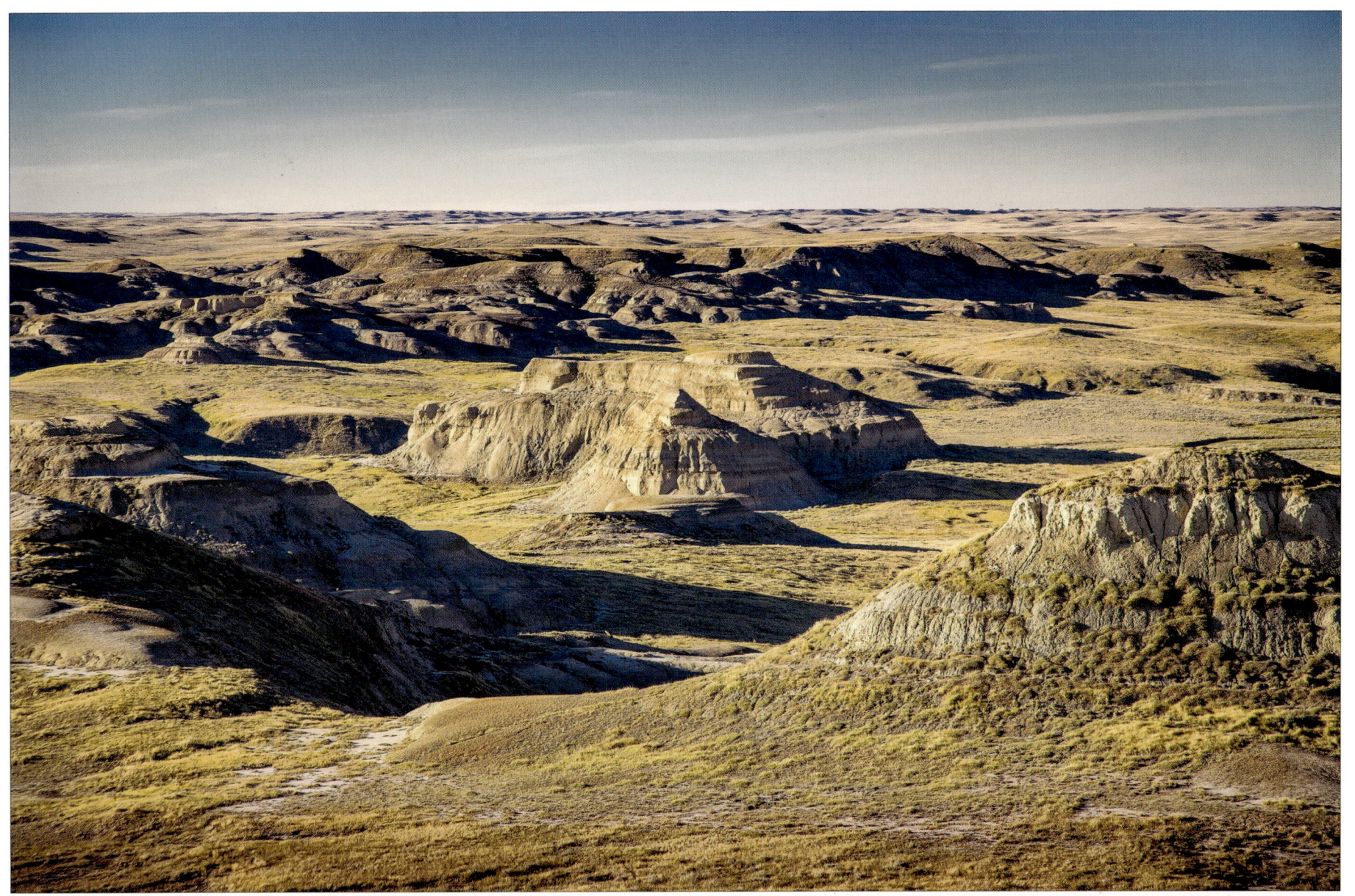

FORMS OF DEVOTION
The low sun setting in the west gives this image its golden glow. The central buttes are framed by the surrounding hills and highlighted by long shadows. While the terrain at Killdeer looks flat and unassuming, in reality it's quite severe.

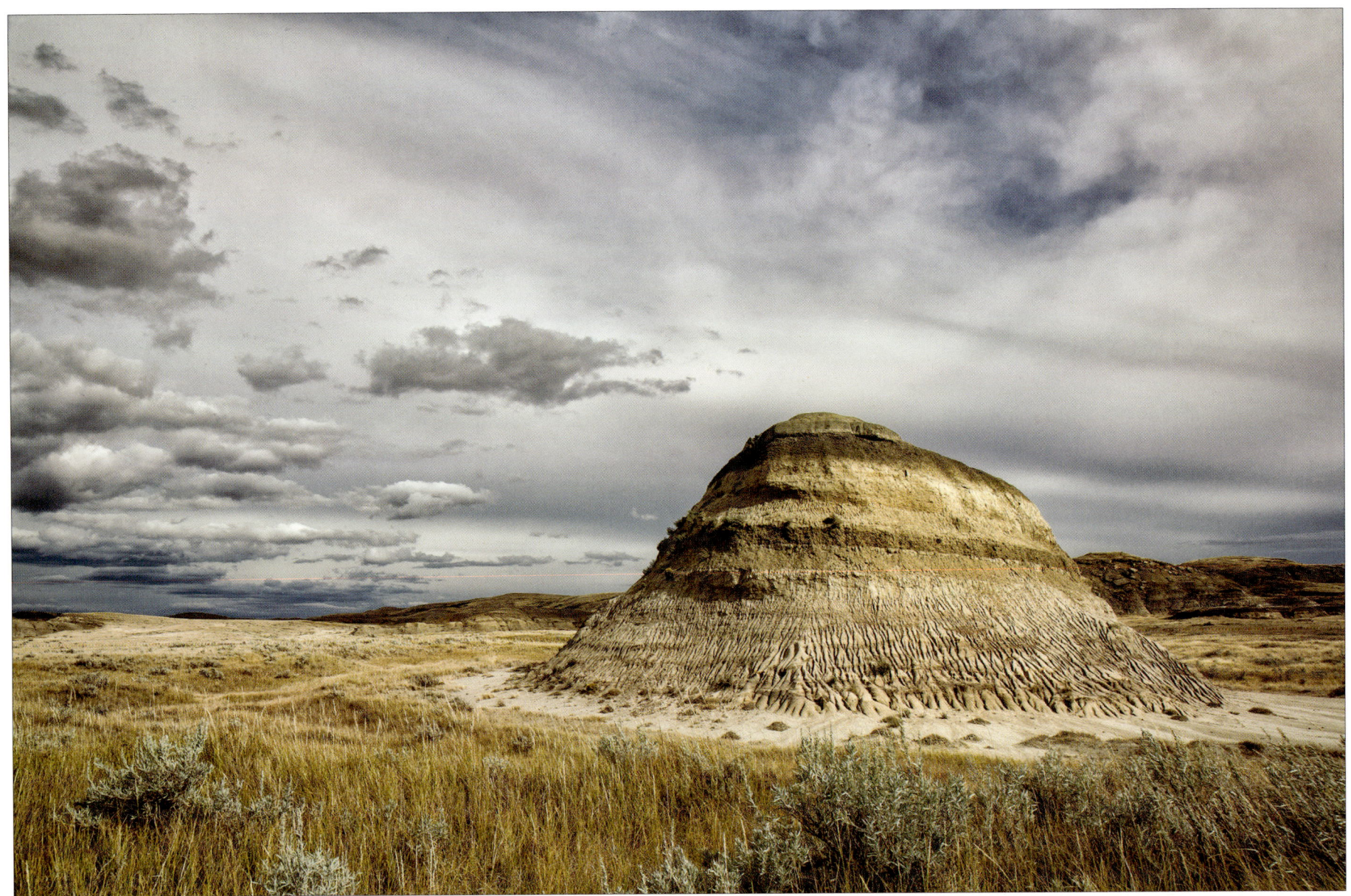

WILD MAN'S BUTTE
This is a great example of a rogue butte, or "dobie" as the locals call them.

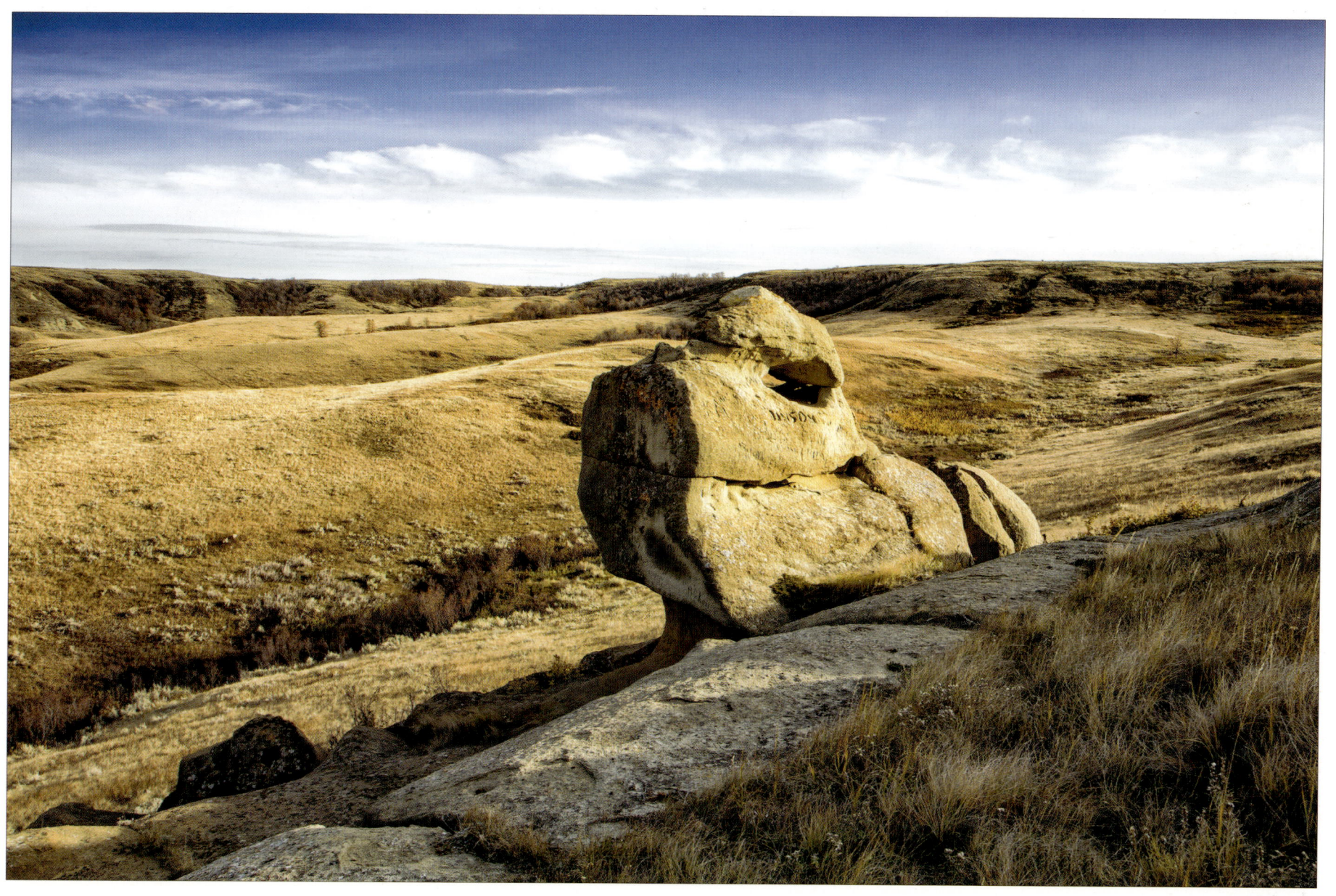

GARGOYLE

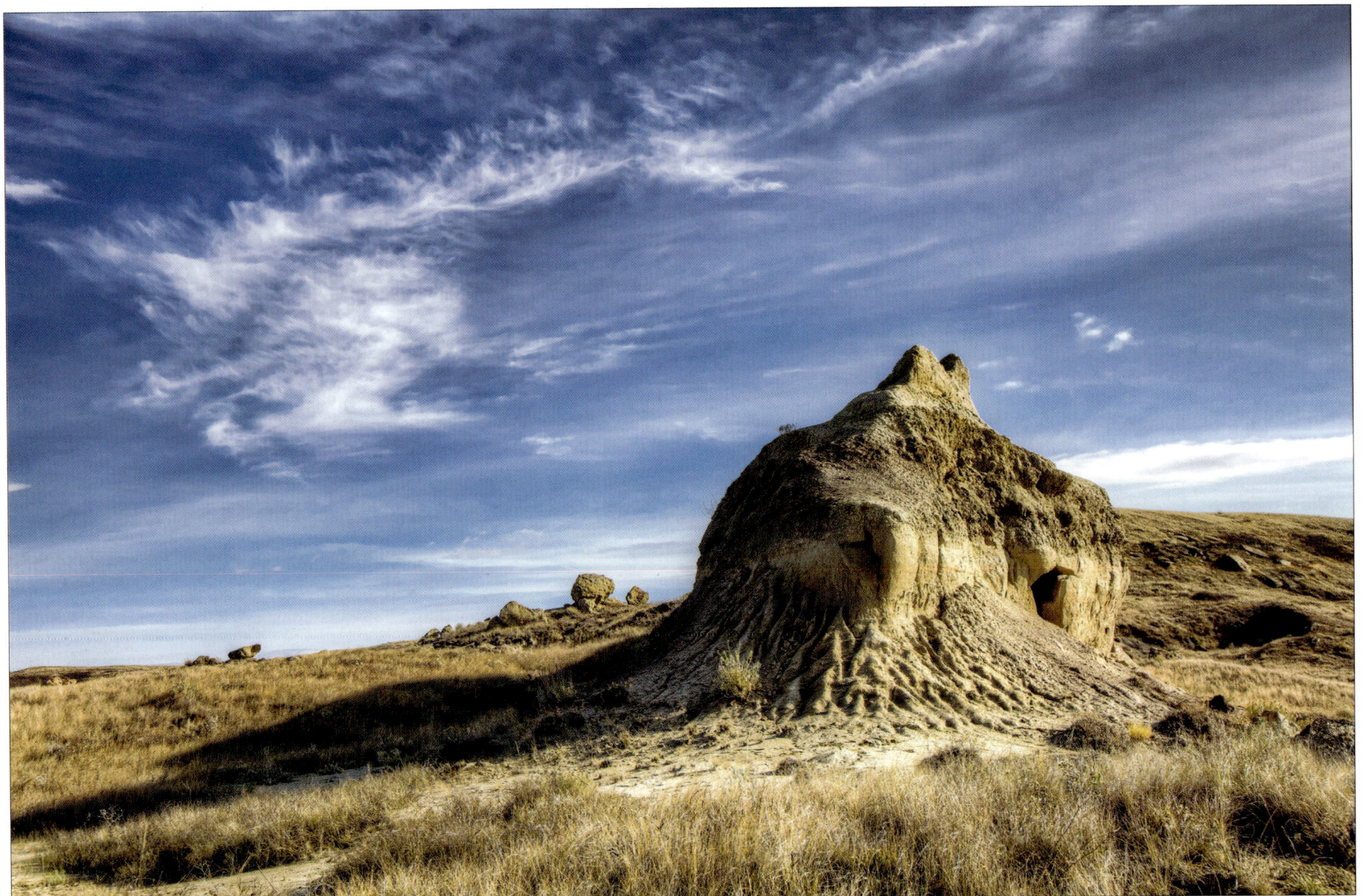

NOT WANTED ON THE VOYAGE
This odd-shaped butte reminds me of a ship that has run adrift. The hatch is about to open and its occupants are about to emerge.

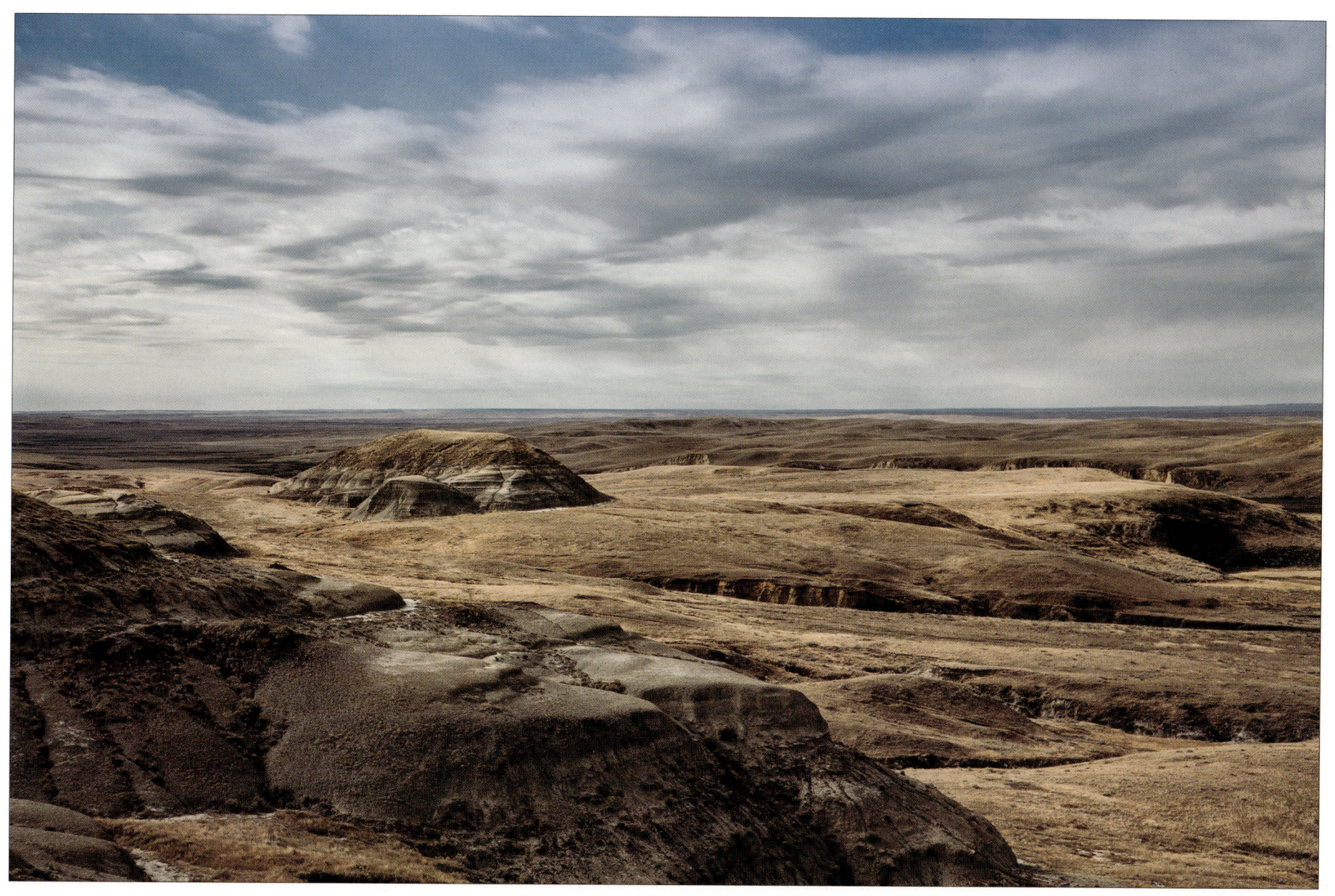

YEAR OF THE FLOOD

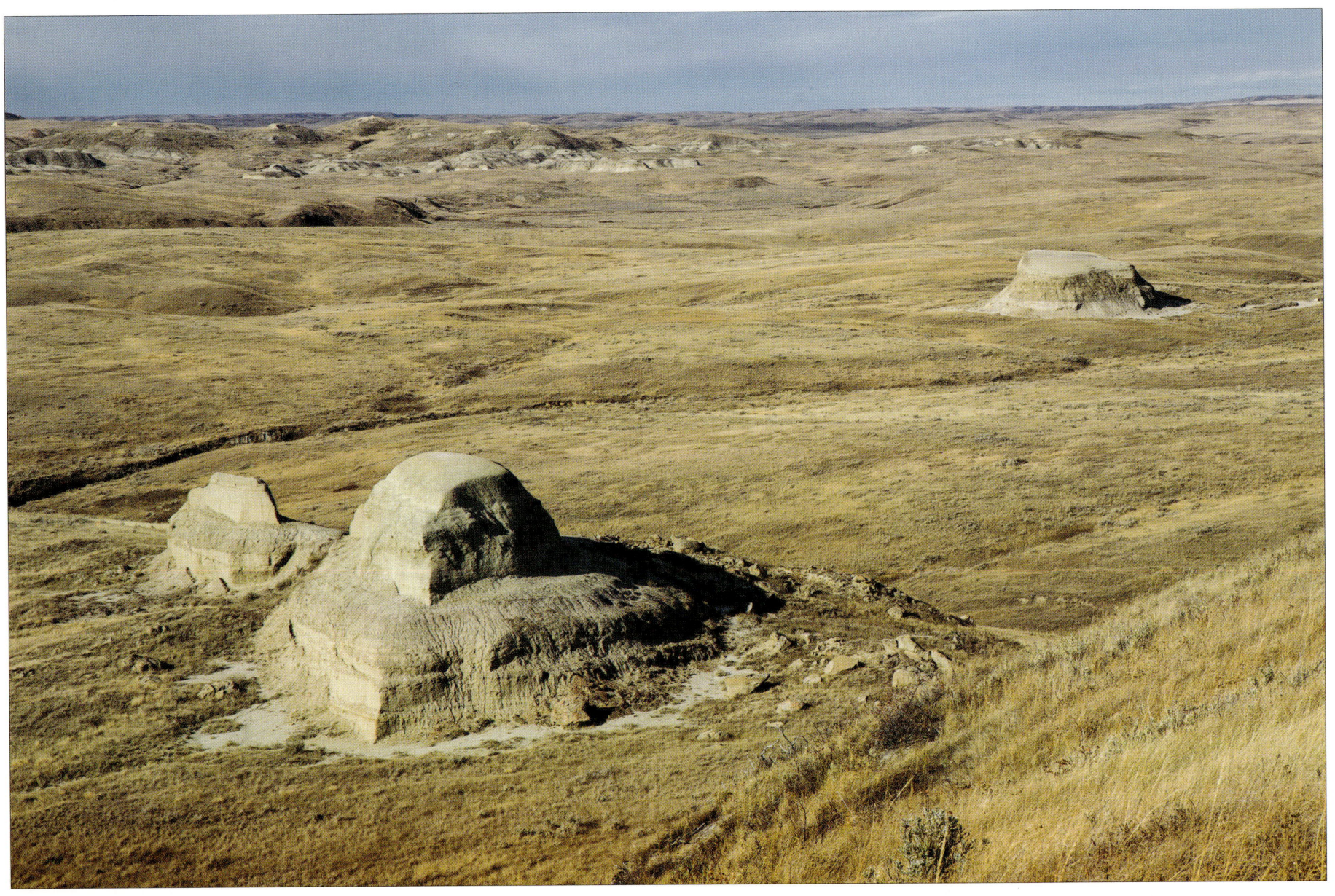

GREAT DEPRESSION

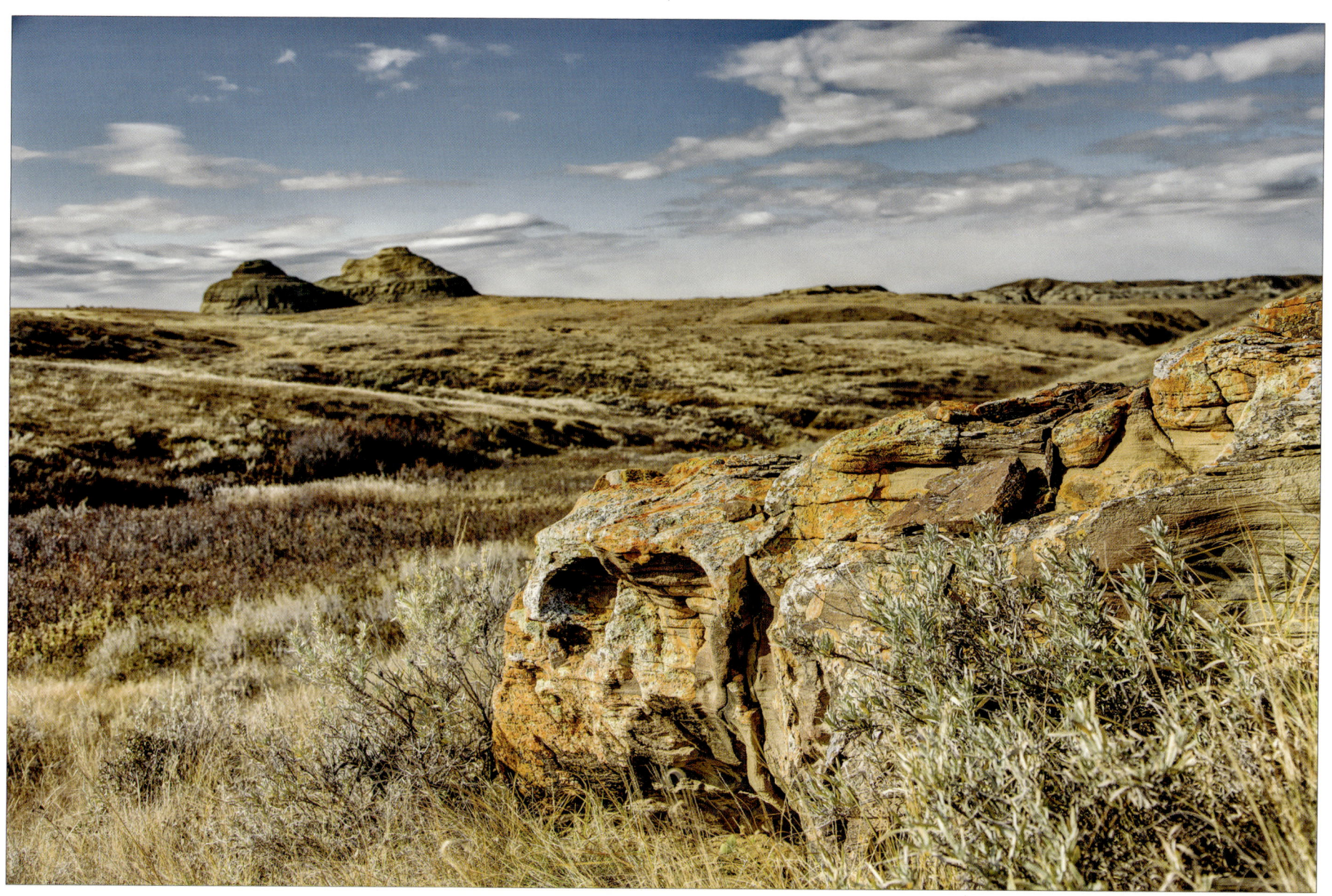

GENERATION X

59

DINOSAUR PROVINCIAL PARK

Rugged is an understatement when describing Dinosaur Provincial Park, located only 50 kilometres north of Brooks, Alberta, along the Trans-Canada highway. Situated within the Red Deer River Valley, it's an extraordinary place where the surreal topography is as complex and diverse as you'll see. Even the twisting road descending into the park offers distinct and contrasting landscapes. To the left of the main entrance road is the Red Deer River, winding its way through a sprawling badlands vista. The scene is rich with semi-arid badlands vegetation, including large cottonwood trees lining the riverbanks. A scenic campground can be seen in the distance. Ochre and sienna hills perfectly frame the view. Glance to the right, however, and you encounter a bleak and grey sun-baked wasteland completely devoid of life. Trench-like crevices are gouged into the ashen hills. The landscape is as barren as the moon.

Further down the road, and located behind the park's Interpretive Centre, the landscape changes yet again. The magic of erosion has formed a number of large and dramatic sandstone sculptures that reach for the sky. The views are breathtaking. One thing you begin to understand about the badlands is that they are full of contradictions. It's almost a given that for everything enchanting about this landscape, there is also something haunting. It's a strange but alluring paradox. To truly experience these contradictions, you must venture forth and explore the terrain on foot.

You will discover that there is as much going on below the surface as there is above. You don't have to explore too far before you find shallow caves and sinkholes collapsing into the earth. These are the product of a series of subterranean water routes beneath the ground. Fortunately, park staff does a good job cordoning off these areas with yellow tape. Still, a park staff member told me that on one occasion a horse and rider were sauntering along a pathway when the earth collapsed under their weight. Heavy equipment had to be brought in to rescue the horse, stuck in the saturated sinkhole. It's not an exaggeration to say that Dinosaur Provincial Park has many secrets.

The badlands at Dinosaur Provincial Park—as its name suggests—are a treasure trove of Cretaceous-period dinosaur fossils. It is one of the richest fossil sites in the world and it was designated a UNESCO World Heritage Site in 1979. Astonishing as it may seem, 70 million years ago, dinosaurs, some gigantic in size, roamed this area. One such dinosaur was the pterosaur, a flying reptile unearthed in these badlands, with a wingspan of an incredible 15.5 metres. In fact, there are over 500 recorded specimens found here, including many of the larger dinosaurs, such as Albertosaurus, Edmontosaurus, Triceratops, and, of course, the main attraction, the carnivorous Tyrannosaurus rex.

RIGHT: DINOSAUR PROVINCIAL PARK

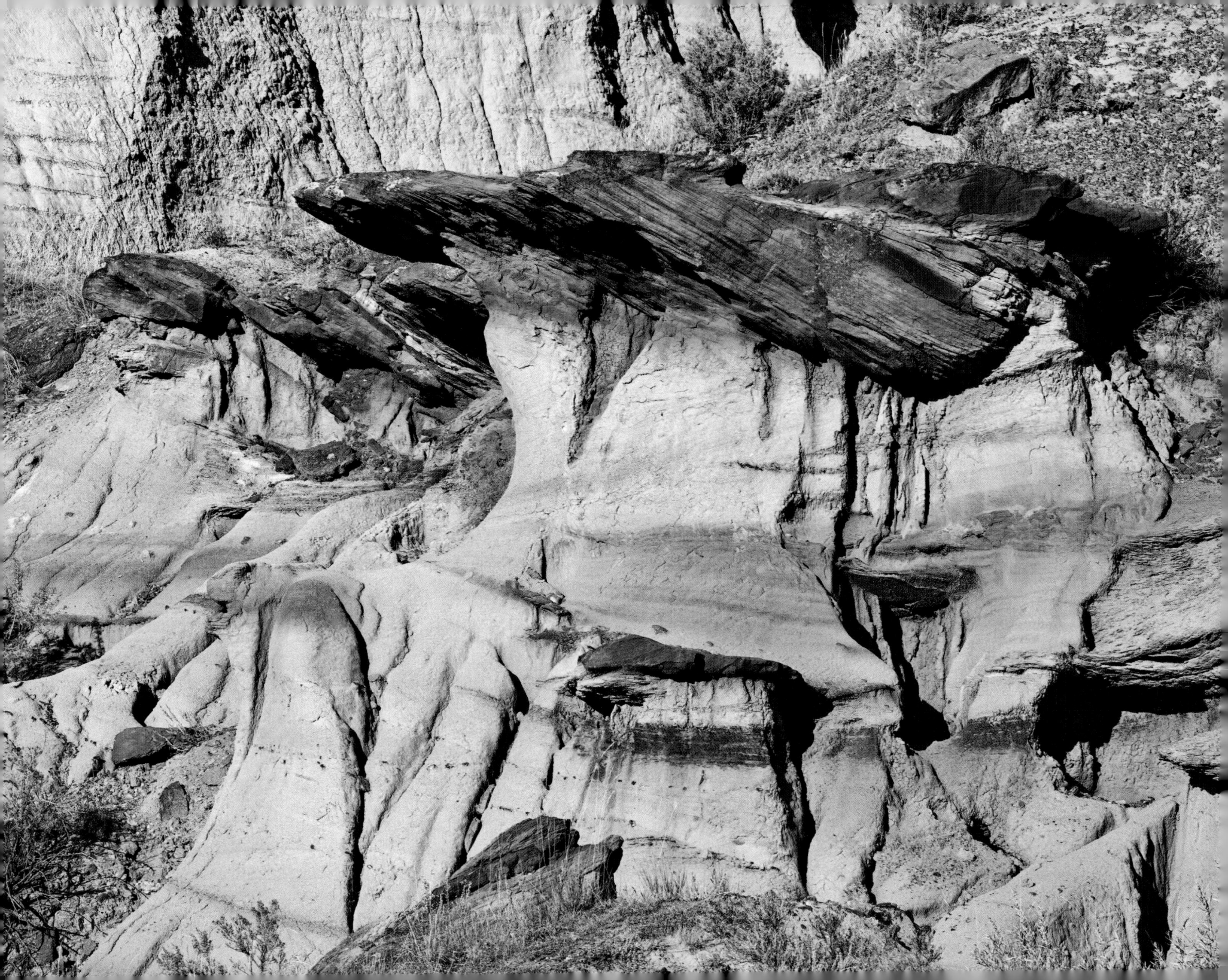

The Canadian Geological Survey was aware of fossils in this area as early as 1881, when geologist George M. Dawson and his assistants explored the area. One of his assistants, Joseph Burr Tyrrell (after whom the Royal Tyrrell Museum in Drumheller is named), later took over the exploration and found the first dinosaur skull in the area in 1883. Tyrrell must have been aware of the significance of this area, but ultimately the timing was wrong to fully expose the fossil wealth. Alberta (then the Northwest Territories) was just being settled and the Government of Canada had other priorities.

It wasn't until 1909, when Barnum Brown from the American Museum of Natural History in New York City overheard a chance remark by a visitor that "bones like these were as common as dirt" where he came from. Brown, no doubt intrigued, further engaged the man, whose name was J. L. Wagner, a rancher near Michimichi Creek, which is situated adjacent to the Red Deer River Valley. Brown travelled all the way from New York to investigate the claim and was astounded to find that Wagner wasn't exaggerating. The Red Deer River Valley was in fact a dinosaur fossil warehouse. Brown immediately returned with a crew, proper equipment, and a 3.6 x 9-metre boat that he used as a floating camp from which he scoured the hills for possible dig sites.

Brown loaded up eight tons of fossils the first year to send back to New York. The Great Canadian Dinosaur Rush had begun. Canadians took notice. Because there was neither a proper museum anywhere in Canada to house dinosaur fossils, nor anyone with the skills or wherewithal to properly excavate them, they hired American Charles Sternberg and his three sons. There was tense competition between the Brown and Sternberg groups, who were both vying for the next big discovery. Since both men used boats as floating camps on the confined Red Deer River, they came in contact with one another on numerous occasions and each kept a watchful eye on what the other group was doing. Ultimately, though, there were more than enough fossils for everyone.

By 1915, the Great Canadian Dinosaur Rush was essentially over. The Great War was affecting funding, and Barnum Brown didn't return because the American Museum estimated that it had a surplus of fossils that would take five technicians two years to prepare and catalogue. The Canadians were also beginning to impose rules and limit the exportation of fossils. Still, during this six-year period, dinosaur bones from what was then called the Deadlodge Canyon badlands, had been sent to museums all over Europe, the United States, and Canada. The

Victoria Museum (now called the Royal BC Museum) was built in 1911 and the Royal Ontario Museum in 1914. Various digs have continued throughout the years, and today they are under the jurisdiction of the Royal Tyrrell Museum of Paleontology in Drumheller.

Dinosaur Provincial Park is an alluring place of open secrets. It's a museum without walls where the terrain changes as dramatically as if moving from a gallery of contemporary sculpture to a gallery of antiquities. It's a graveyard where memories have come to die, a sanctuary where echoes are resurrected, and an oasis where the lost becomes found.

Dinosaur Provincial Park has five self-guided interpretive trails to explore. The Prairie Trail and Coulee Viewpoint Trail are located behind the park's Interpretive Centre. The Badlands Trail, Trail of the Dinosaurs, and Cottonwood Trail are located past the campground and along the Public Loop Road. All of these are highly recommended.

They offer a wide variety of great views and interesting information. Also highly recommended are guided tours, family programs, backcountry excursions, and actual dinosaur digs.

But remember that you are in rattlesnake country. Scorpions and the occasional black widow spider can also be found.

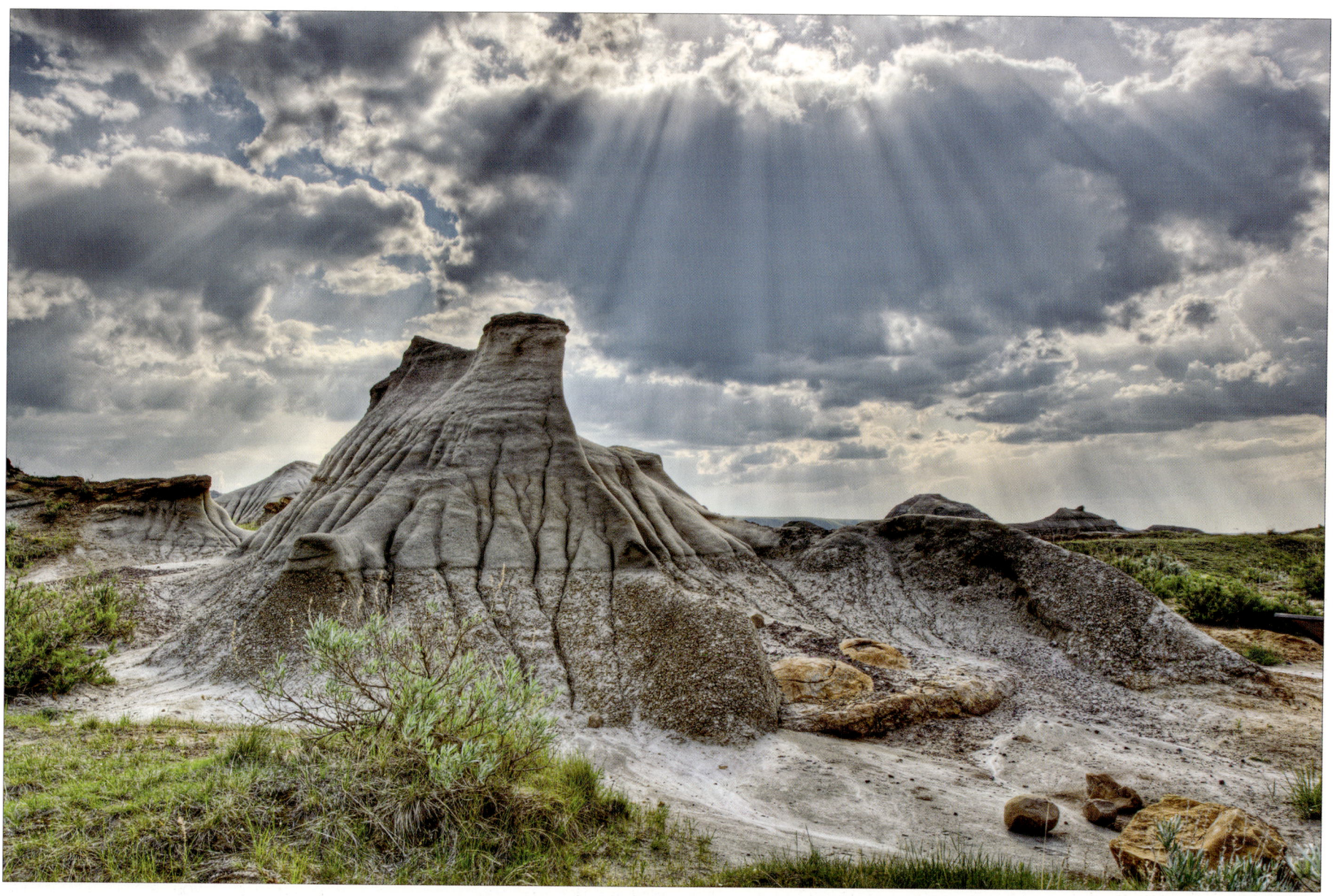

SCARHEAD
A curious sandstone structure photographed one evening when this dark and foreboding sky suddenly opened up and showered it with rays of light.

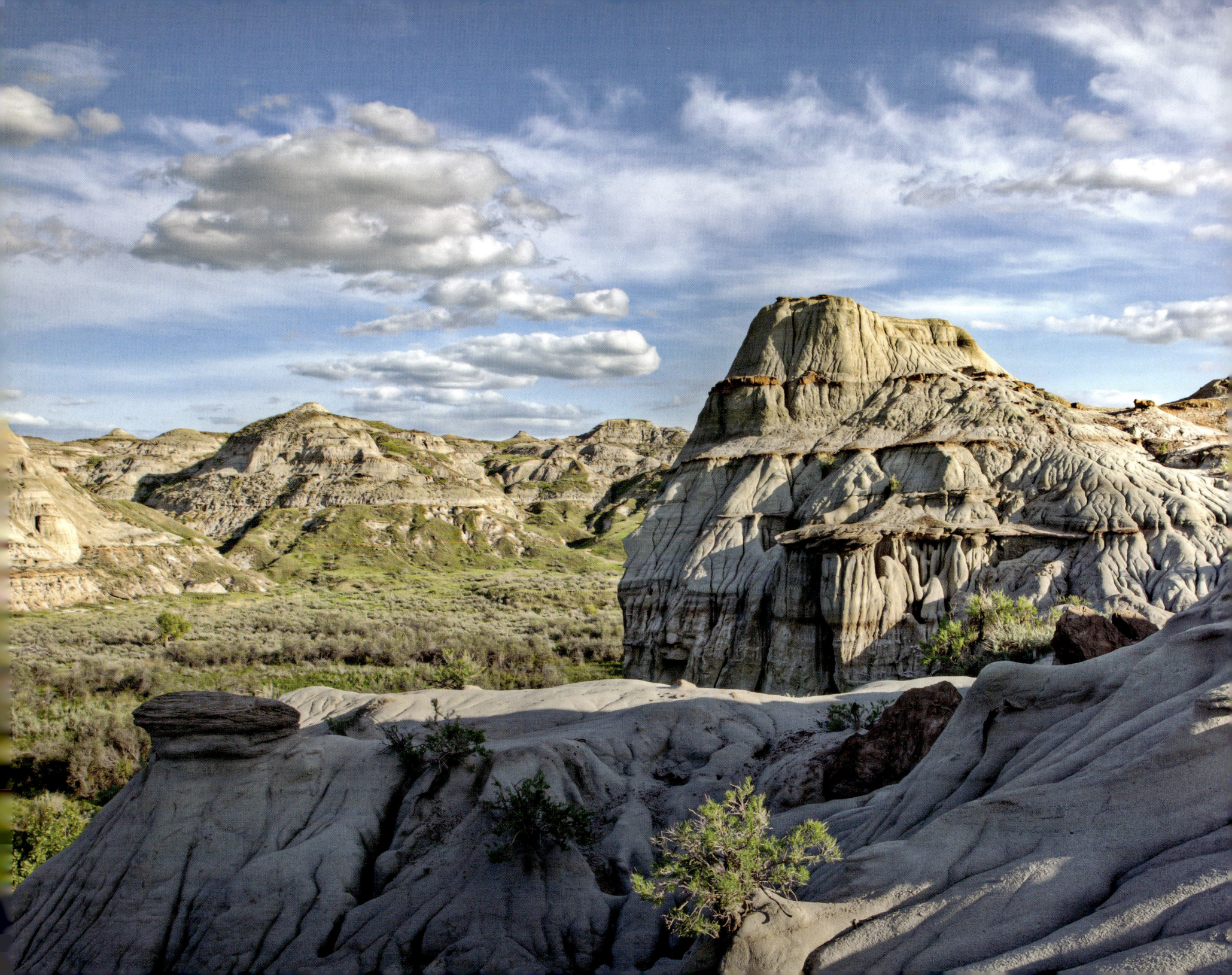

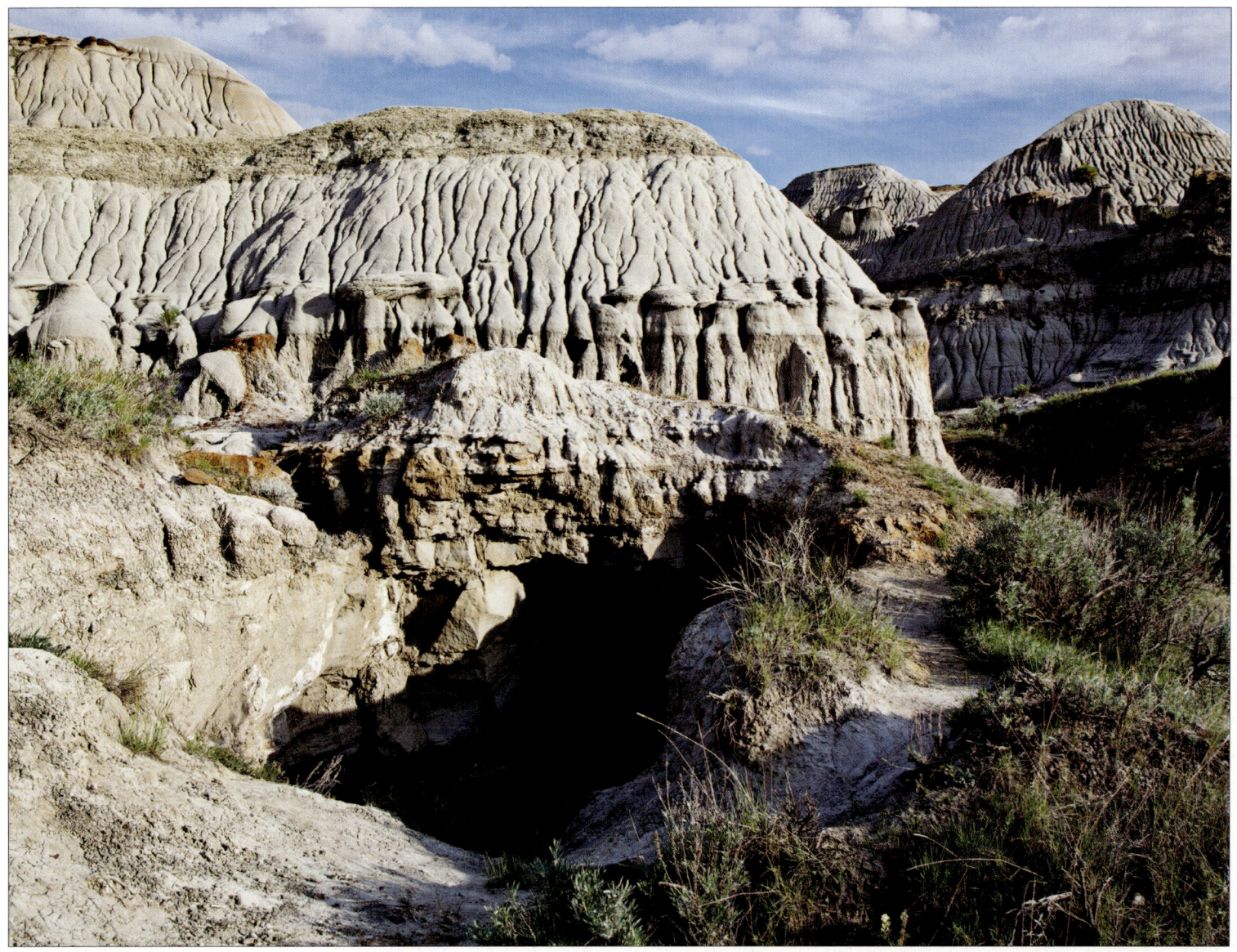

LEFT: OPEN SECRETS
Sunrise along Coulee Viewpoint Trail.

ABOVE: GRAVE SECRETS
With the densely rilled slopes in the background it is easy to see where the water has been, but it's not overly apparent where it has gone. Subterranean piping and sinkholes are common at Dinosaur Provincial Park and most other badlands.

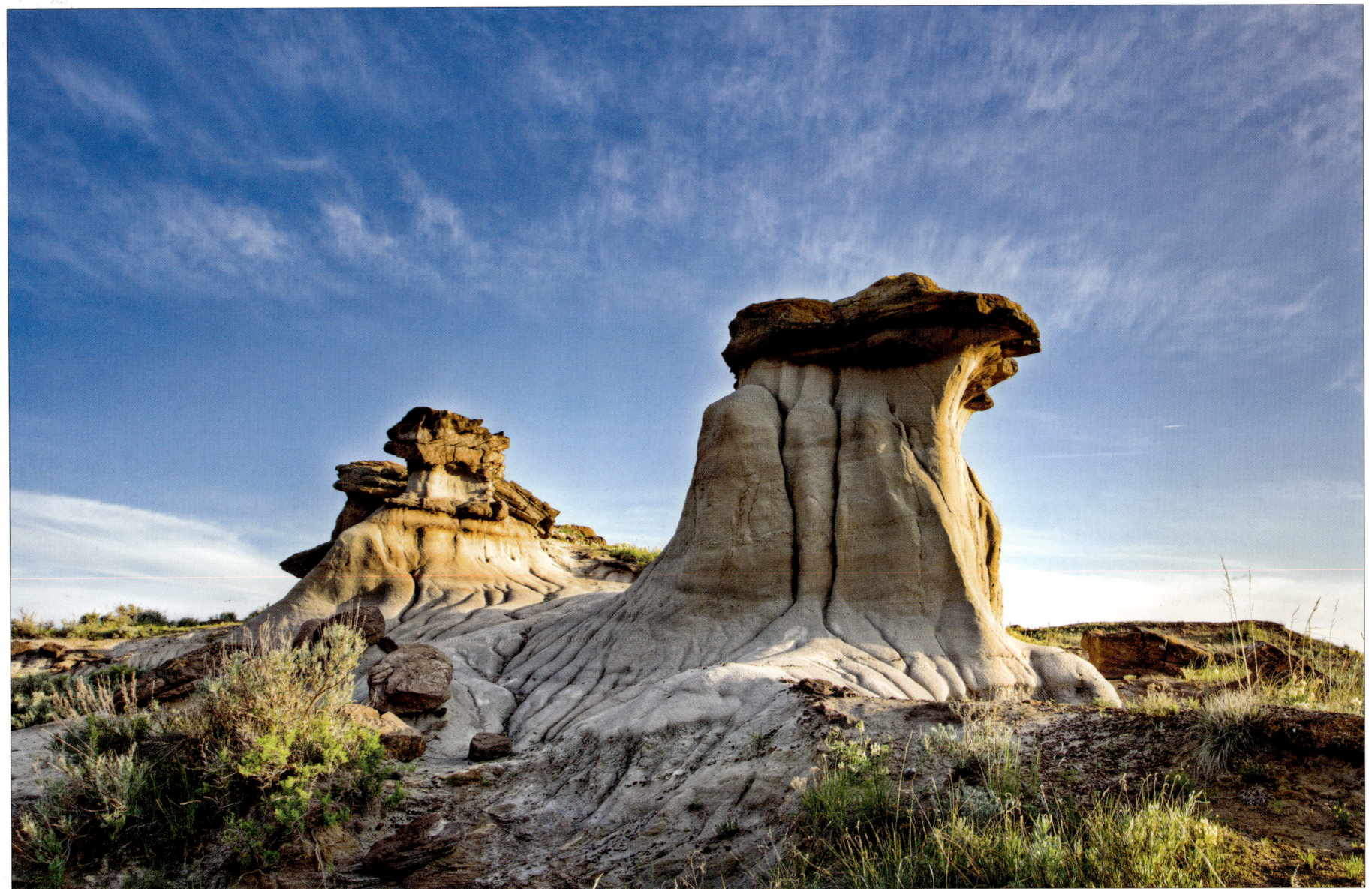

ABOVE: DEFIANT SPIRITS
This powerful hoodoo with large caprock endures the erosive forces of nature and defiantly occupies a prominent position along the Public Loop Road.

RIGHT: PROCEDURES FOR UNDERGROUND
You can see in this photograph the beginnings of a subterranean waterway.

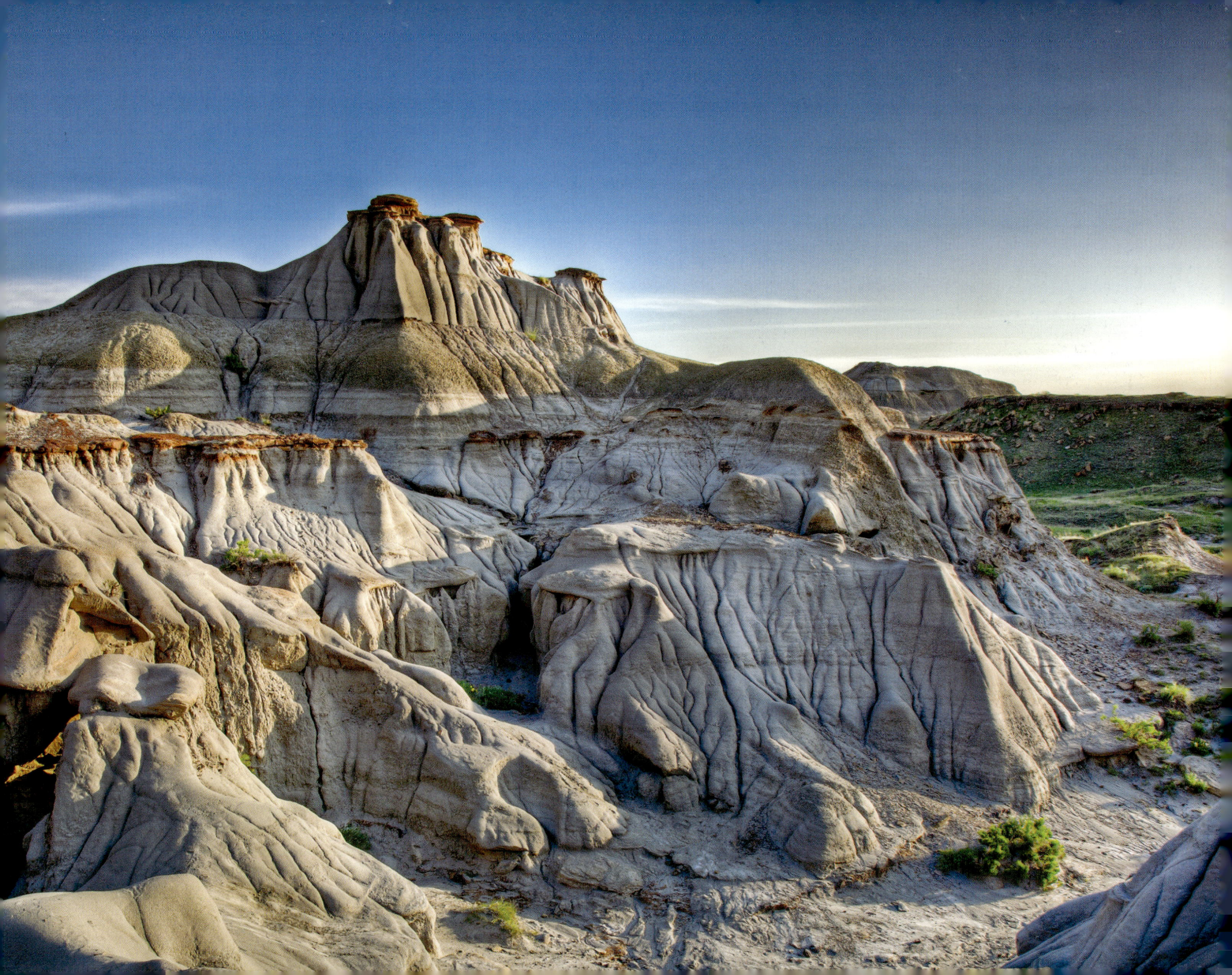

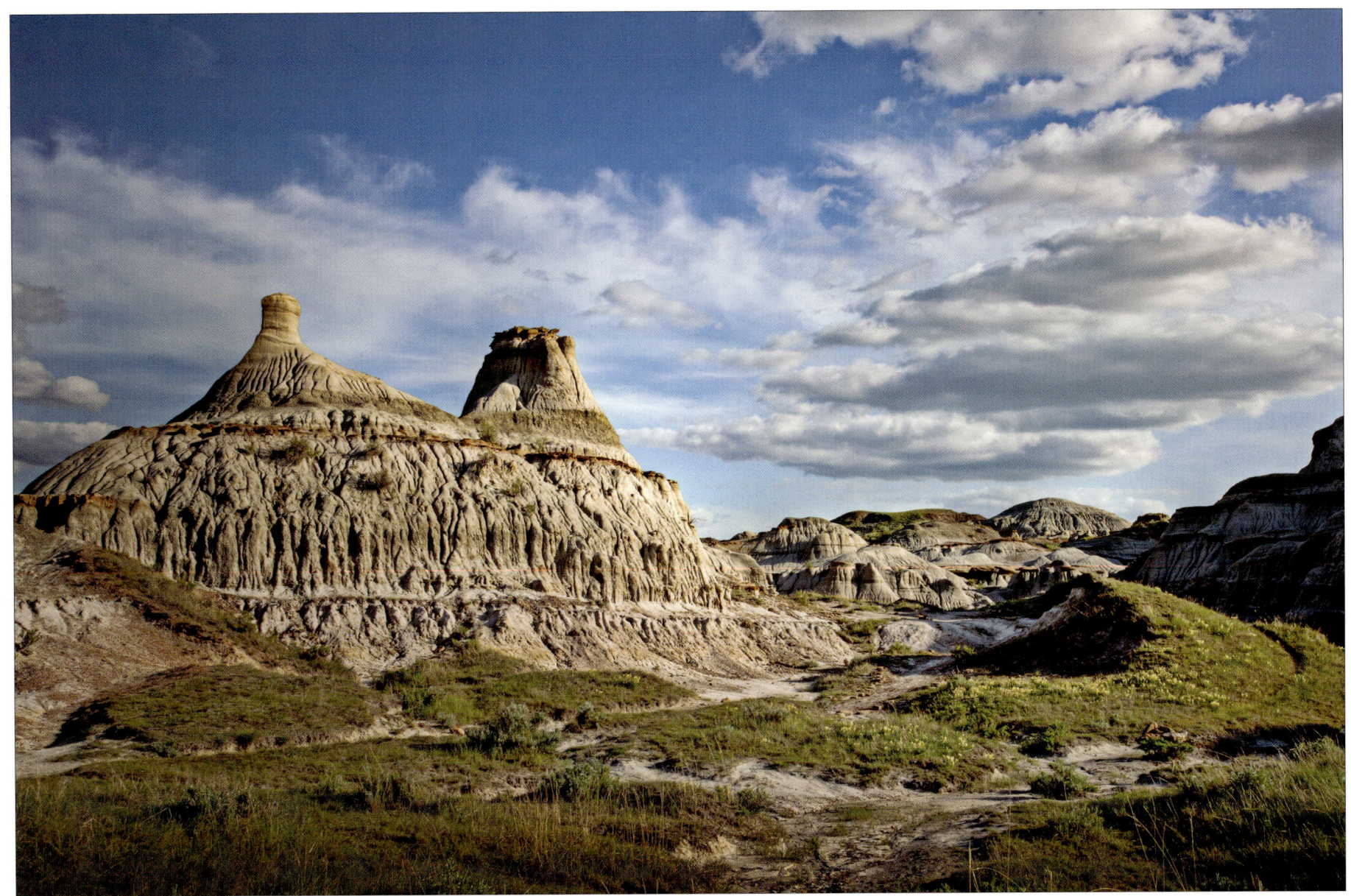

VALLEY OF CASTLES
Classic scenery at Dinosaur Provincial Park with spectacular light along the Coulee Viewpoint Trail.

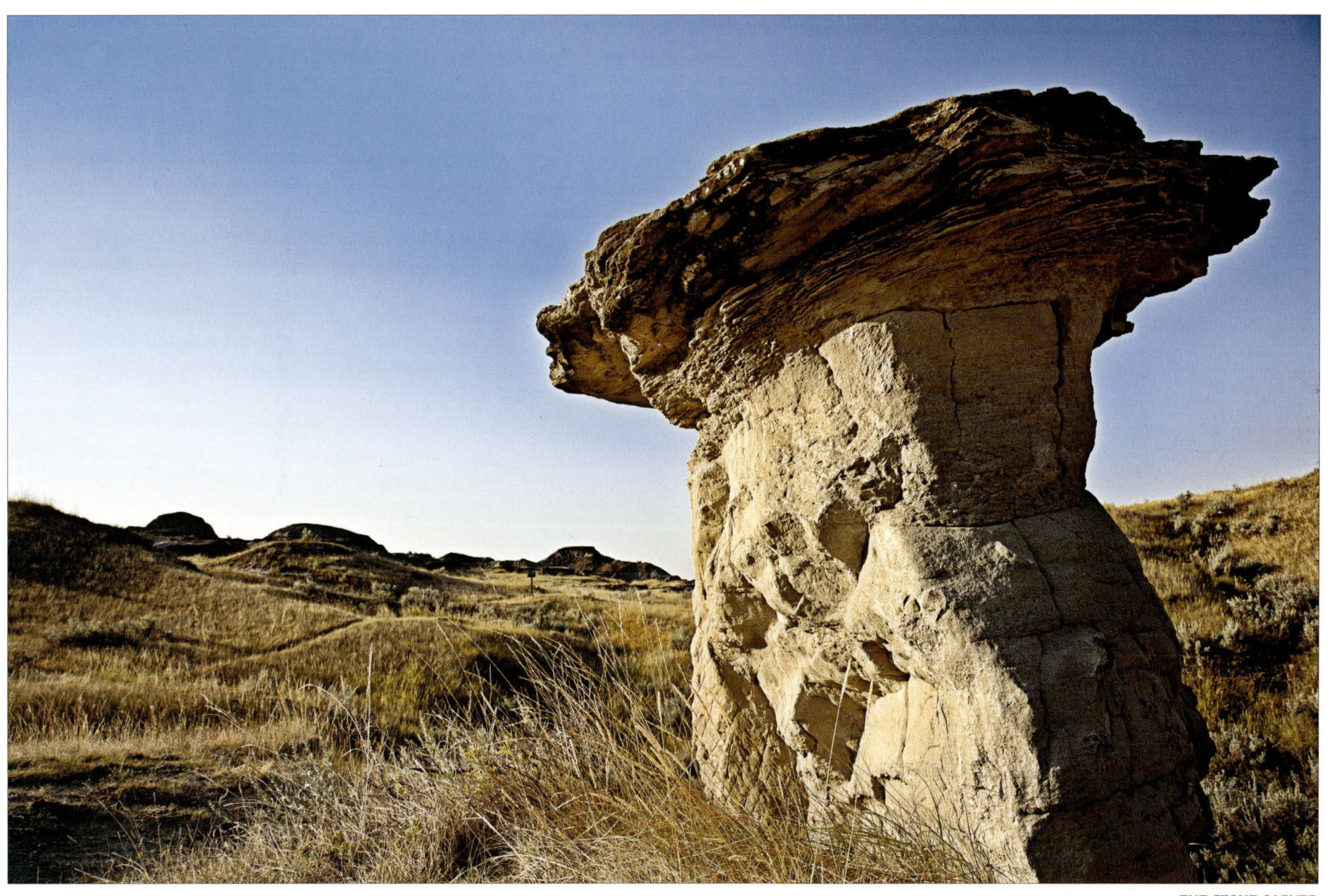

THE STONE CARVER

"I saw the angel in the marble and carved until I set him free." *Michelangelo Buonarroti*

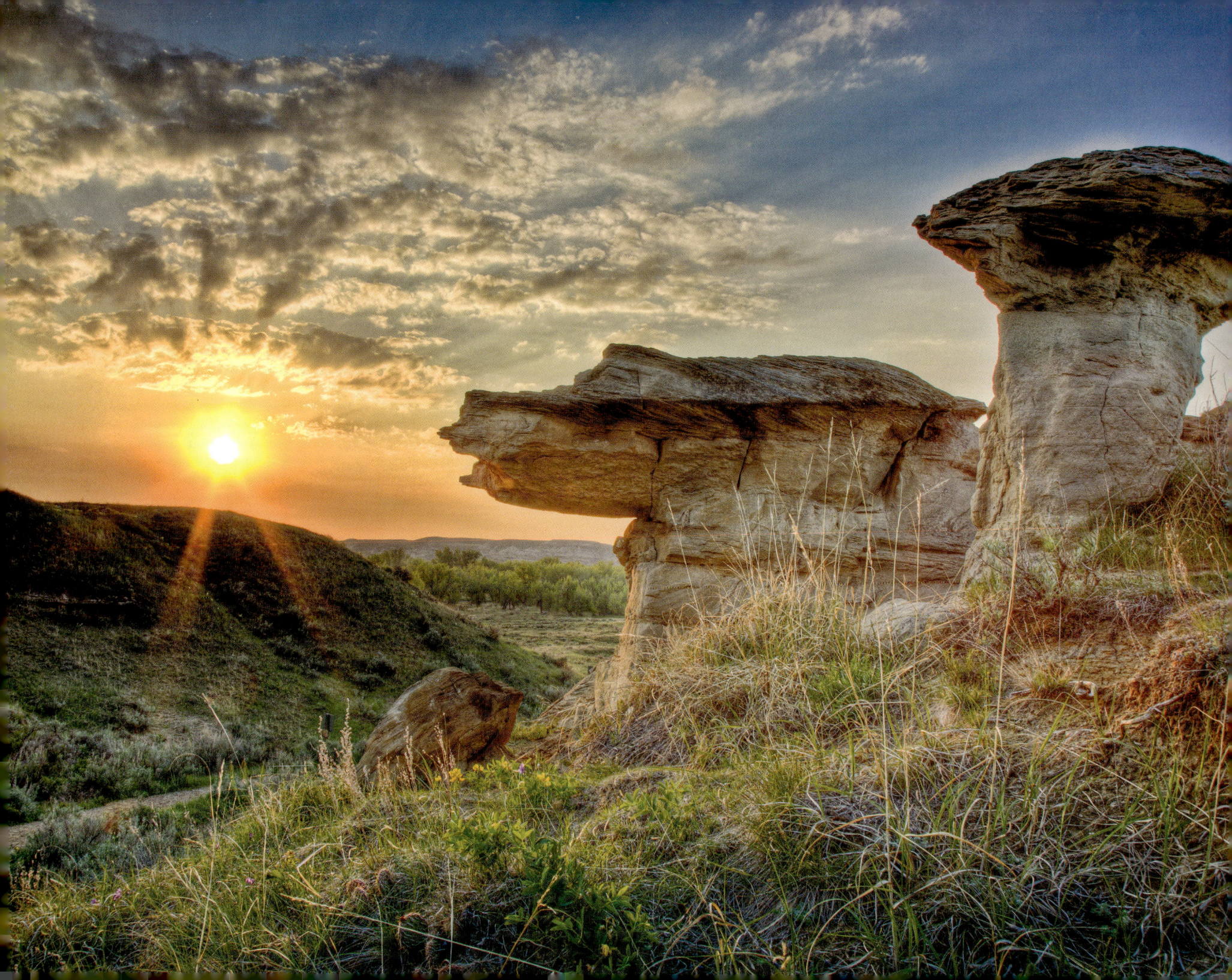

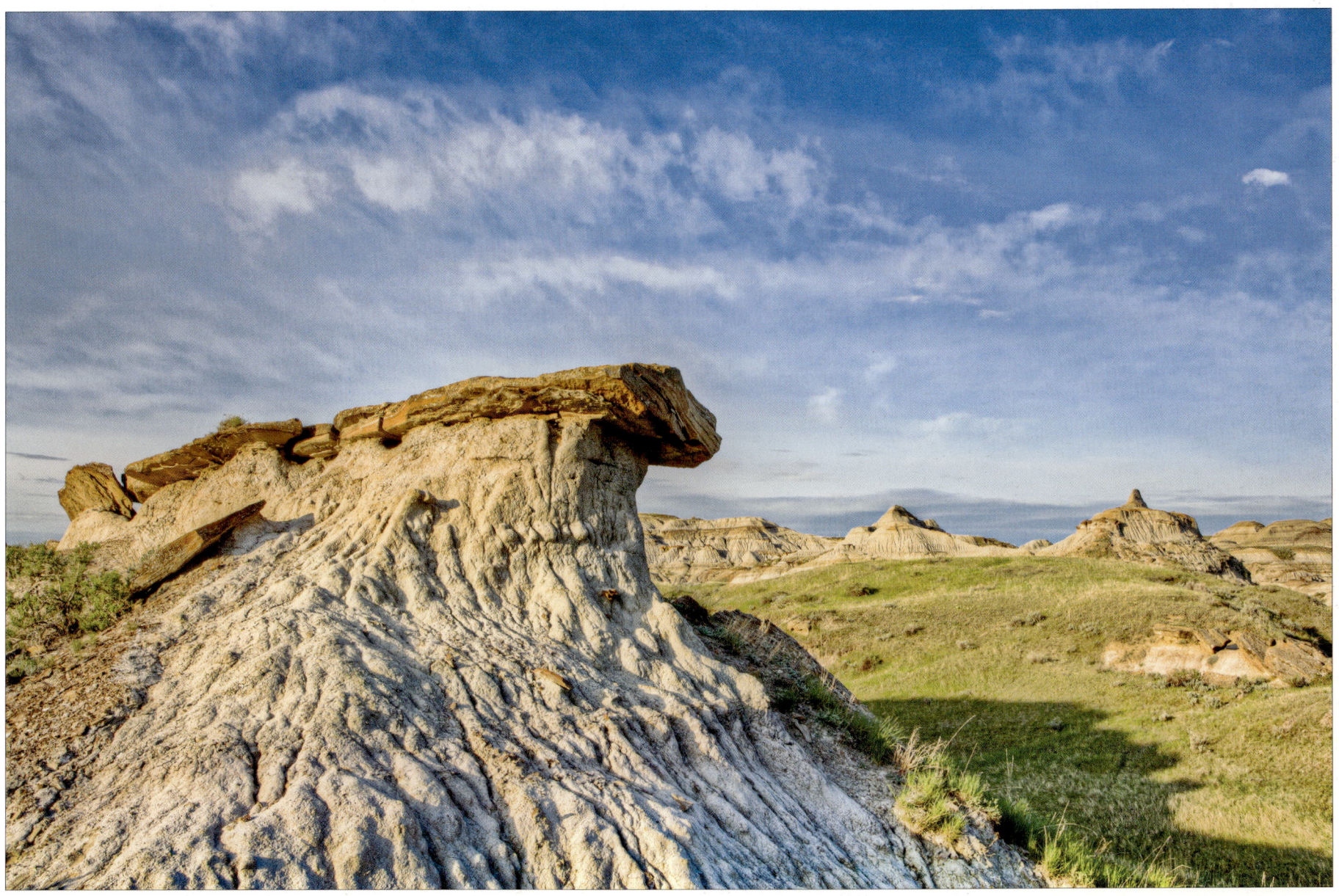

LEFT: TWO SOLITUDES (SUNSET)

ABOVE: THUNDER FIST
Dinosaur Provincial Park has a fantastic library of badlands vistas. Here is another fine view with the Valley of Castles in the distance.

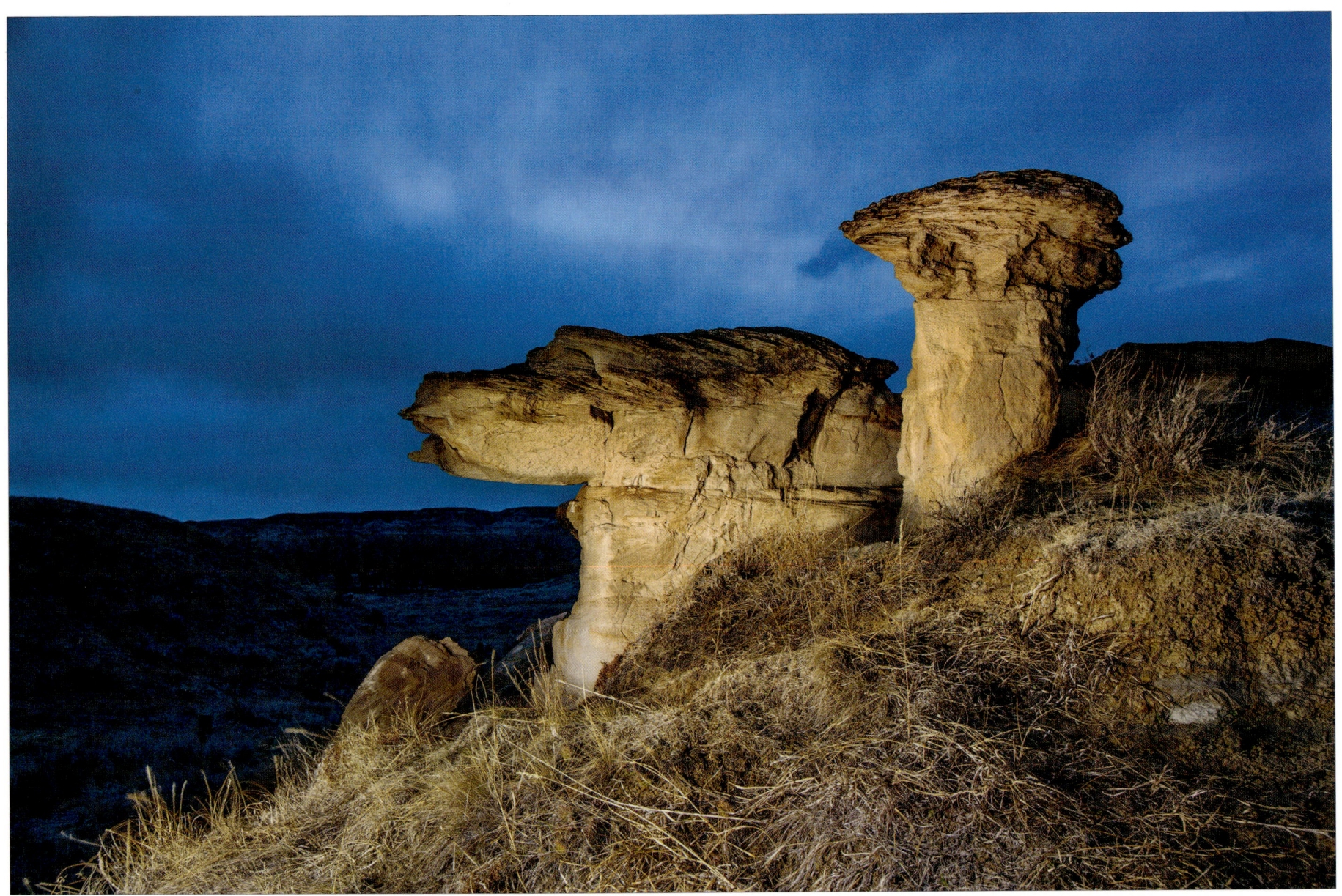

TWO SOLITUDES (NIGHT)

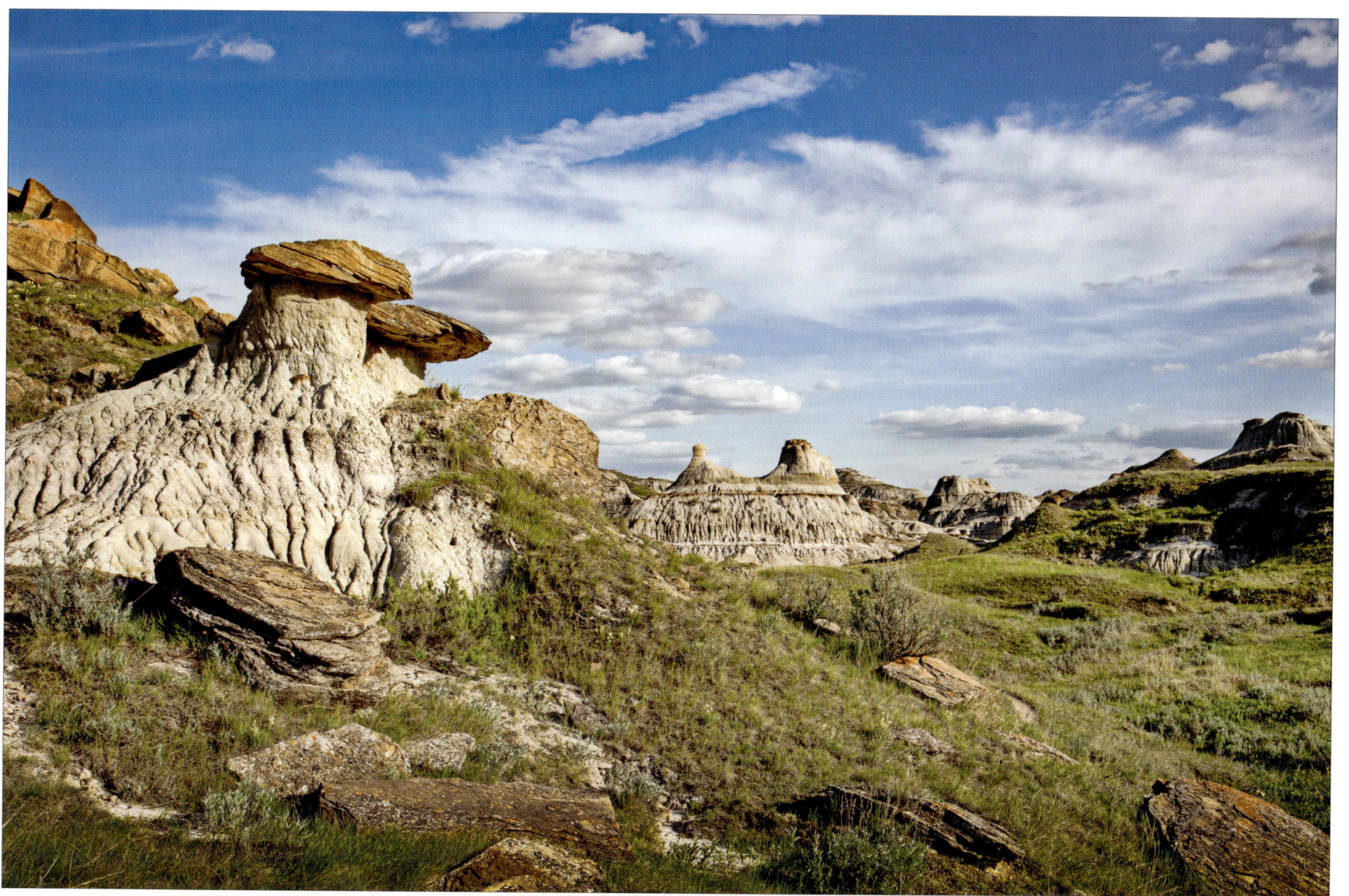

ABOVE ALL THINGS
Despite mornings like this, the winds around here seem to carry memories of the region's dark past. I find myself thinking of explorer David Thompson who, in the 1780s, witnessed the aftermath of a smallpox epidemic that nearly wiped out a Blackfoot tribe in this area. He describes a "stench" of death where those still alive "were too weak to move."

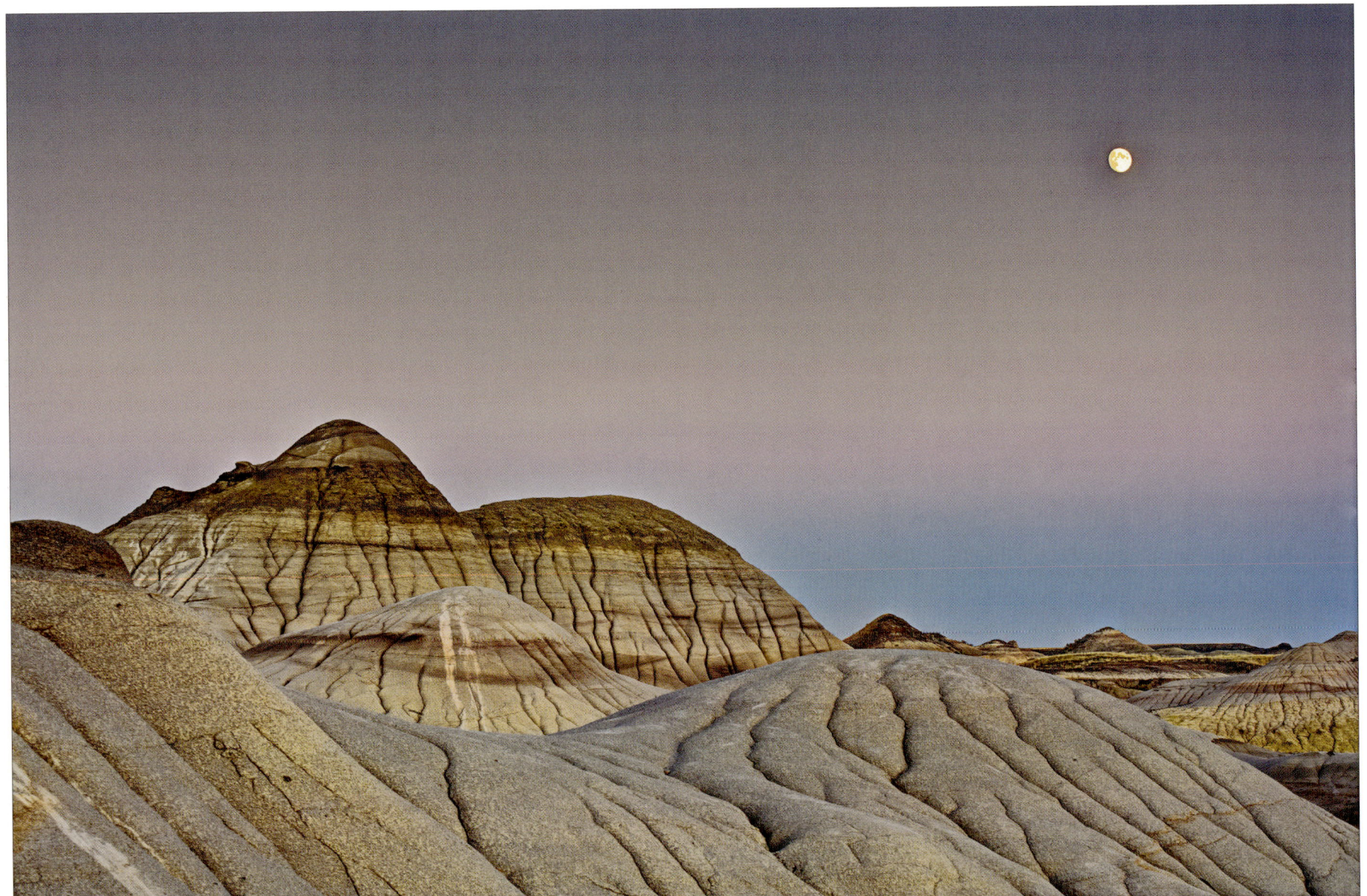

NIGHT GAMES
While the badlands are a complex and diverse ecosystem, this moon-like section
is about as barren and desolate as you can find.

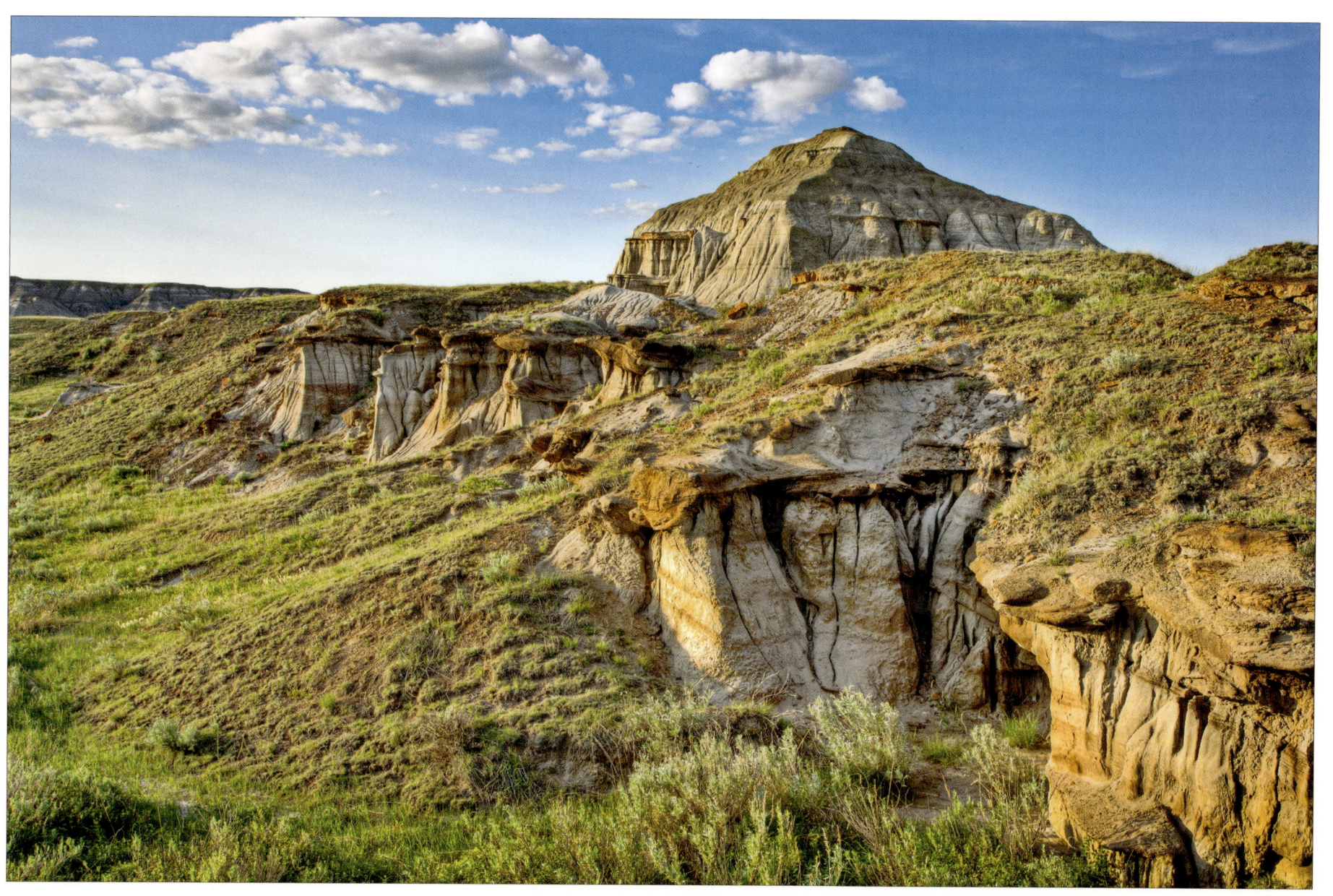

RATTLESNAKE RIDGE
This is rattlesnake country. Highway signs outside the park urge you to slow down and avoid driving over snakes.

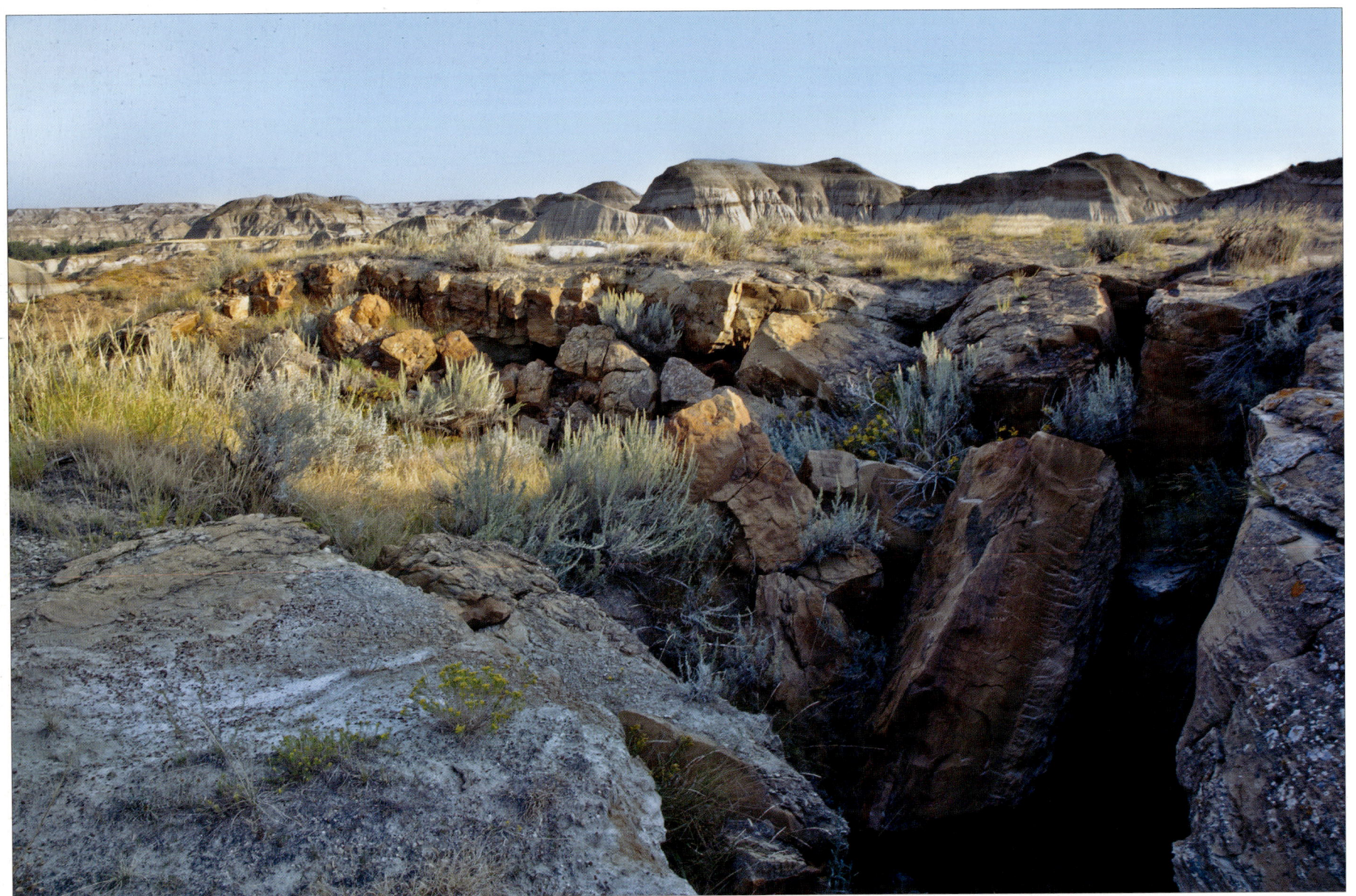

ABOVE: THE BONE CAGE
Large rocks crumbling into an abyss hint at the often-violent geologic
wrestling match taking place beneath the badlands.

RIGHT: TO THE BARRICADES

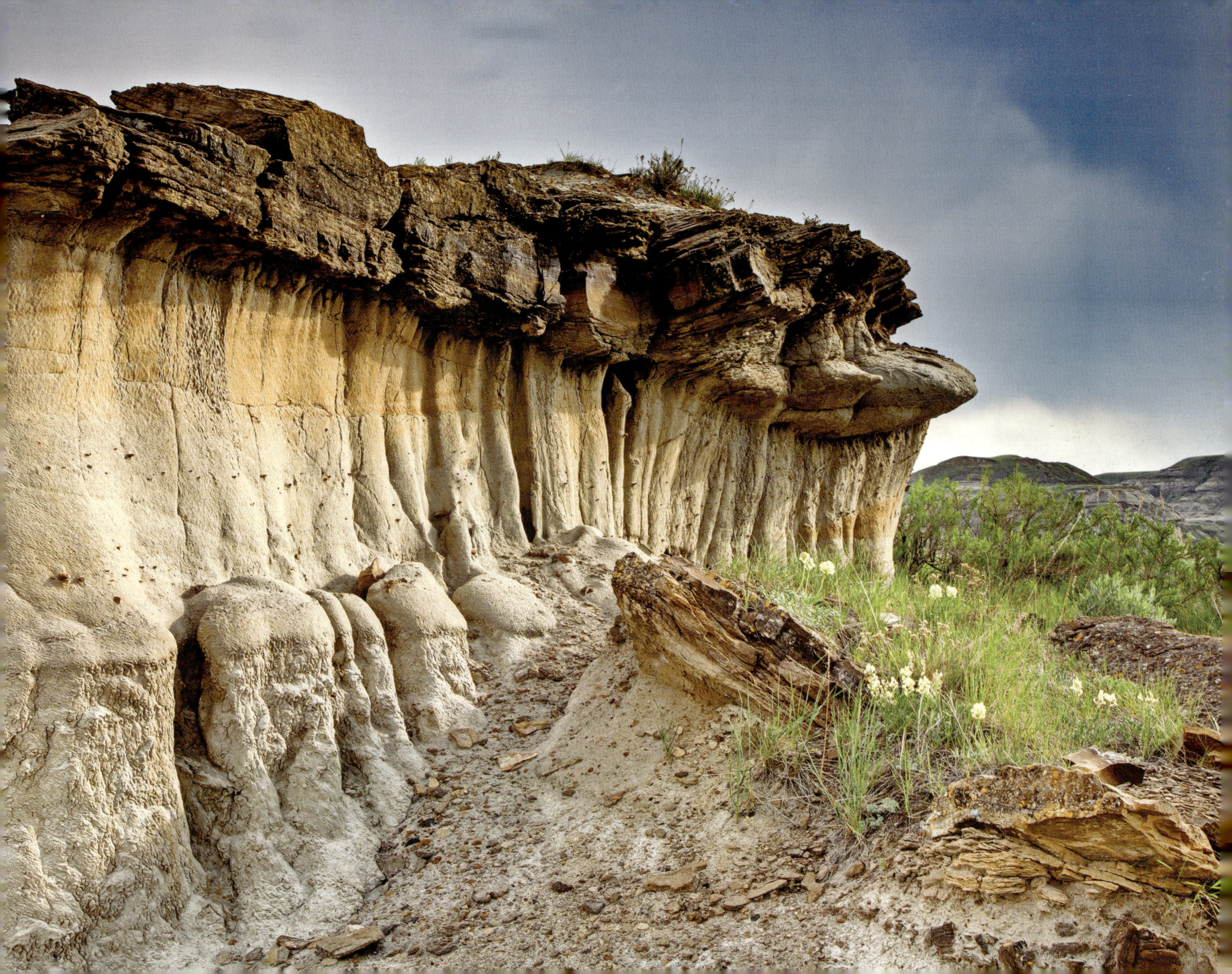

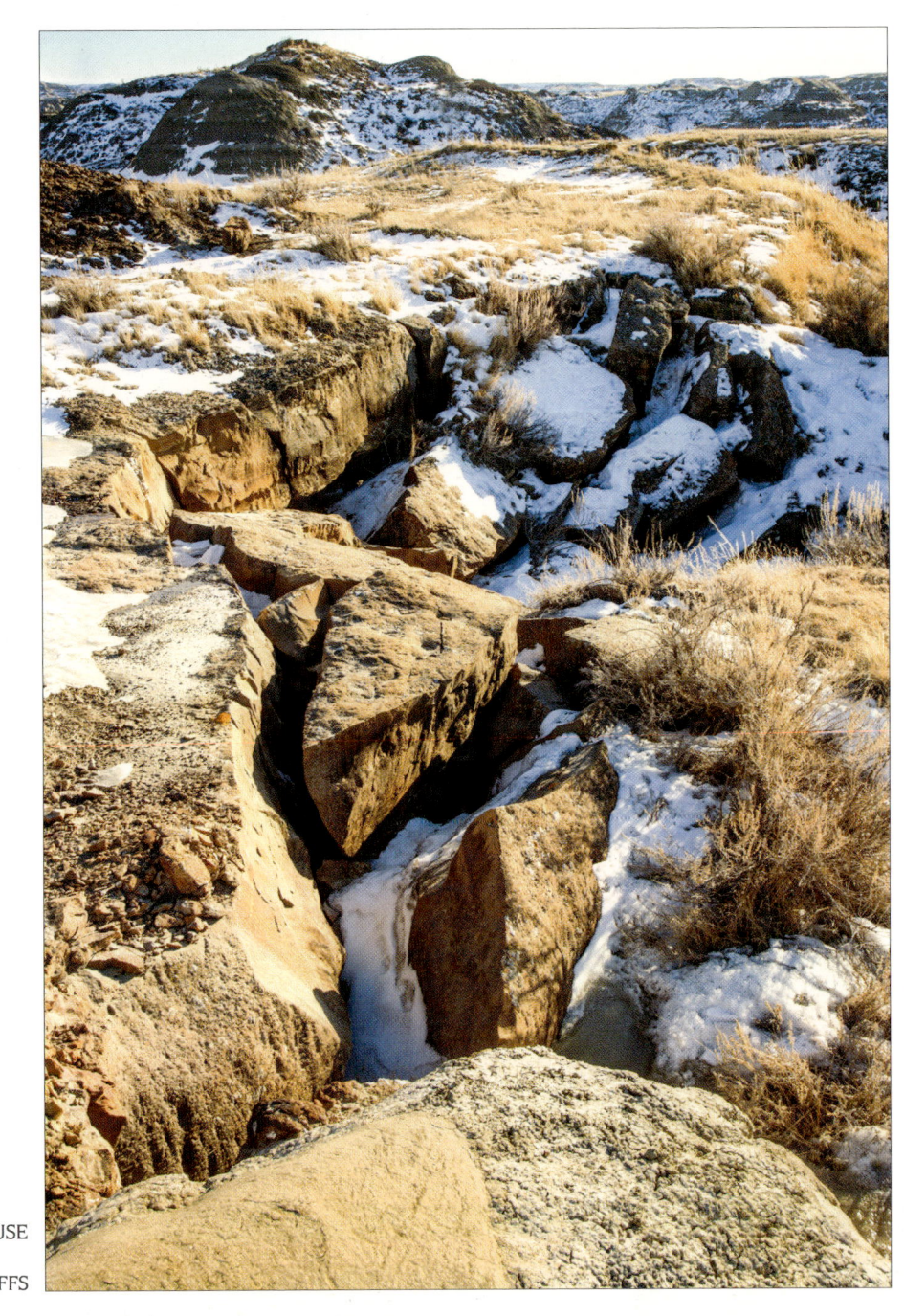

THE BIRTH HOUSE

FAR RIGHT: COPPER CLIFFS

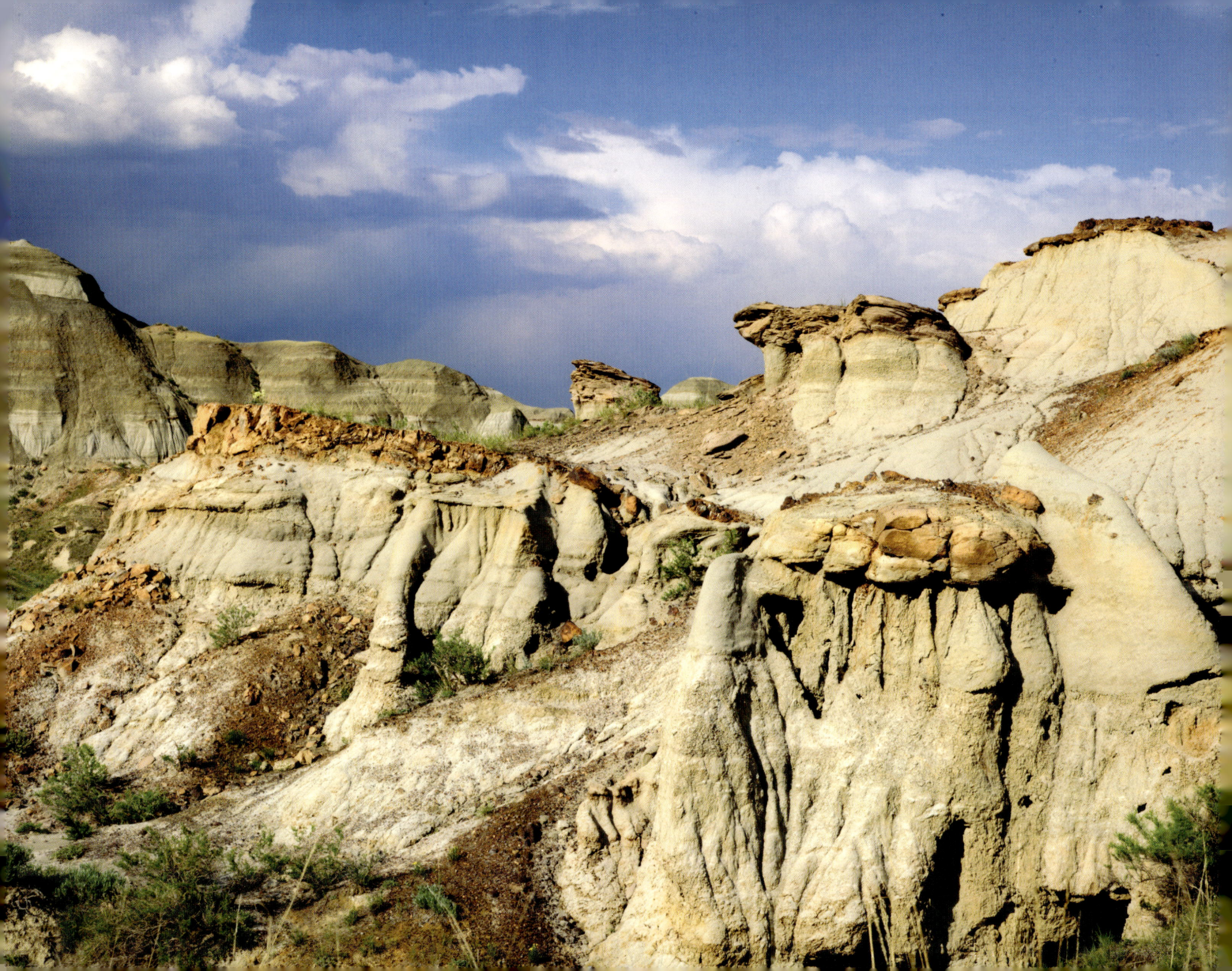

COTTONWOOD TRAIL

I had visited and explored the magnificent badlands of Dinosaur Provincial Park numerous times, and in various seasons, before finally deciding to hike the visually barren Cottonwood Trail. The way the park is set up, there are three trails along what is called the Public Trail Loop Road. This is a one-way dirt road that I would hesitate to tackle in wet weather without a four-wheel drive vehicle. The Cottonwood Trail is the third hike along this road, so it will likely be your last option of the day. This flat path cutting through a field of thorny buffaloberry always looked to me like a waste of time. What cottonwood trees I could see were clumped together far off in the distance lining the riverbanks and were at least a kilometre from the road. As an artist and photographer, I was focused on visual inspiration, and exploring the vast geography of metaphors contained within these badlands was always rewarding. I was consistently finding some new stone sculpture to photograph, or even revisiting an old one in the hope that I would discover a new angle to showcase, or, better yet, be rewarded with a dramatic sky and spectacular light. So after spending hours hiking and photographing the open-air art gallery called Dinosaur Provincial Park, I would usually drive by the cottonwood trailhead, pause briefly, look out over the underwhelming vista, and continue onward. This, I discovered, proved to be a huge mistake.

The Cottonwood Trail has some of the most spectacular trees I have ever seen. While most trees seem to be fairly uniform and blend together, cottonwoods tend to have their own unique growth pattern that gives them an individual character. In Dinosaur Provincial Park, there are a number of cottonwoods that, quite frankly, are completely mesmerizing. In fact, two of them could easily be included in Thomas Pakenham's book, *Remarkable Trees of the World*. The first cottonwood you encounter, the "Tree of Remembrance," about 500 metres into the trail, is one of them. You notice immediately that this solitary tree has presence. It's also growing quite far from the river where most of the other cottonwoods are growing. Cottonwoods typically flourish close to water because their roots always need to be in contact with the water-table or they will die. This 200-year-old cottonwood seems to defy that notion. As you approach the tree, the second thing you notice is its massive trunk and remarkable branch pattern, one of which has somehow contorted to form the letter "R." And if that isn't strange enough, this particular branch appears to be dead while the rest of the tree is thriving. Fortunately, a sign beside the trail offers an explanation. During drought conditions, cottonwoods are able to prune themselves by

RIGHT: THE MATRIARCH

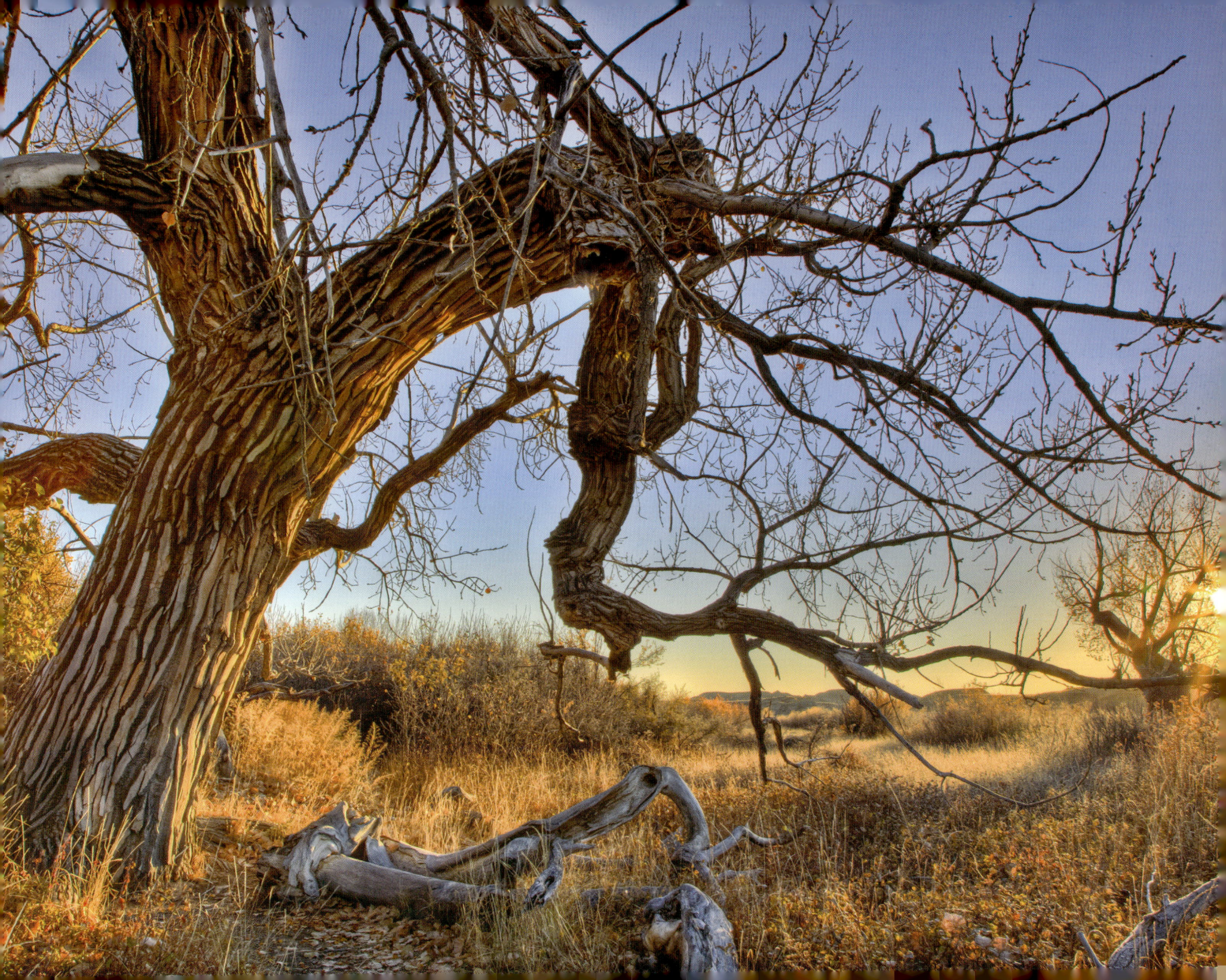

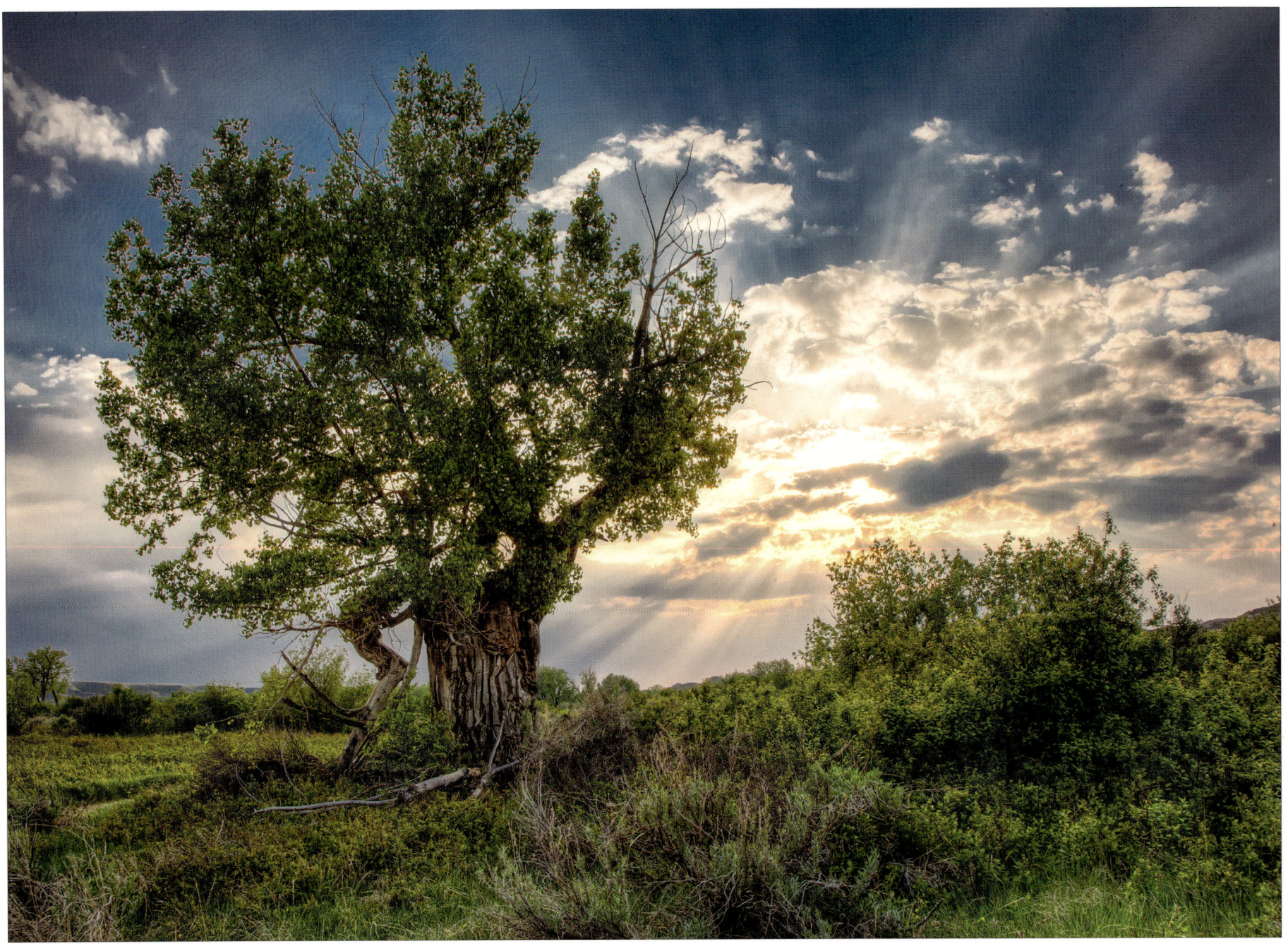

cutting off the water supply to some of their branches in order stay alive. There is a bizarre deformity to this tree but it manages to retain an odd symmetry. "Tree of Remembrance" appears quite statuesque and regal—much like an ambassador welcoming you to a foreign land.

If the "Tree of Remembrance" welcomes you to the trail, the next large and imposing cottonwood, only 50 metres further along the path, accosts you. One approaches this tree with apprehension. This cottonwood looks like a mythical serpent, the hostile gatekeeper of the trail. It stops you in your tracks and dares you to pass. This tree is called "The Matriarch," and, like the "Tree of Remembrance," it too, is a solitary 200-year-old cottonwood, seemingly growing too far from a water source. For the record, the average lifespan of a plains cottonwood tree is between 100 and 150 years. "The Matriarch" is battle-scarred and menacing. She gives the overwhelming impression that her long tentacle-like branches might spring to life and grab you at any moment.

What makes both of these trees riveting is the extraordinary life-and-death struggle manifesting itself within their spirit. While both trees are alive and thriving, it is clear that sections have been sacrificed or condemned, and they are not going gently. Instead of withering and falling off, they writhe, twist, and contort downward as if frantically searching to re-root themselves into the earth. The result, though, is inevitable. The ground surrounding these trees is littered with large rotting branches that have lost this battle. One can only sit back and wonder how immense these cottonwood trees would have been had these vanquished branches not been sacrificed.

These mystical cottonwoods emanate a distinct spirit-like aura. The Plains Indians revered these old trees and would place their dead on platforms among the branches. If you look closely, some of these cottonwoods seem to be growing out of themselves, like a phoenix reborn from its own ashes. Where else but amidst the visual drama of the hostile badlands would you expect to find this kind of surreal battle?

LEFT: TREE OF REMEMBRANCE

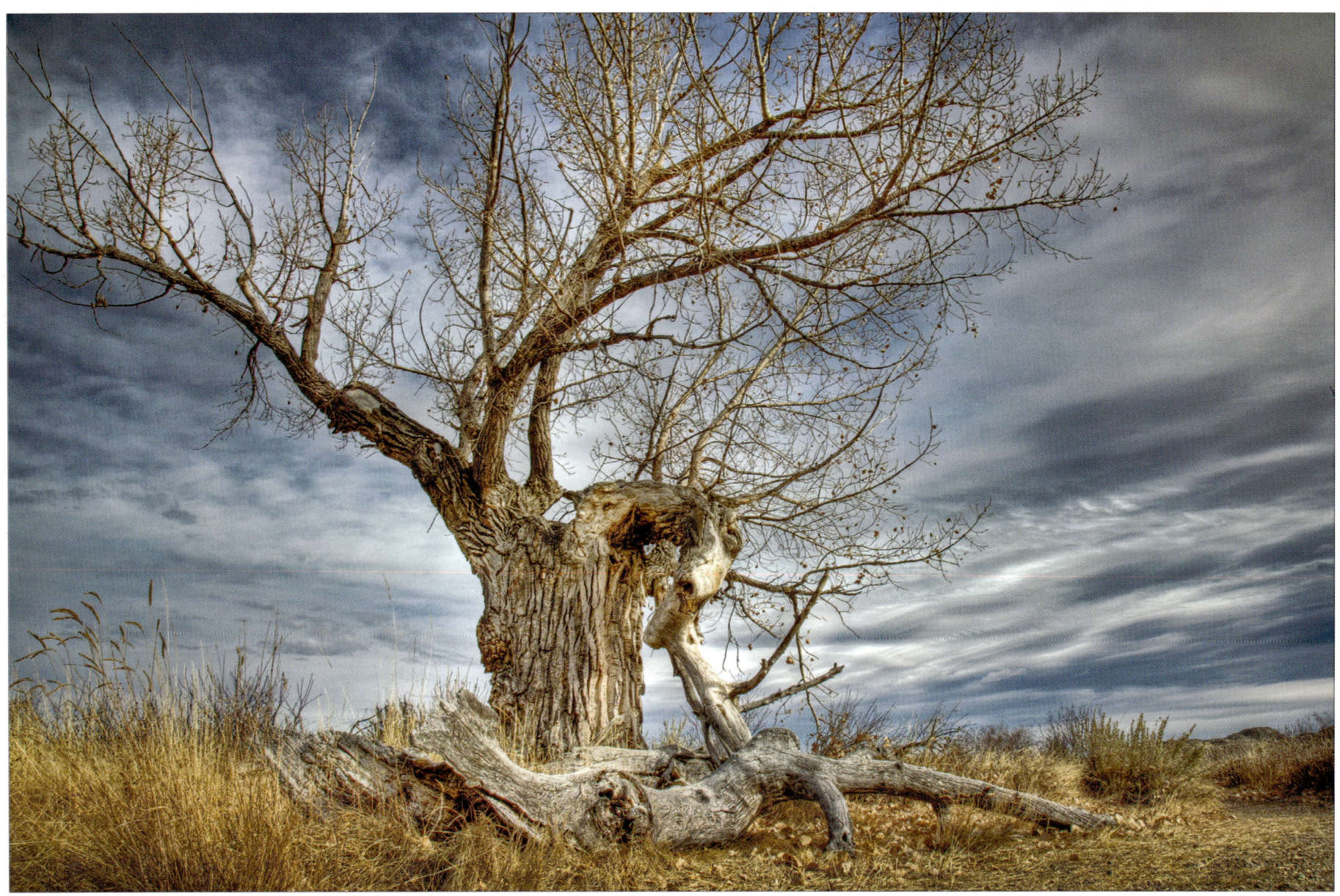

TREE OF REMEMBRANCE

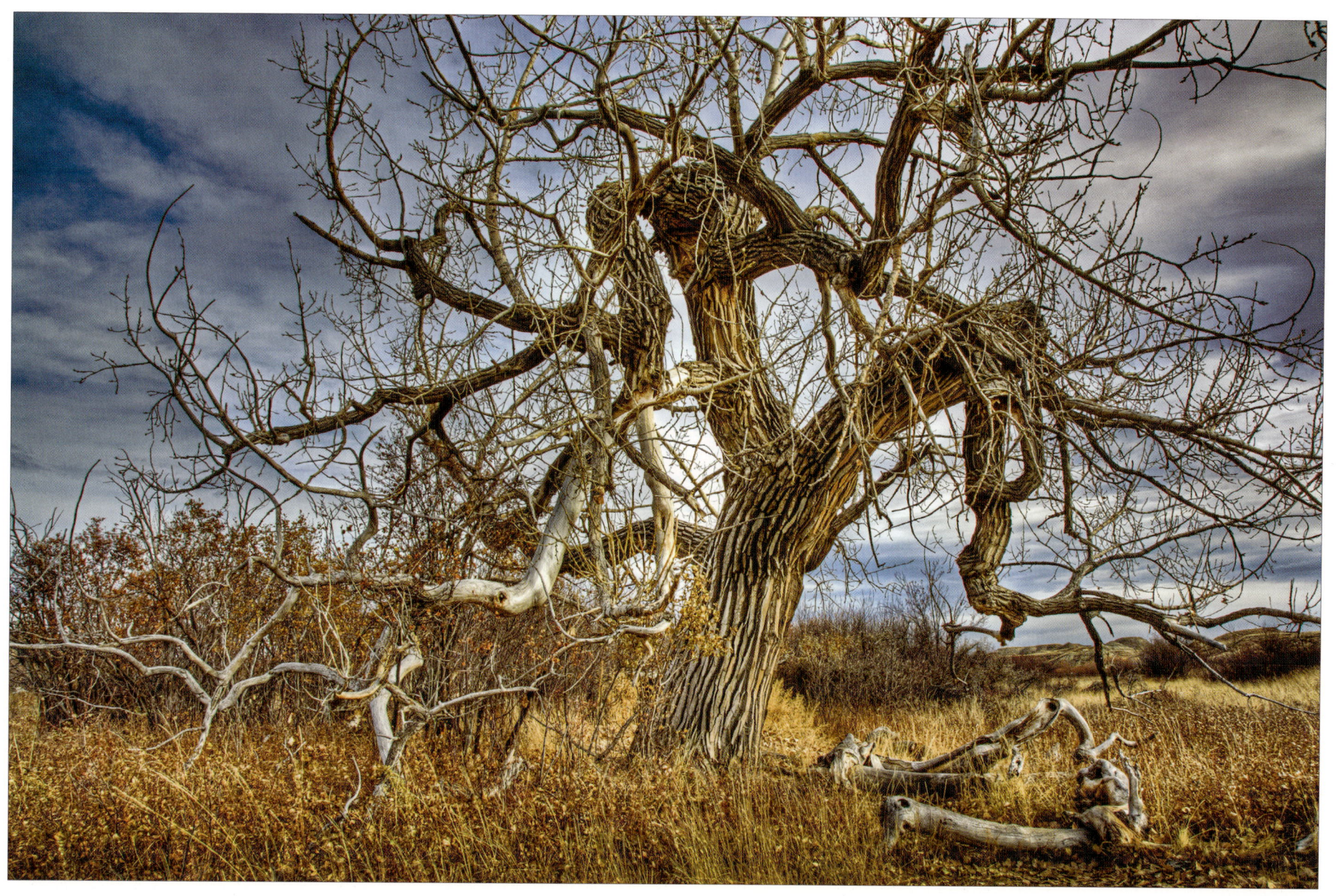

THE MATRIARCH

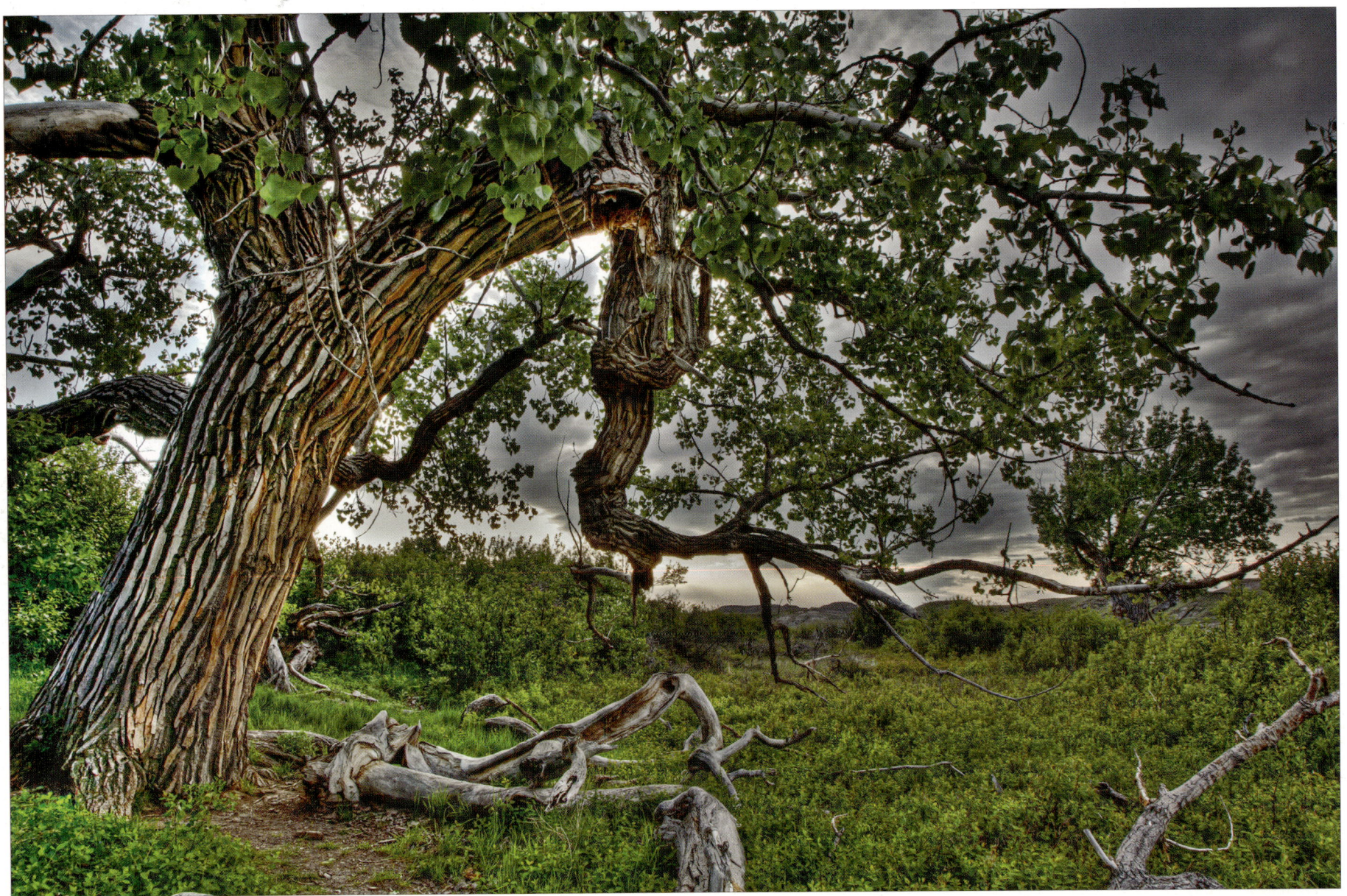

ABOVE: THE MATRIARCH

RIGHT: TREE OF REMEMBRANCE

This photograph was one of three images in this book included in the show "Contemporary Indigeneity" at the Great Plains Art Museum at the University of Nebraska.

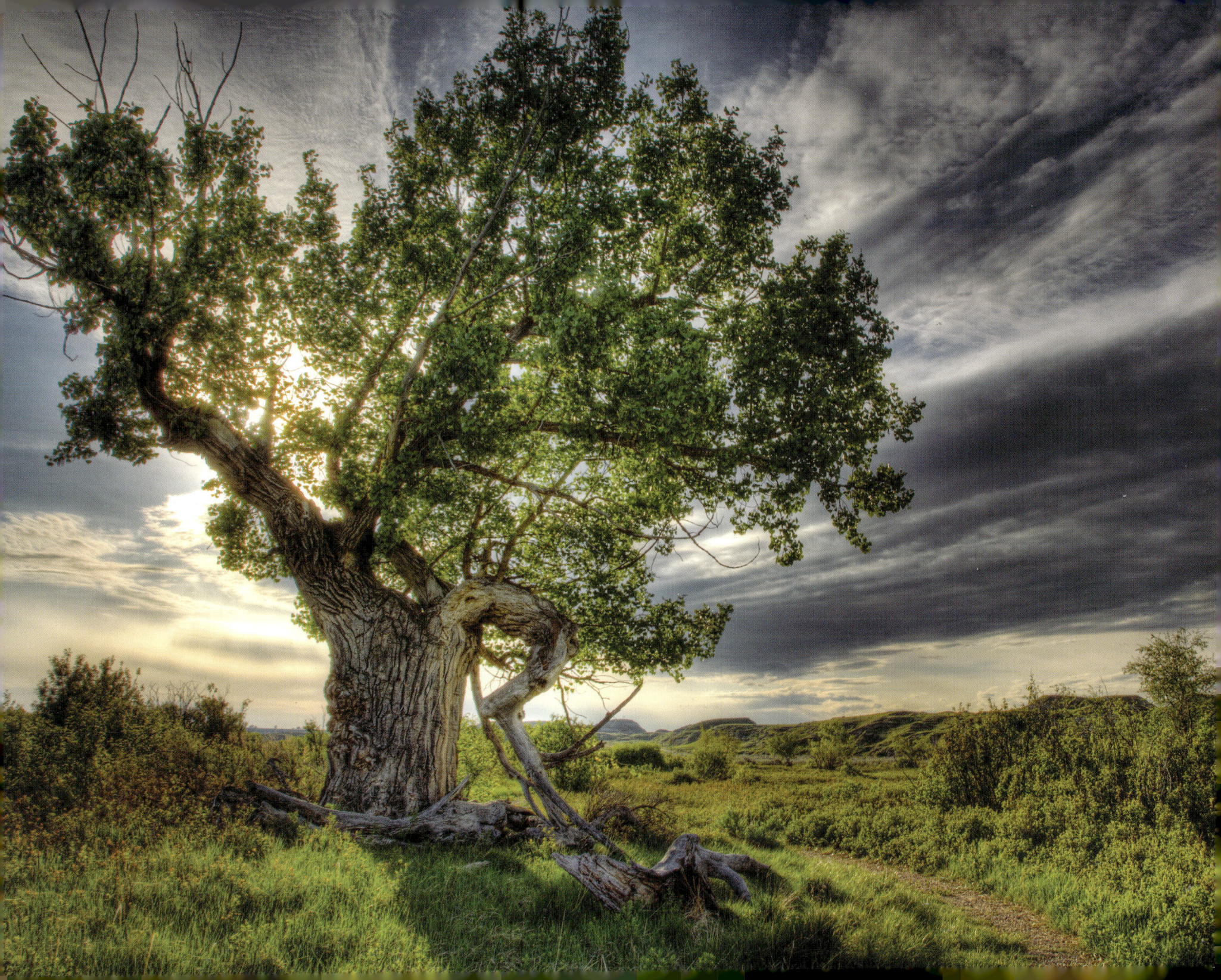

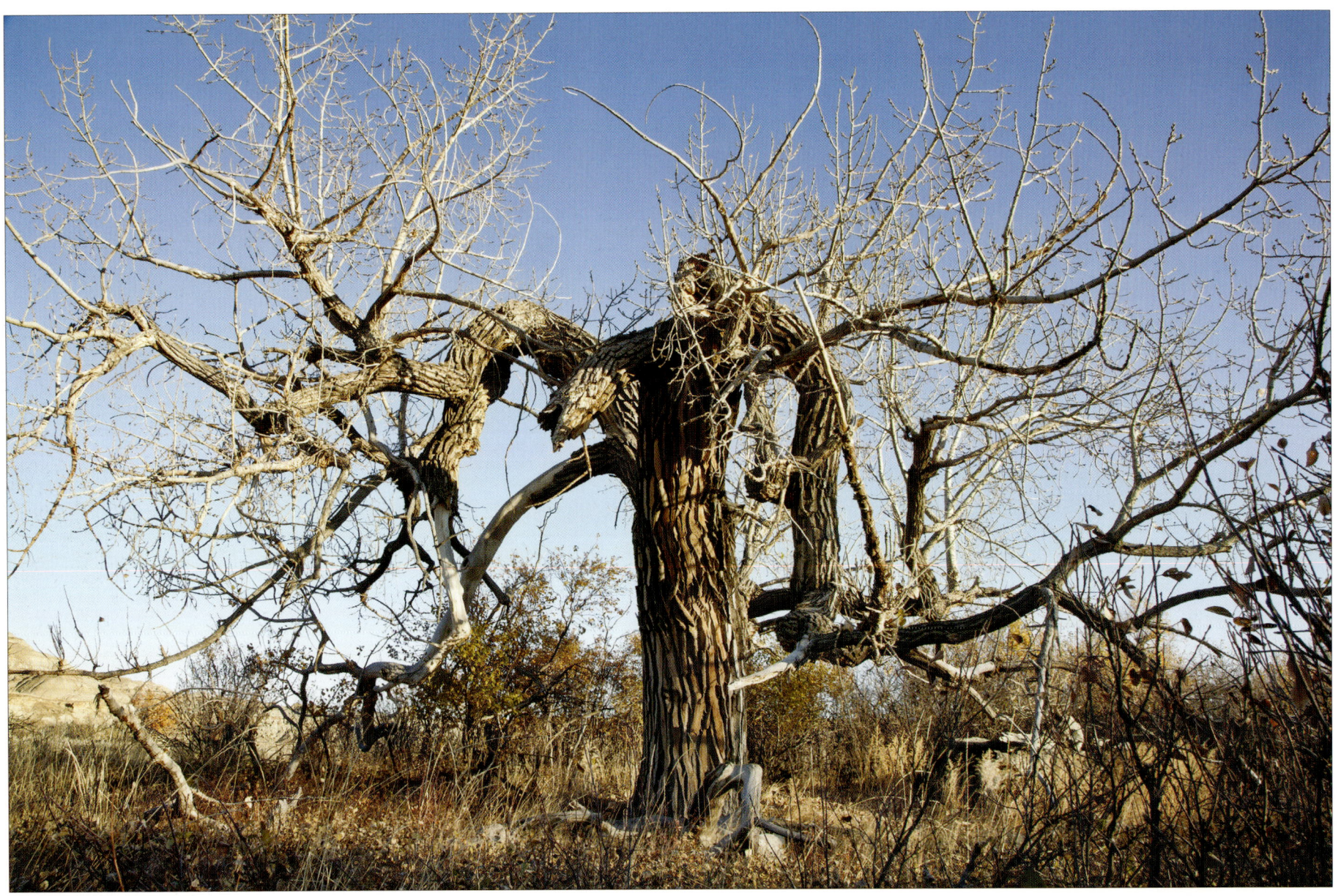

MORAL DISORDER

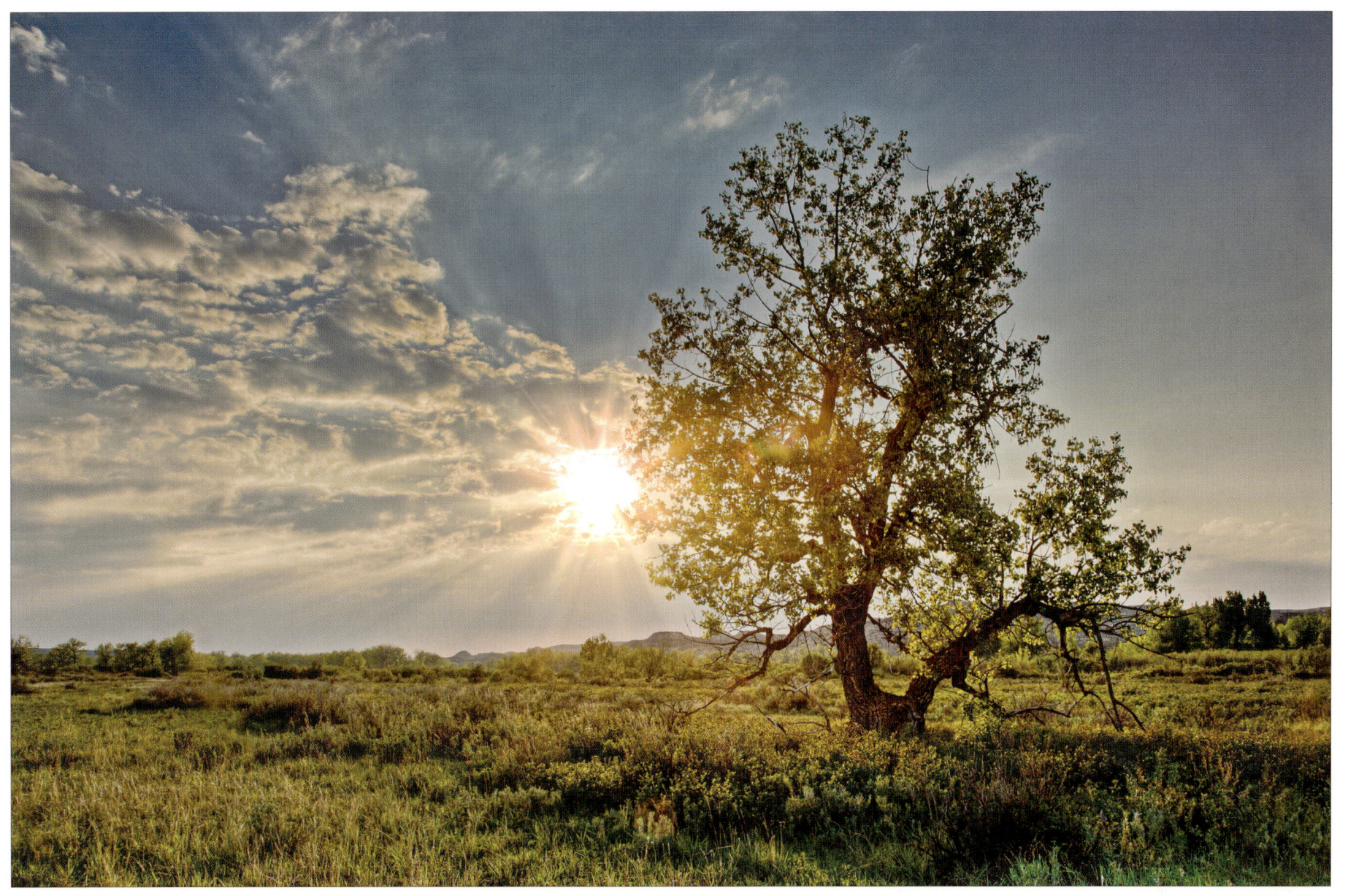

TREE OF HOPE

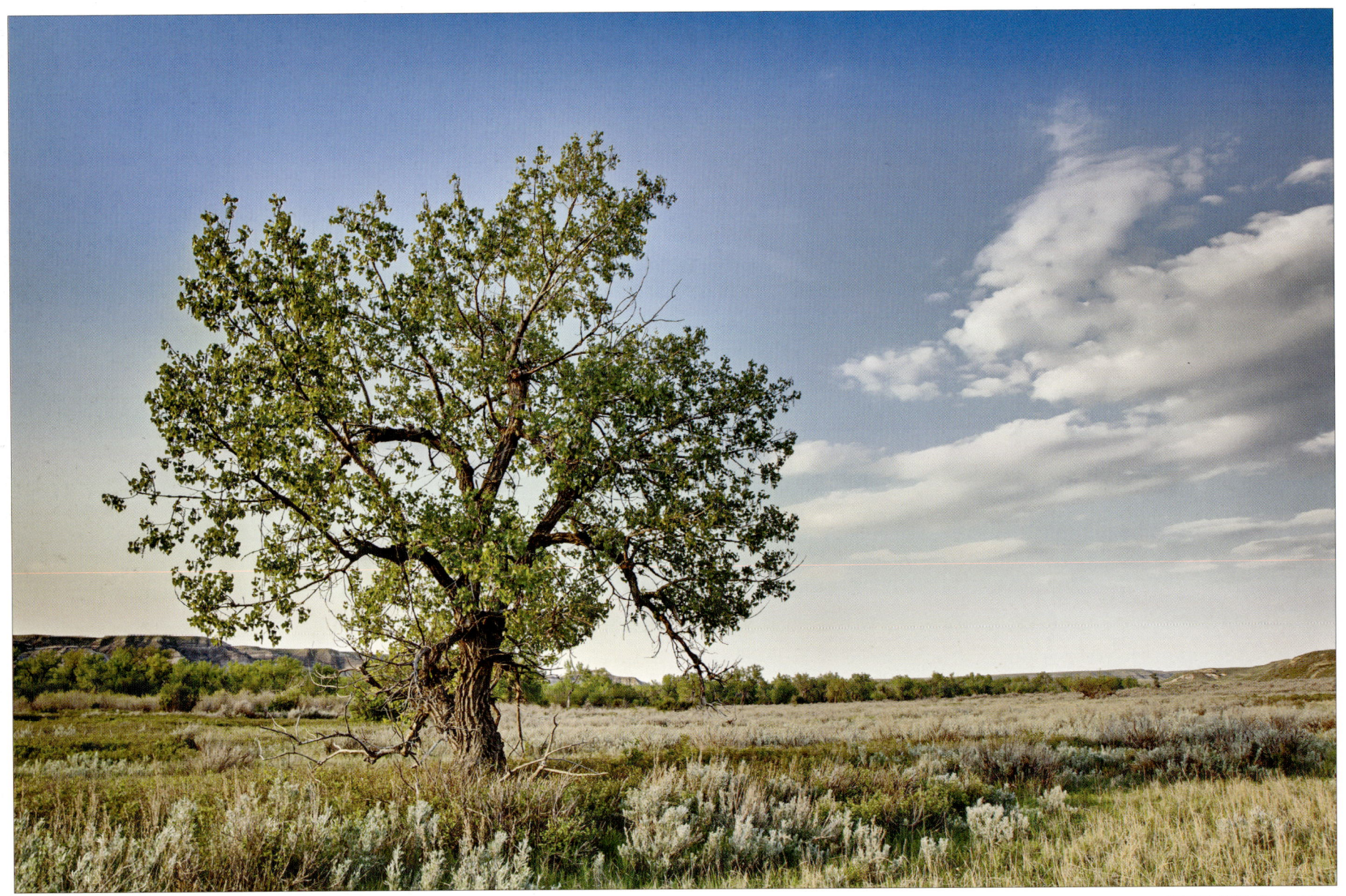

ABOVE: RUNAWAY

RIGHT: THE MATRIARCH

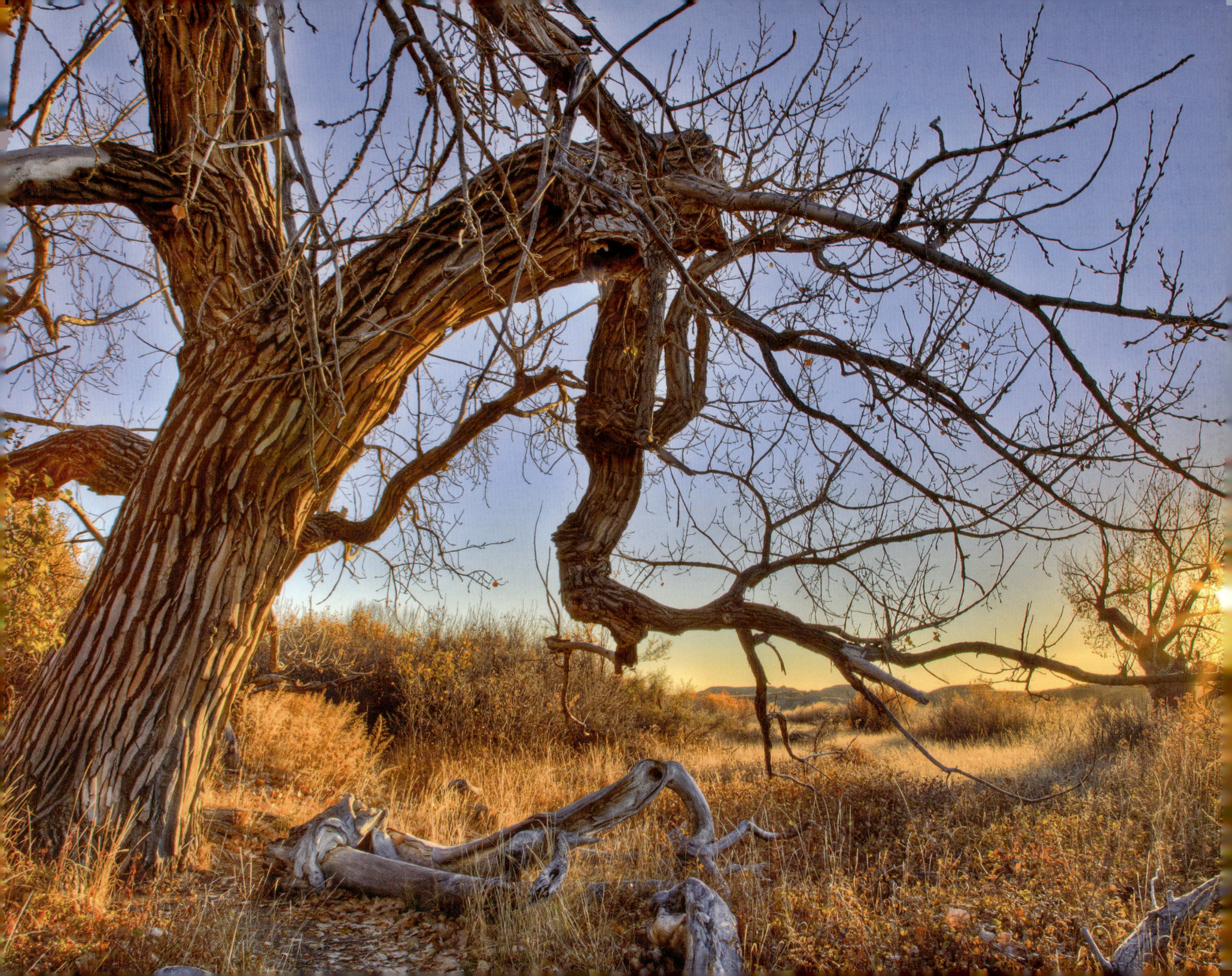

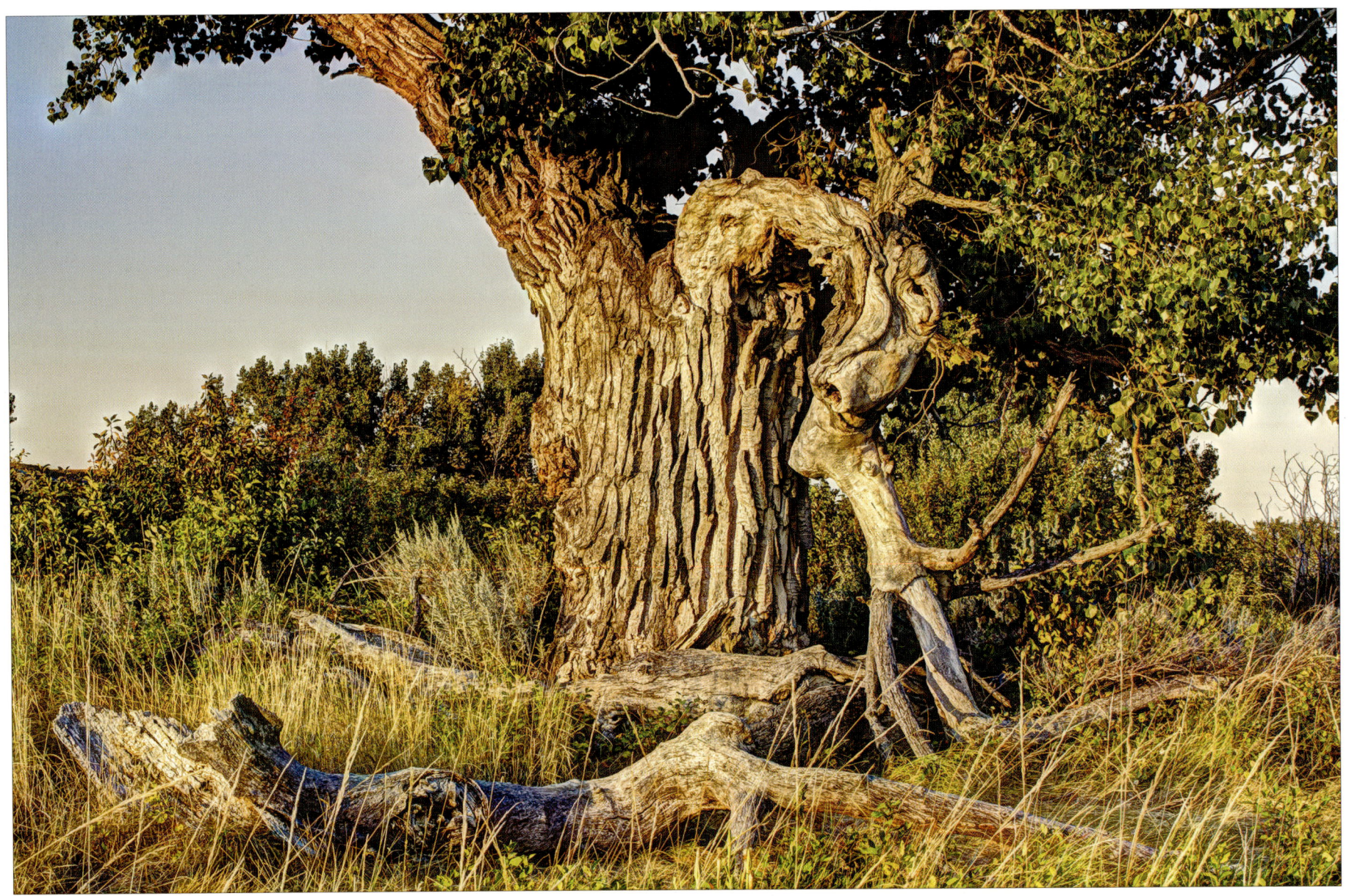

TREE OF REMEMBRANCE

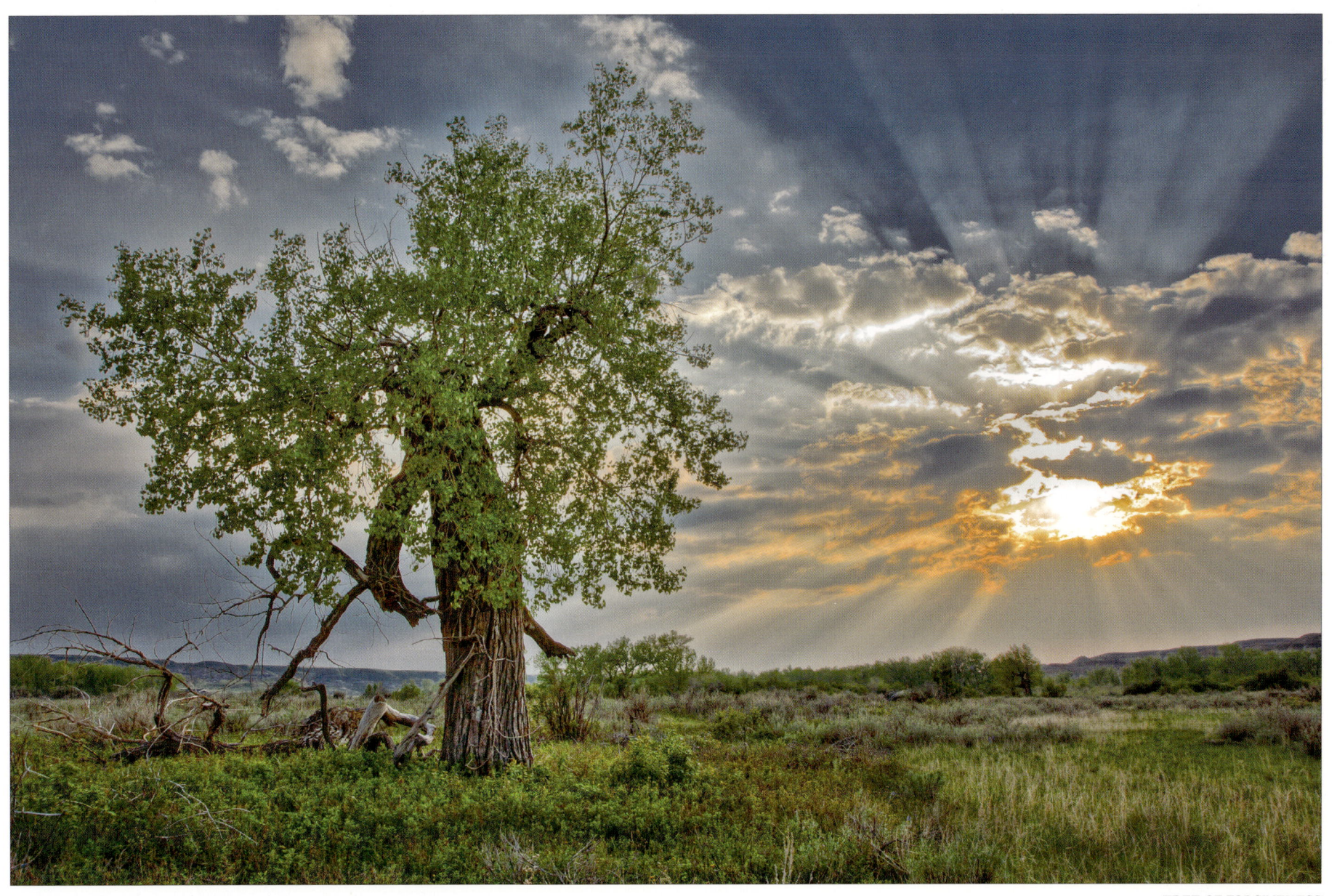

TREE OF FORGIVENESS

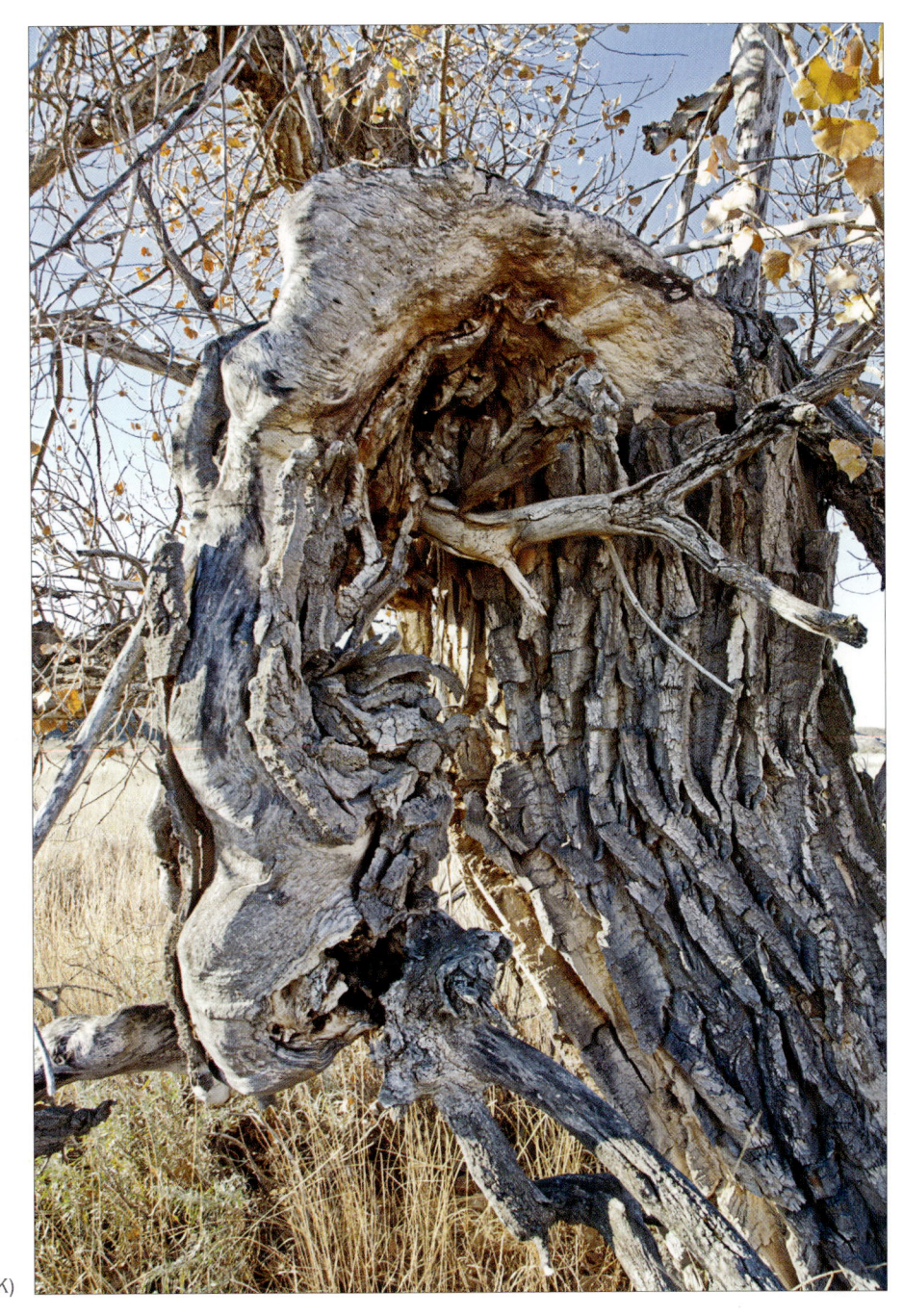

COTTONWOOD (DETAIL OF BARK)

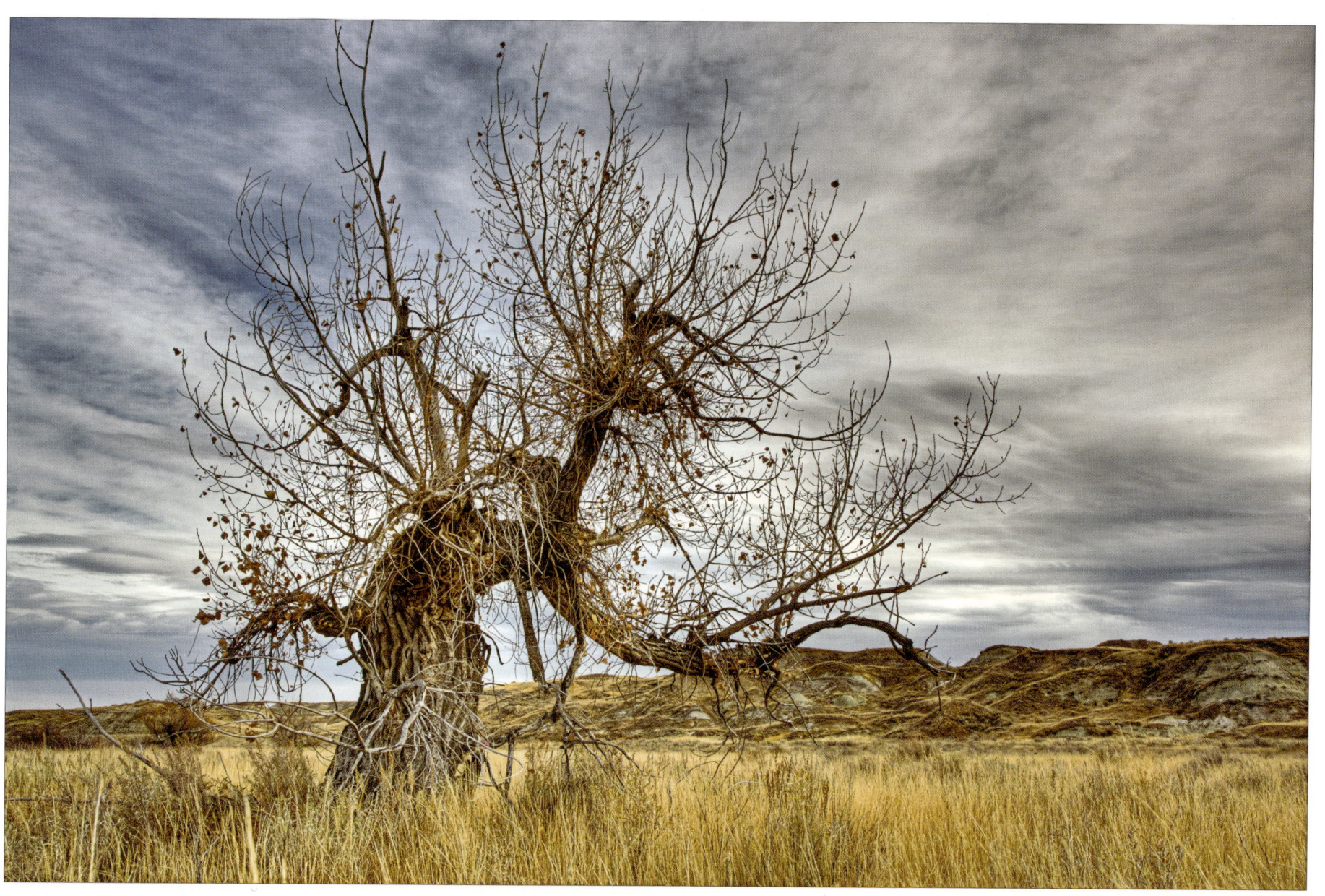

BODILY HARM

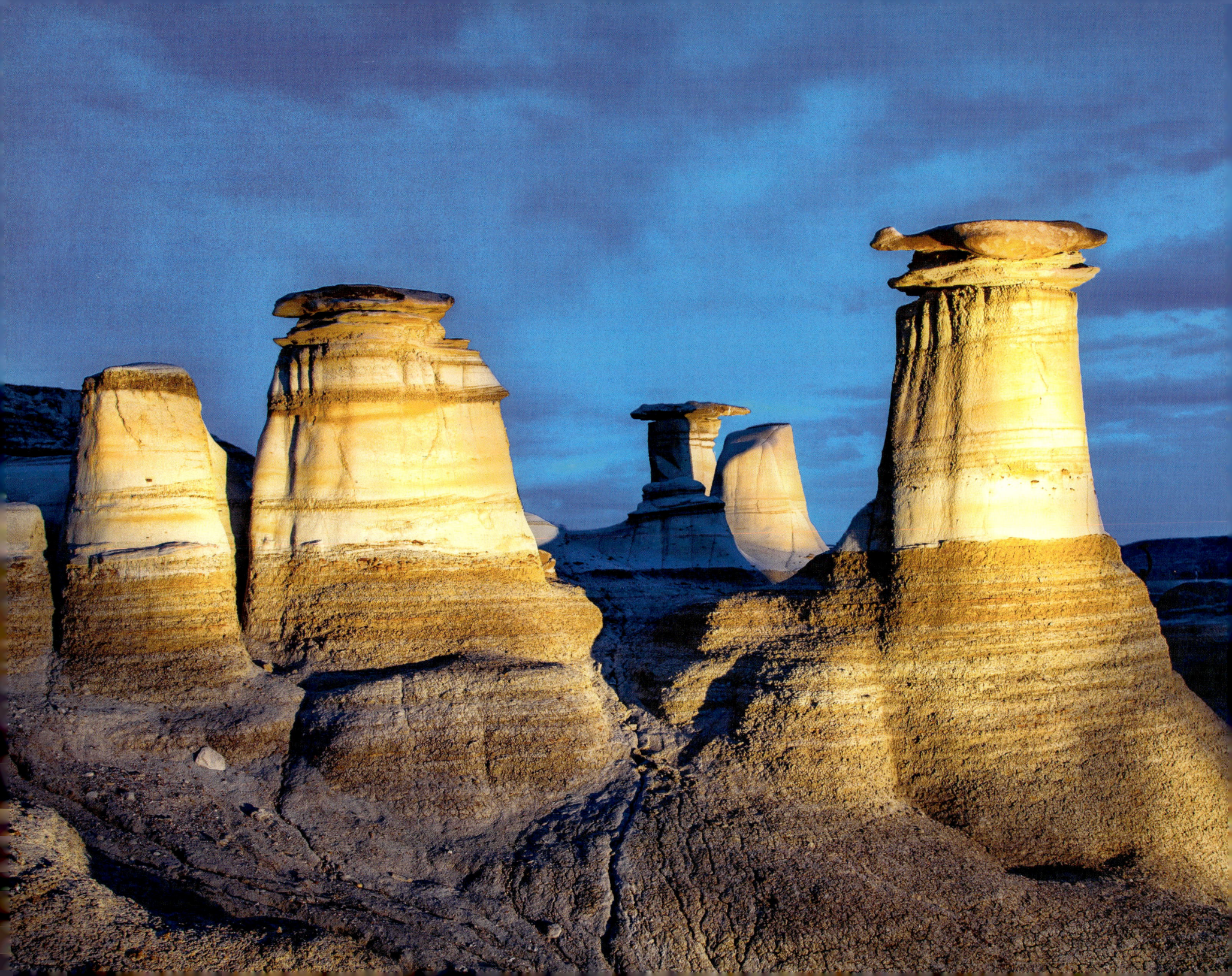

DRUMHELLER

Located one hour northeast of Calgary, Alberta, along Highway 9, and nestled perfectly among the picturesque badlands of the Red Deer River Valley, the Town of Drumheller boasts a number of prominent attractions. Two very special ones stand out from the rest. First and foremost, Drumheller is home to the world-class Royal Tyrrell Museum. What the Louvre is to art, the Royal Tyrrell is to paleontology. The exhibition space alone is a monstrous 4,400 square metres. The main building is 11,200 square metres, and the ATCO Tyrrell Learning Centre is an additional 1,500 square metres. This facility is the only Canadian museum dedicated exclusively to the study of paleontology.

The jewel of the museum is Dinosaur Hall, which contains one of the world's largest displays of fully reconstructed dinosaur skeletons. There are 40 in total, including all of the big names, like Albertosaurus, Stegosaurus, Camarasaurus, Triceratops, and, of course, the King of the Tyrant Lizards, Tyrannosaurus rex. The other must-see is the Lords of the Land exhibit. This exhibit showcases some of the museum's most scientifically significant and rare fossils in the collection. Included here is the renowned blackened T-rex skull called "Black Beauty." Also included is likely the most prized specimen in the entire collection—a skeleton of a juvenile Gorgosaurus in a dramatic death pose, with its head snapped back so far that it almost touches its raised tail. This specimen is the most complete of its kind in the world. Also of significance is the initial find of the museum's namesake, Joseph Burr Tyrrell (pronounced TEER-uhl), who discovered an Albertosaurus in 1884. At the time, Tyrrell was a geologist working with the Geological Survey of Canada.

One attribute that makes the Royal Tyrrell unique among museums is its location. While the world's big-name dinosaur museums are located in Berlin, Brussels, or Chicago, paleontologists at the Royal Tyrrell in Drumheller need only walk out their back door to explore one of the best dinosaur fossil sites on the planet. The Alberta badlands, extending from Drumheller all the way southeast to Dinosaur Provincial Park, are literally a fossil warehouse for dinosaur bones. In fact, the museum seems to make important "recent discovery" announcements at least twice a year. In October 2012, researchers found the first feathered dinosaur in North America. It was located amidst 75-million-year-old rocks almost on their doorstep.

Secondly, Drumheller is also home to a world-class stand of sculpted hoodoos that even have their own parking lot to accommodate the thousands of visitors they attract each year (see "The Stone Diaries"

LEFT: THE STONE DIARIES (NIGHT)
In order to showcase these popular hoodoos in a unique way I decided to "light-paint" them. This type of night photography involves opening the shutter long enough to paint the scene with a light source.

page 96). The parking lot even has a covered picnic area. To their advantage, these chiseled rock pillars, located approximately 16 kilometres southeast of Drumheller, are both clearly visible and easily accessible from Highway 10. Like a work of art, they are also perfectly showcased. Free-standing and tall, this grouping of hoodoos sits perfectly on a natural rock pedestal. Surrounding them is an impressive and natural badlands amphitheatre complete with deep-fissured hills and streaks of black coal. It's a stunning vista. For better or for worse, a metal staircase with three observation platforms has been constructed around these rock monuments, offering the hoodoos protection and sightseers a full 360-degree viewing perspective.

There is ample badlands scenery to explore in and around Drumheller. Horseshoe Canyon, situated along Highway 9 toward Calgary, is easily the busiest. It's an isolated pocket of badlands located approximately 17 kilometres southwest of Drumheller that has been described as a miniature Grand Canyon. The lesser-known Horsethief Canyon, located to the northwest along Highway 838 (North Dinosaur Trail), has just as many breathtaking panoramas. Local legends suggest that Horsethief Canyon was so named because it was the ideal place for thieves to rebrand stolen cattle and horses.

All of these features and more add to the allure of Drumheller and help make it the epicentre of the Canadian badlands.

LEFT: STILL LIFE (NIGHT)
Isolated by painted light, this large statuesque hoodoo, part of "The Stone Diaries" collection, glows against the night sky.

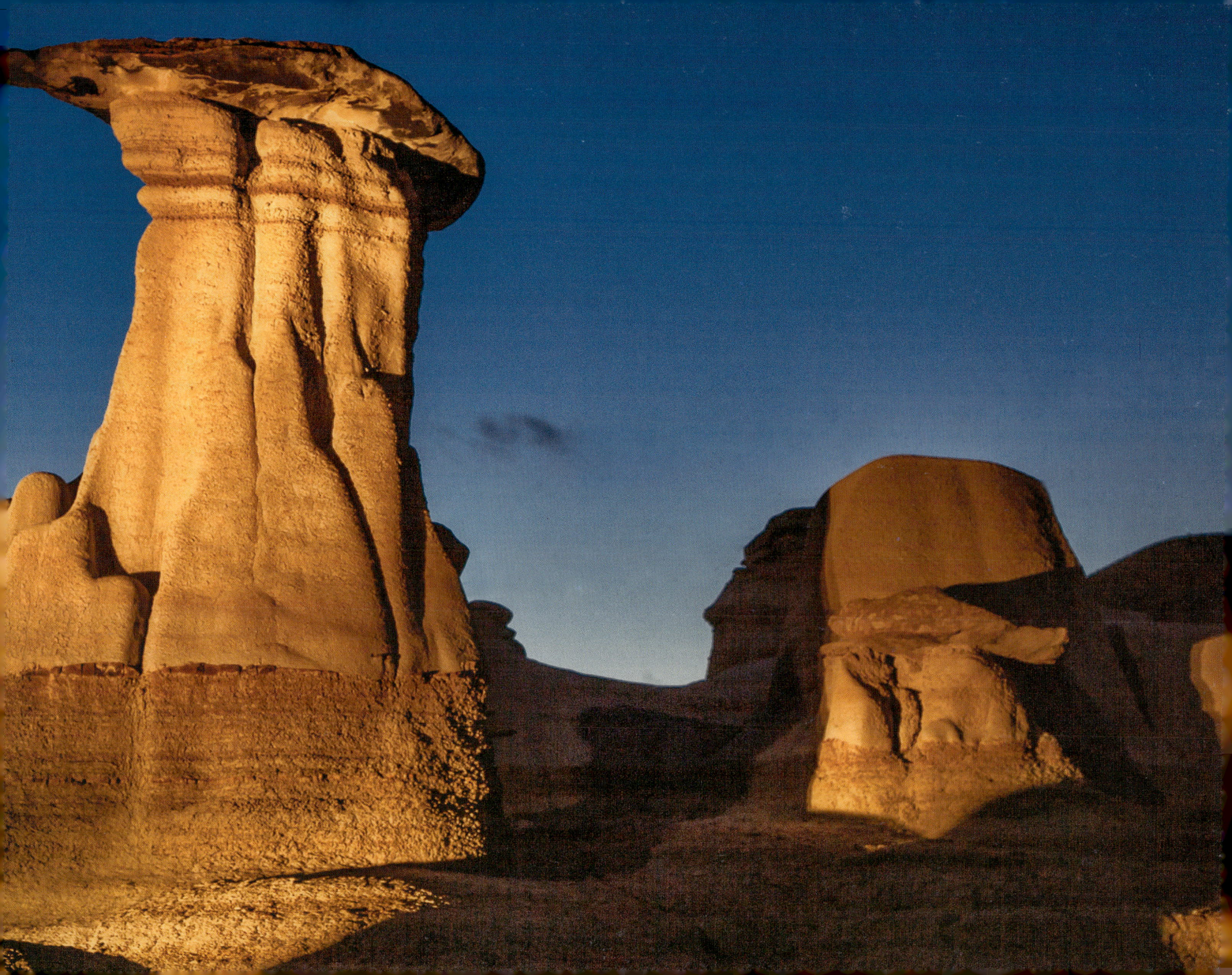

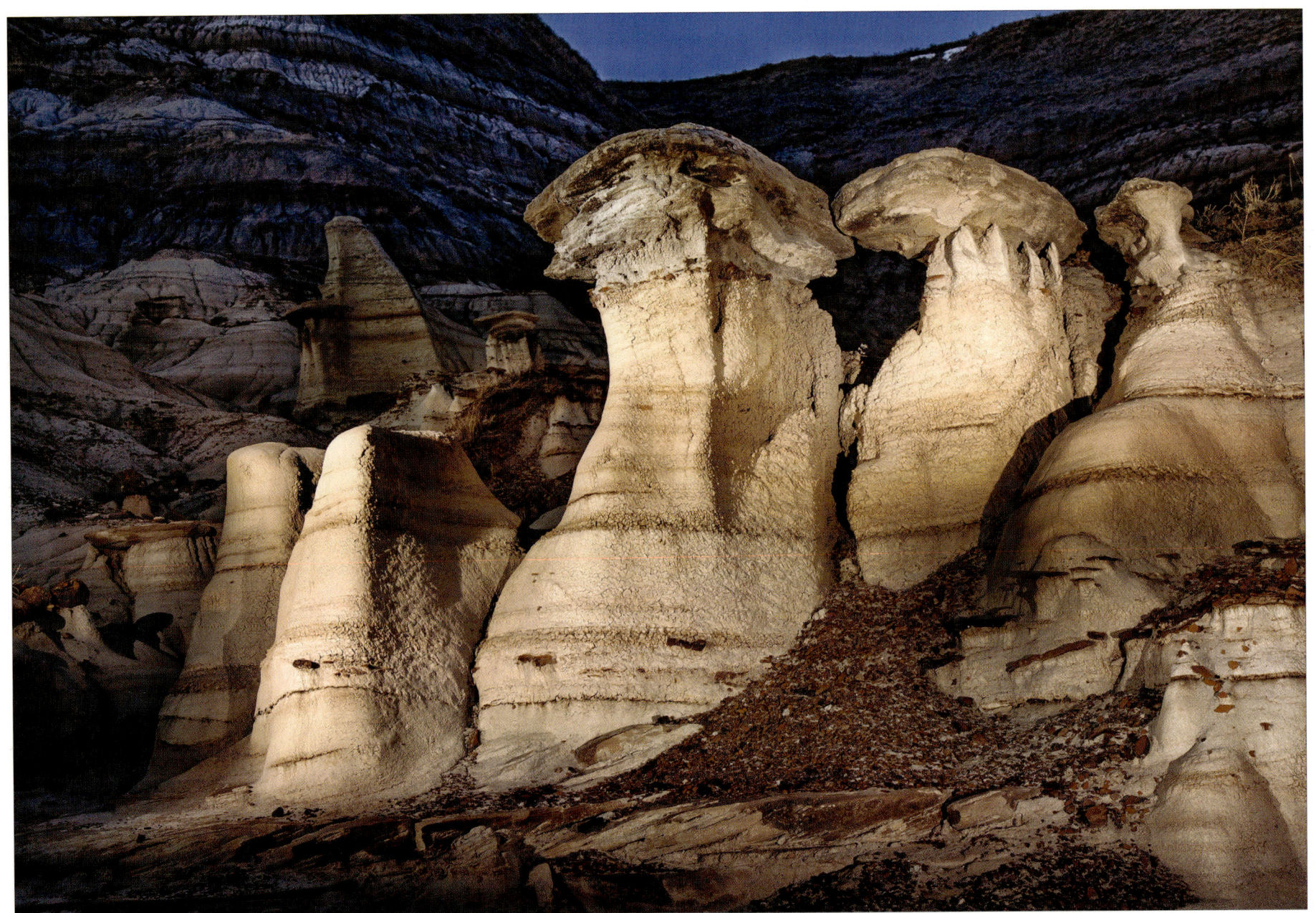

STRANGE THINGS
A light-painted collection of striking hoodoos in Drumheller's Dinosaur Valley.

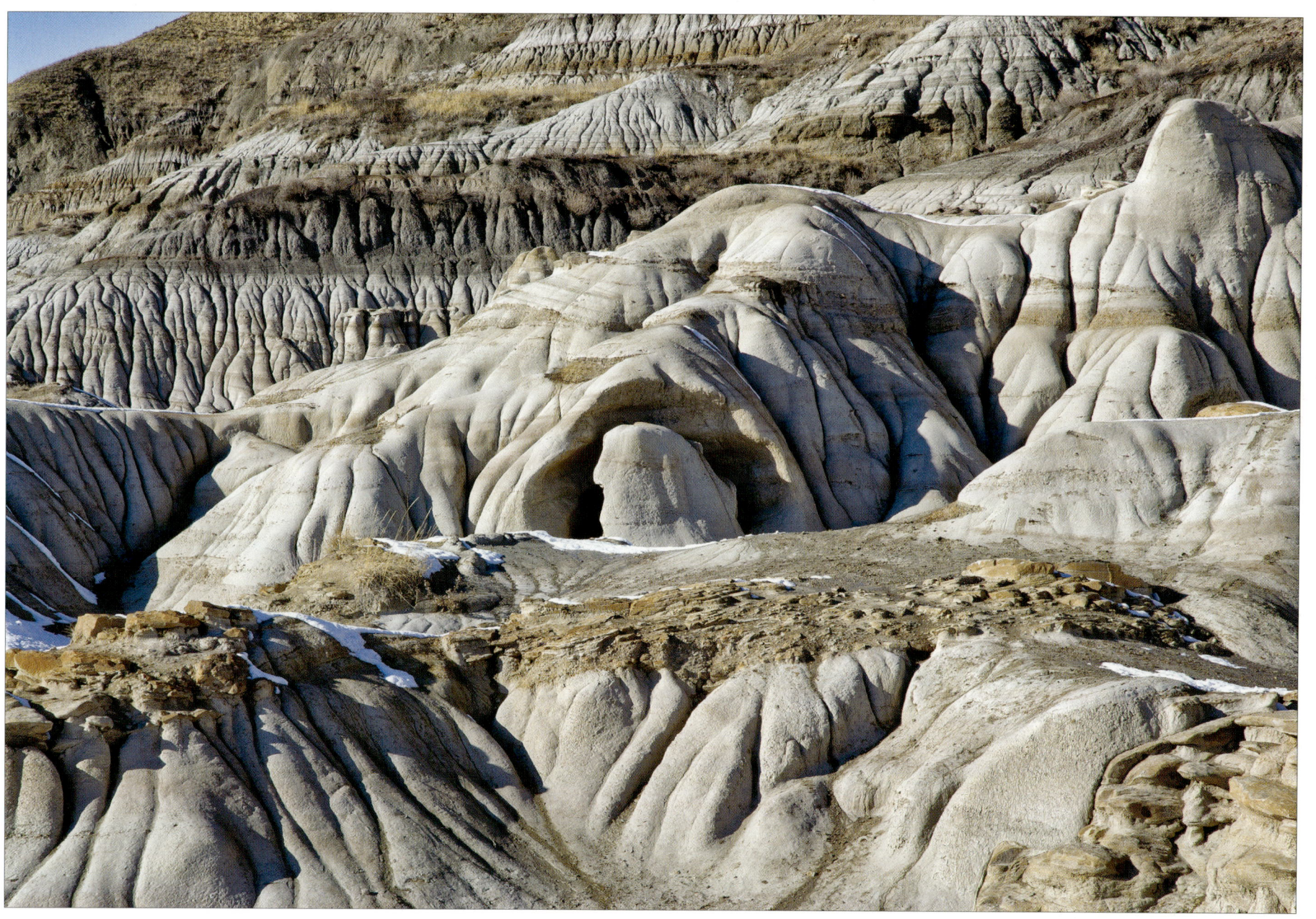

LIVES OF SHADOWS
This secret cave interrupts the ordered chaos of the deep-grooved fissures carved into the hills.

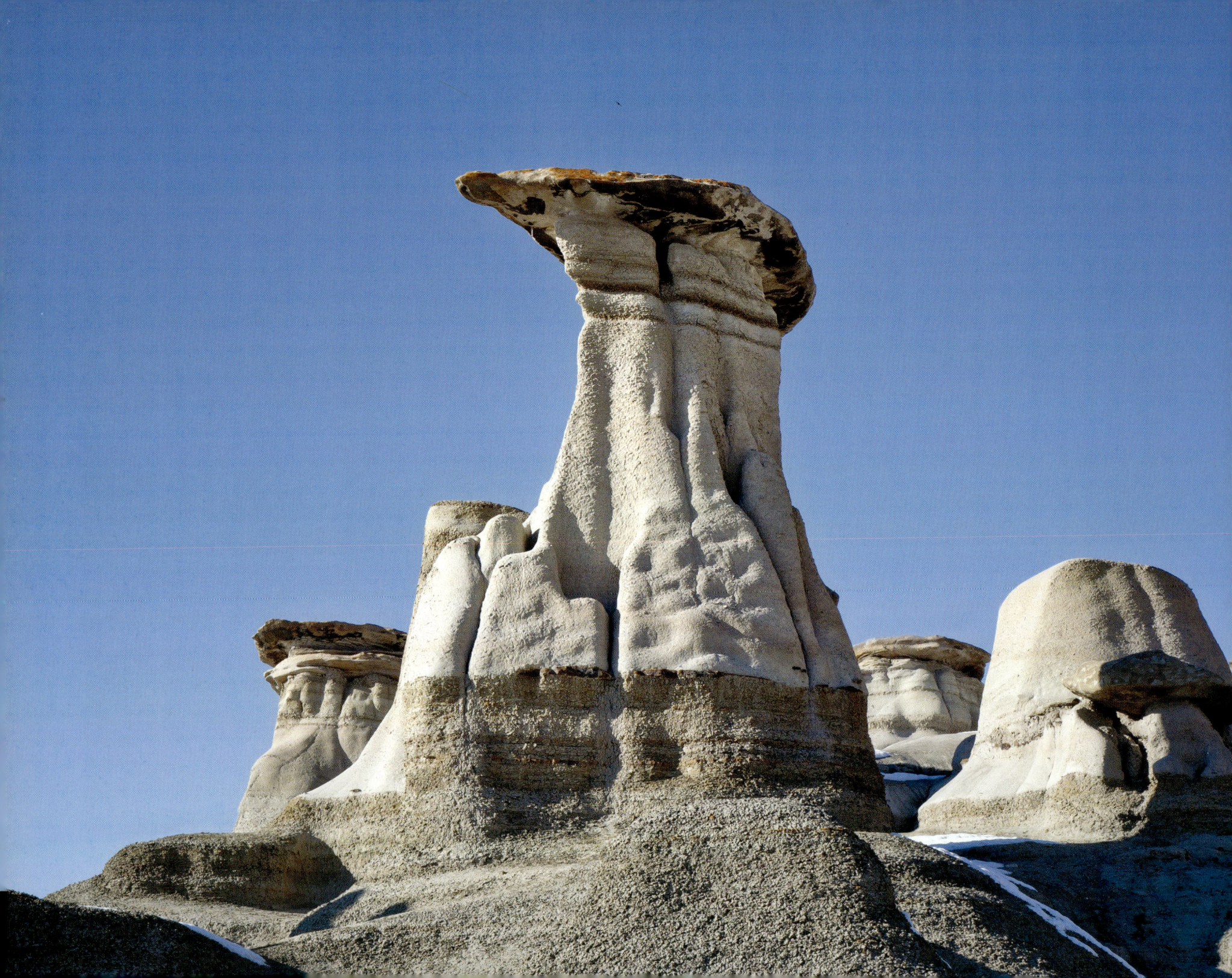

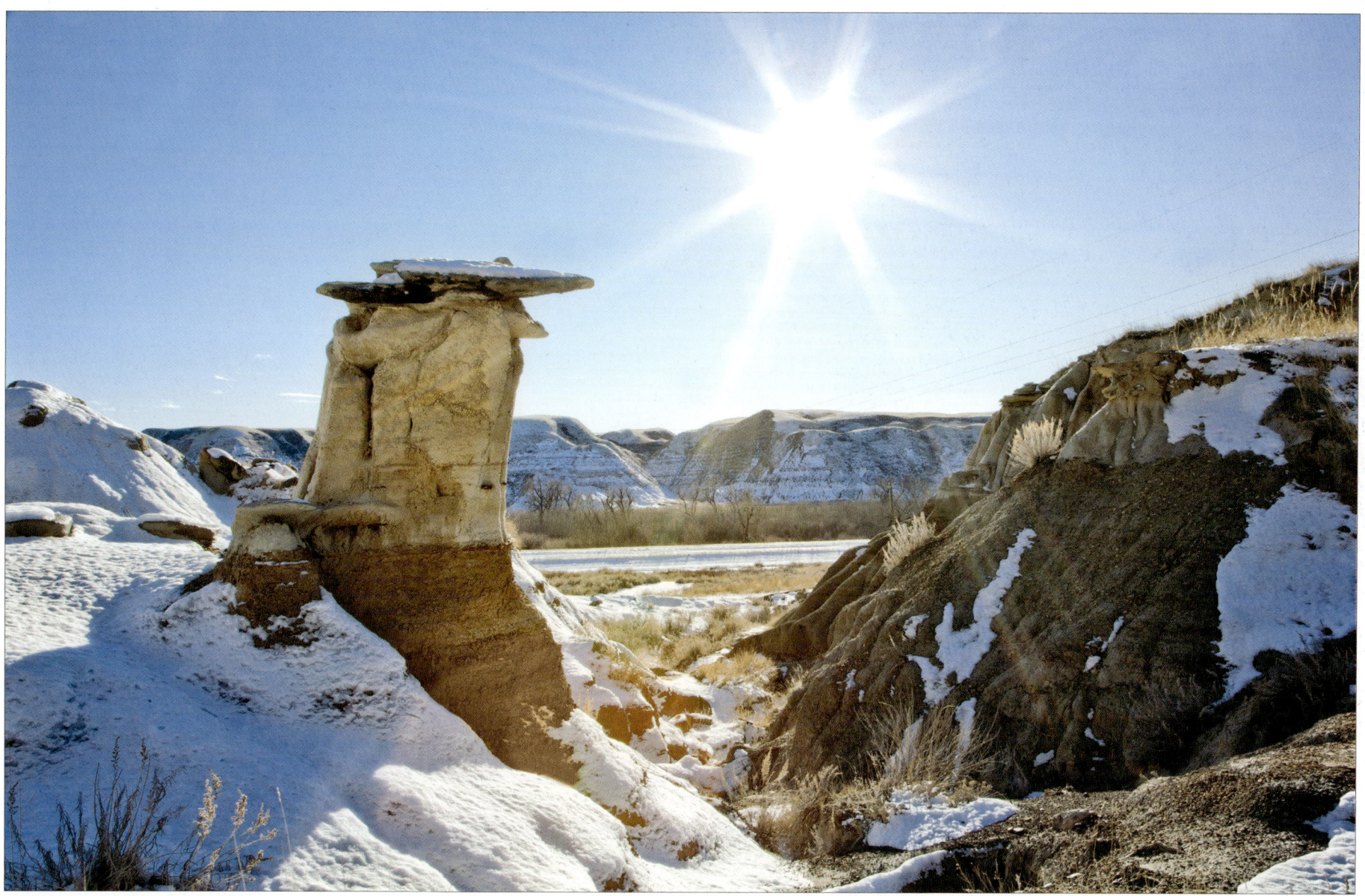

LEFT: STILL LIFE (WITH ROCK AND SANDSTONE)

RESURRECTION
On this winter day the snow was melting and the wet clay was treacherous to walk on. With camera and tripod in hand, I slipped and fell on more than one occasion. Needless to say, I was covered in mud.

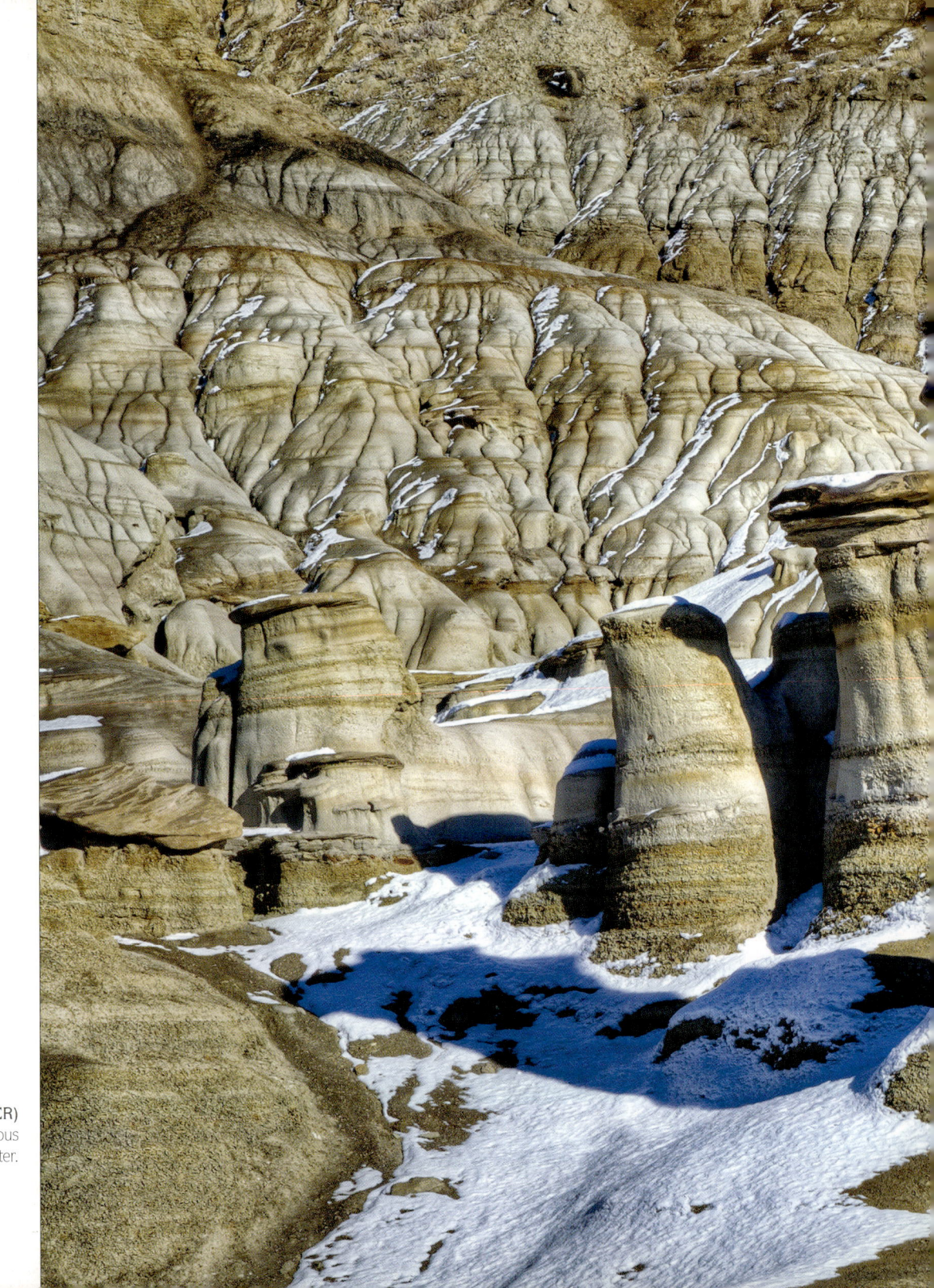

THE STONE DIARIES (WINTER)
With a skiff of snow to highlight the rills, these fabulous
hoodoos look even better in winter.

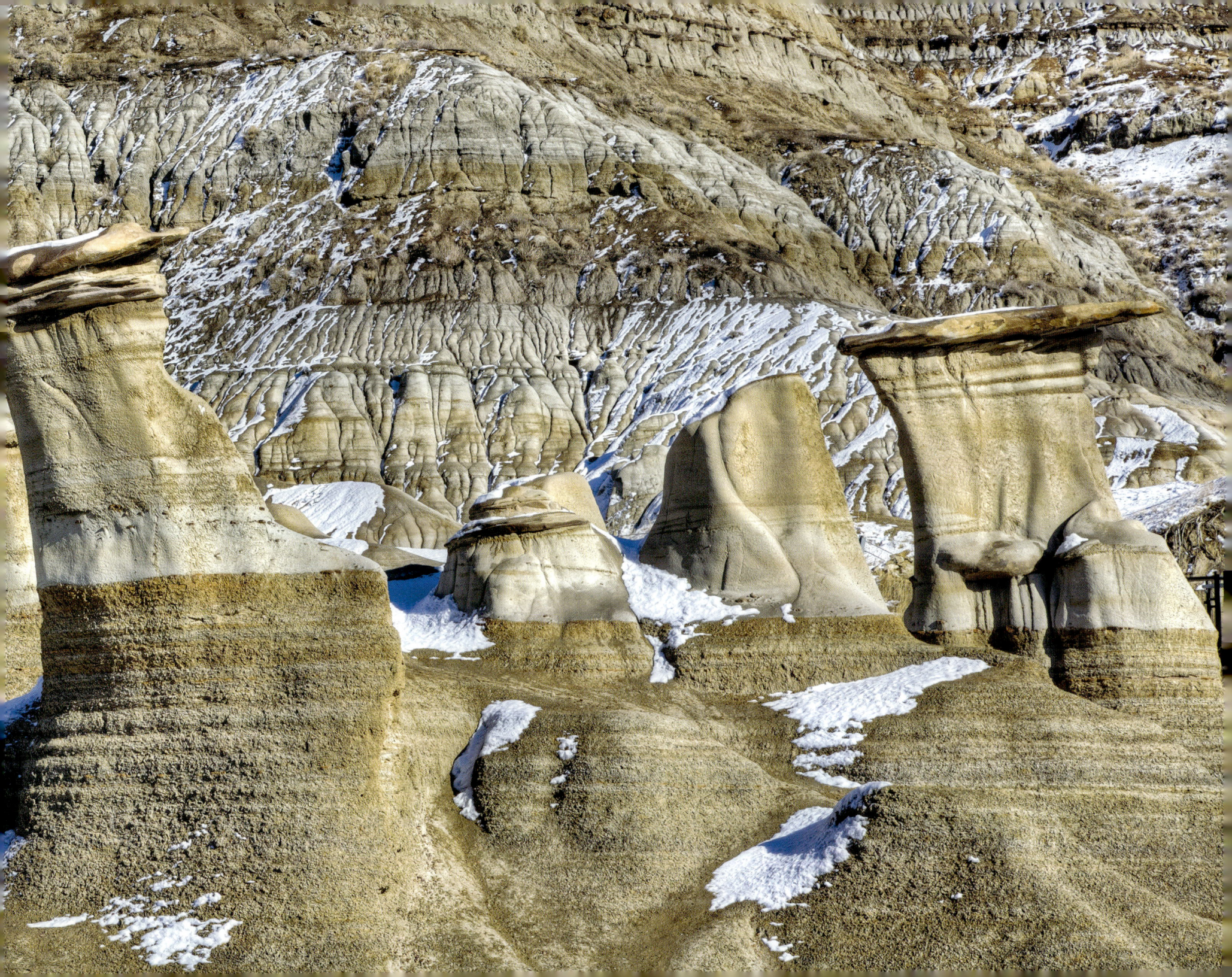

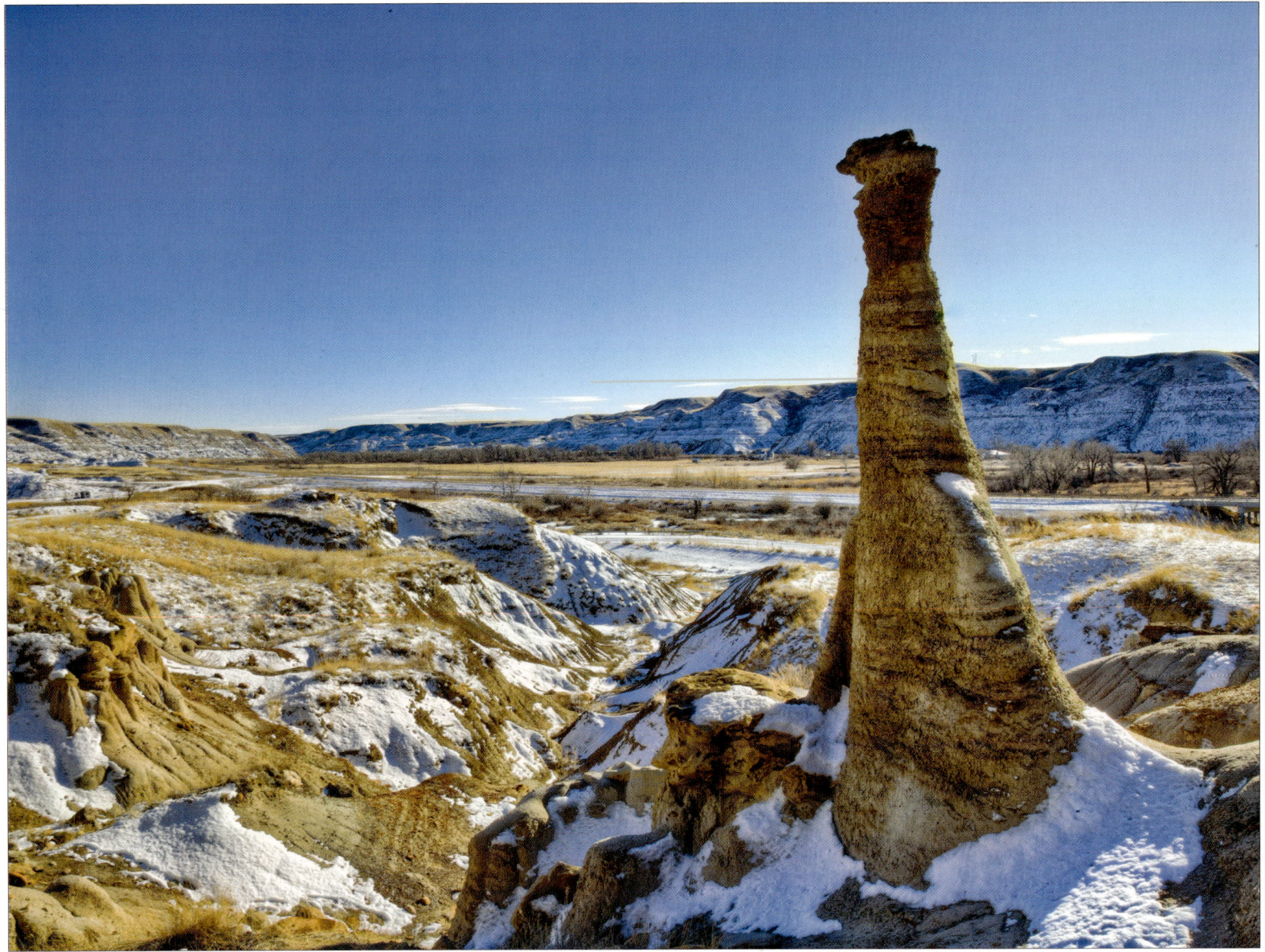

ABOVE: STRANGE FUGITIVE
I saw this spire (a hoodoo without a caprock) from Highway 10 (seen in the background) and hiked into the hills for a better view.

RIGHT: HORSETHIEF CANYON
This canyon has a spectacular Grand Canyon feel to it. Needless to say, it was once a great place to hide stolen horses.

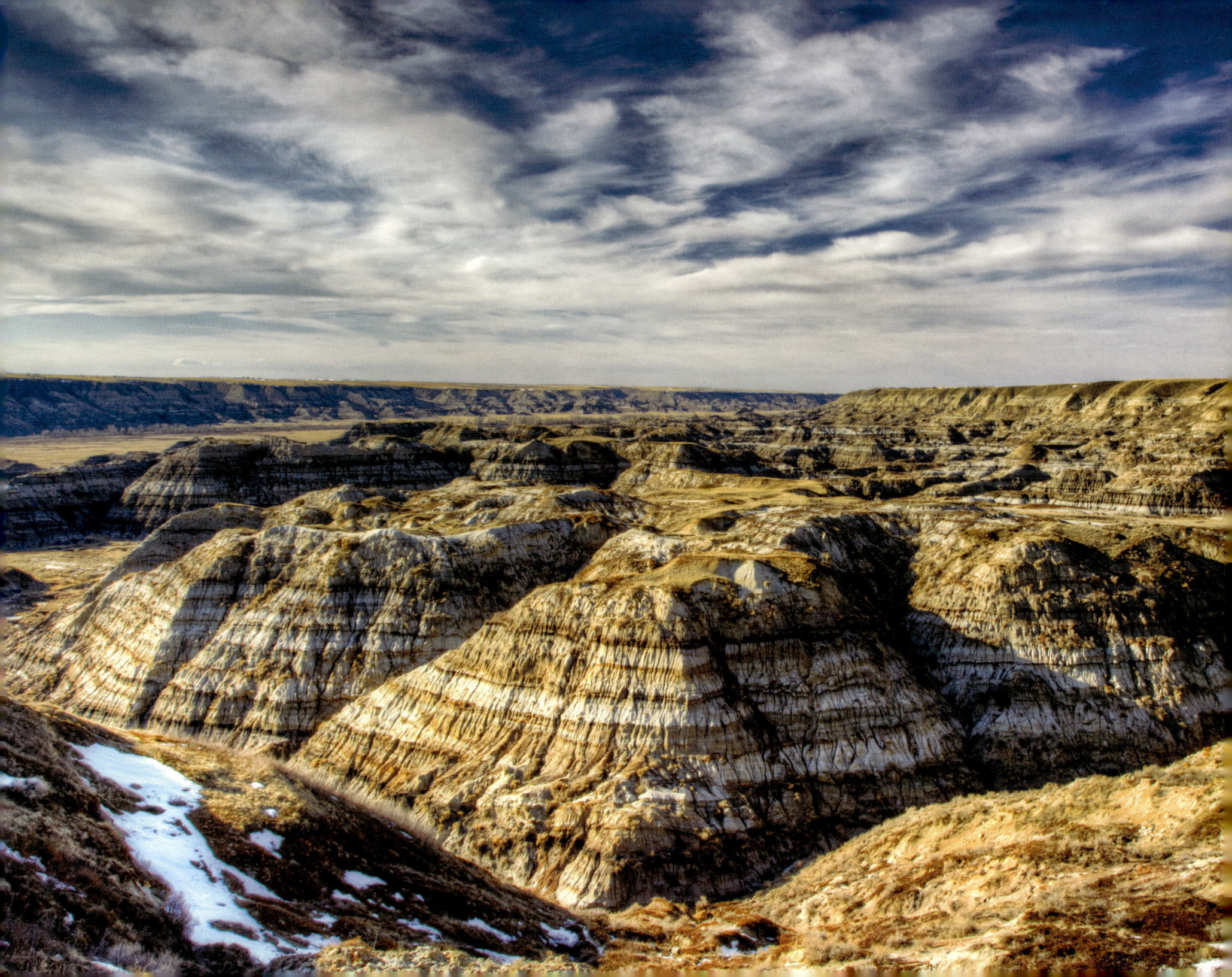

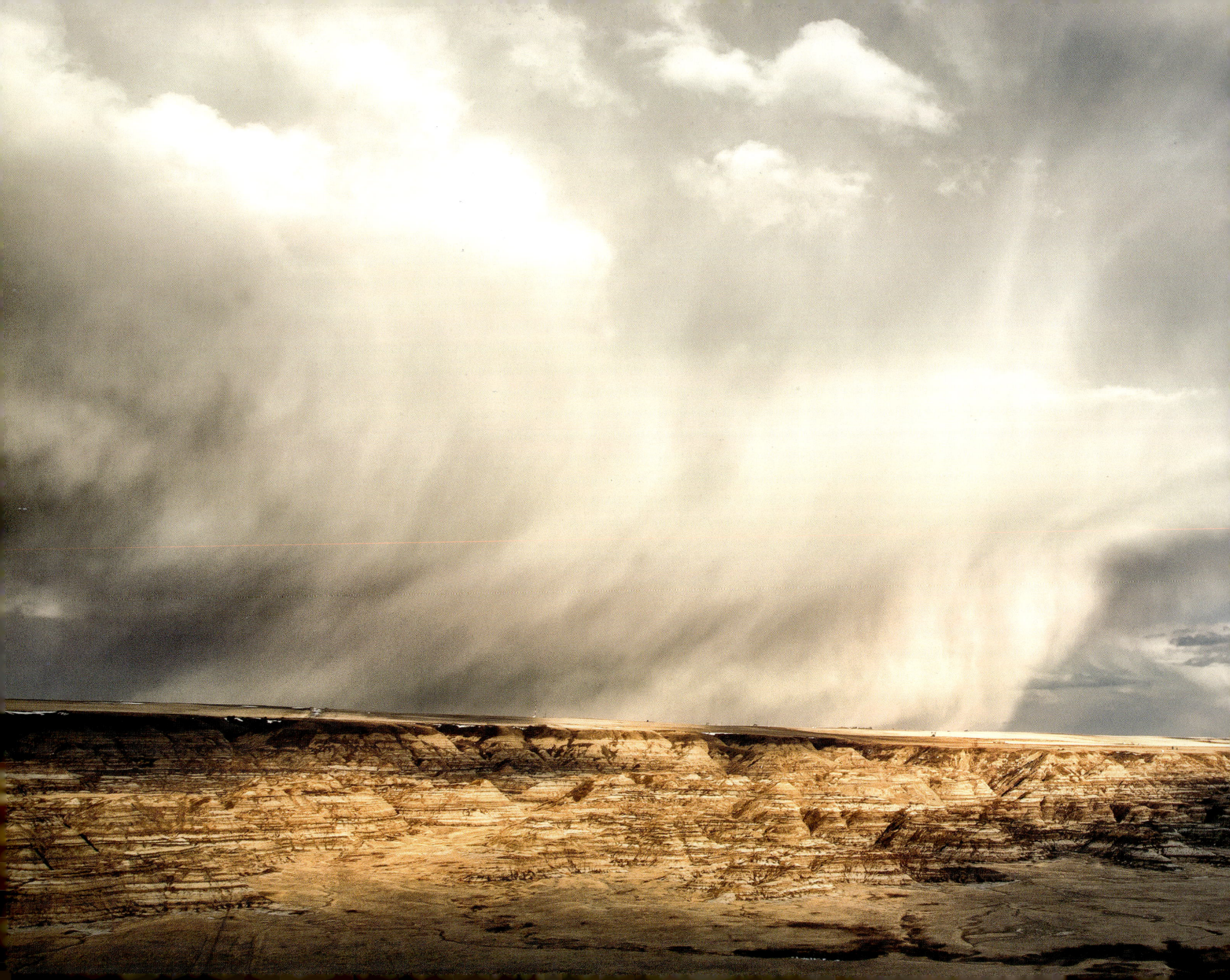

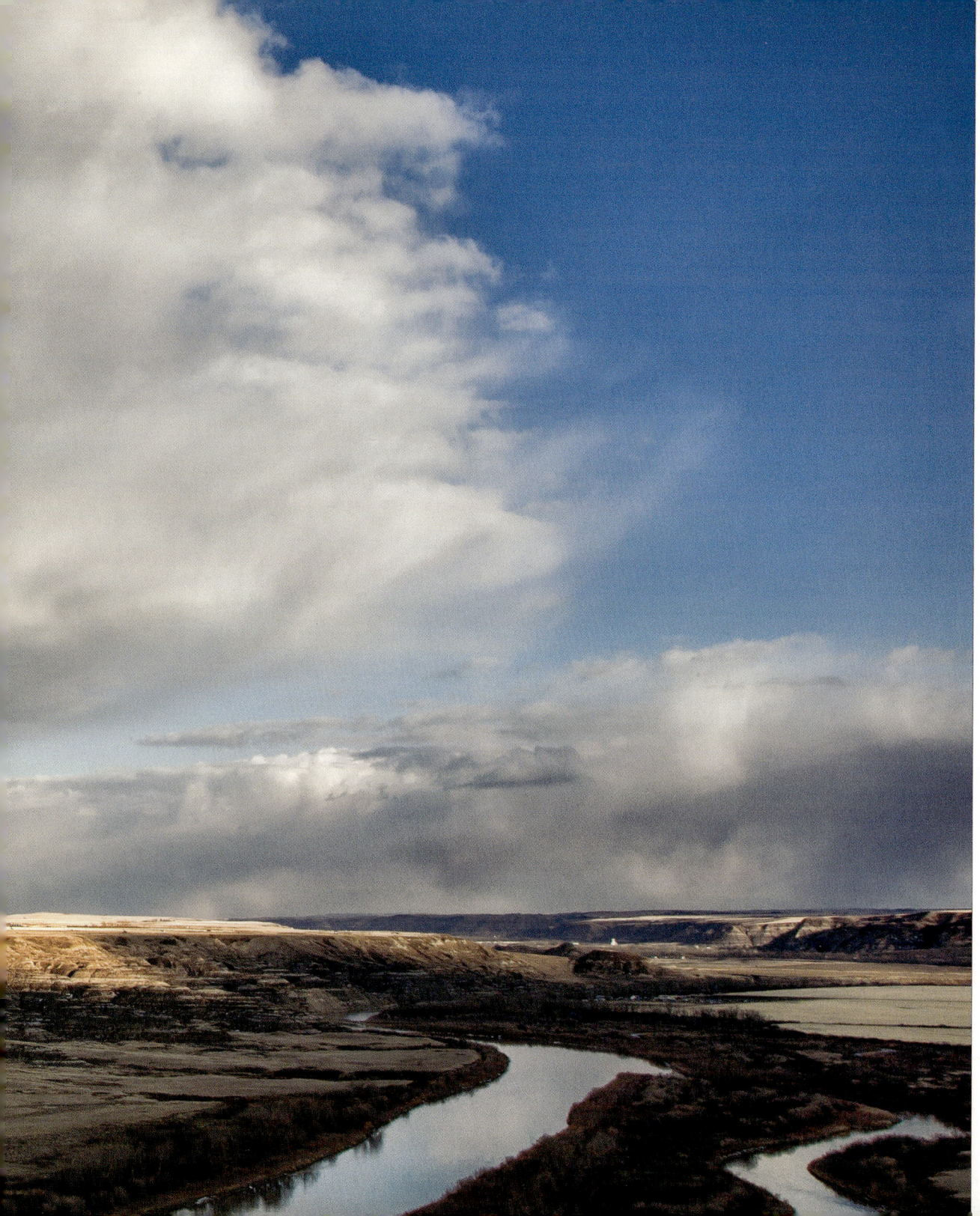

WHO HAS SEEN THE WIND (RED DEER RIVER VALLEY)

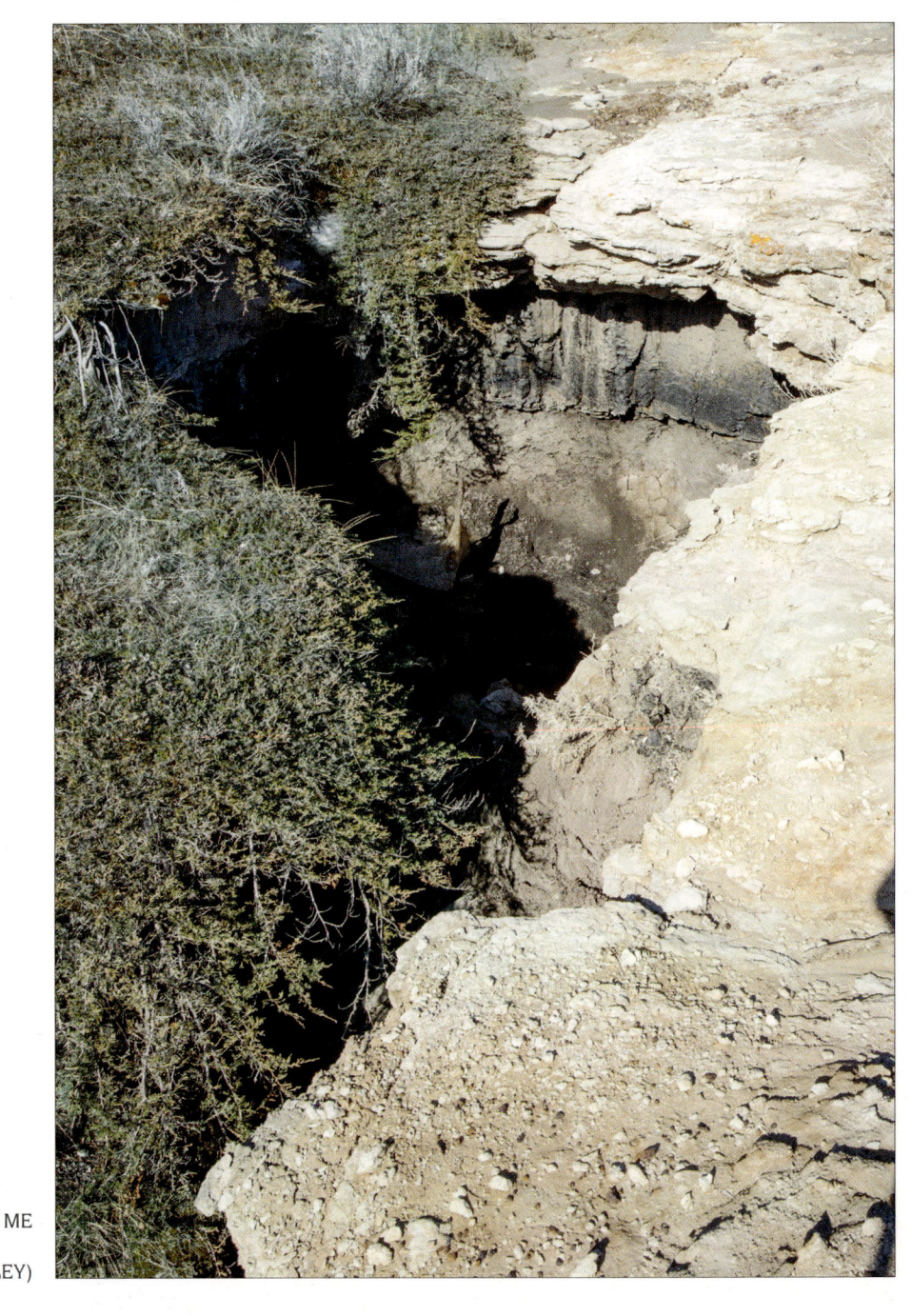

YOUR SECRETS SLEEP WITH ME

RIGHT: LIGHT LIFTING (RED DEER RIVER VALLEY)

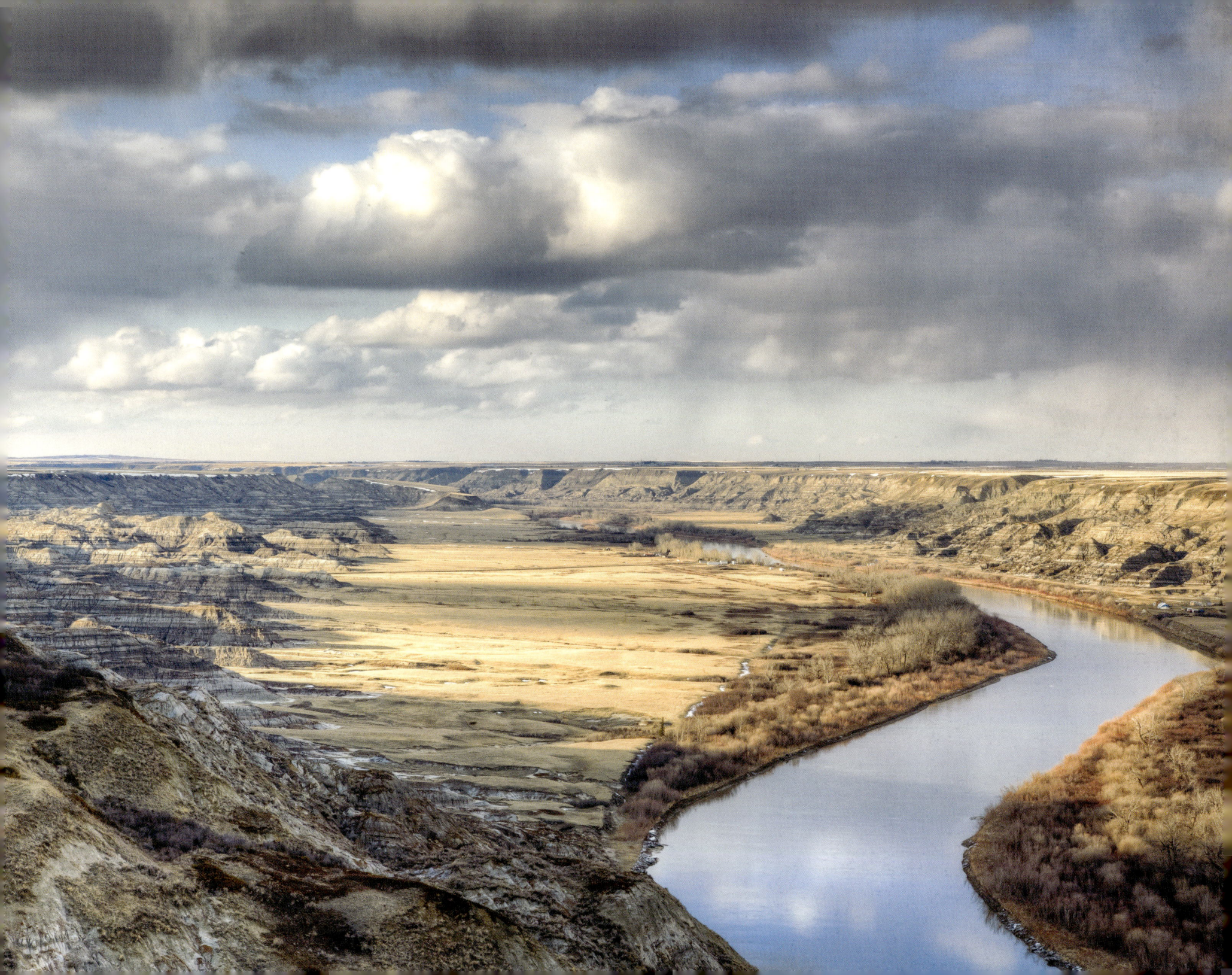

RED ROCK COULEE

Experiencing Red Rock Coulee is like wandering into a field of dreams. The terrain magically shifts from wheat fields to badlands, after which you find yourself staring at hundreds of enormous spherical concretions lining the folds and recesses of these gently rolling hills. With some rocks up to 2.5 metres in diameter, almost the size of a small car, these concretions are among the largest in the world. You forget to blink when you see them for the first time.

According to the book *A Traveller's Guide to the Geological Wonders of Alberta*, concretions can vary quite a bit in terms of size and shape. Some can appear like stems or nodules, millimetres in length, while others are shaped like logs and often mistaken for petrified wood. Mostly concretions are round or oval-shaped. Usually they are symmetrical. Scientists believe that a concretion often begins with an organic nucleus of some kind. This could be a leaf or a bone. The nucleus is then buried in homogenous sediment, which, in the case of Red Rock Coulee, is likely bentonite clay. As large amounts of water pass through the clay, dissolved minerals, such as red iron oxide and silica, accumulate and harden into the pores of the nucleus, thus encasing it. Once this chemical precipitation of minerals begins, the nucleus will grow until a dense rock is formed. So the concretion (as the derivation of the word suggests) is formed over time, similar to a pearl. Often, growth rings and various bands of colour are visible in the concretion. One of the key characteristics of a concretion is this concentric layering. It's unclear why the concretions at Red Rock Coulee have grown so huge. Whatever the reason, the resulting formations, unearthed by erosion, are extraordinary.

In the half-light at dusk or dawn, the concretions appear like a discovery of strangers, wandering souls, exiled here from another world. In the daylight, these giant rust-coloured boulders are just as captivating. For many, they resemble large dinosaur eggs or alien pods in various states of decay. Some of these concretions are broken and cracked, others are sliced and fractured. Most are impressively textured with multicoloured lichen.

These aptly named cannonball concretions dotting the landscape are a curious oddity.

They would attract attention anywhere, but part of their allure at Red Rock Coulee is that they are completely out of context with their surroundings. Red Rock Coulee is actually a badlands microcosm, extending only about one kilometre west. There is plenty of ashen clay and barren hills, but there are no large rugged rock formations, hoodoos, or jagged cliffs like those at Dinosaur Provincial Park or

RIGHT: WE ARE EXILES
Large cannonball concretions dot the unassuming hills like dinosaur eggs at Red Rock Coulee.

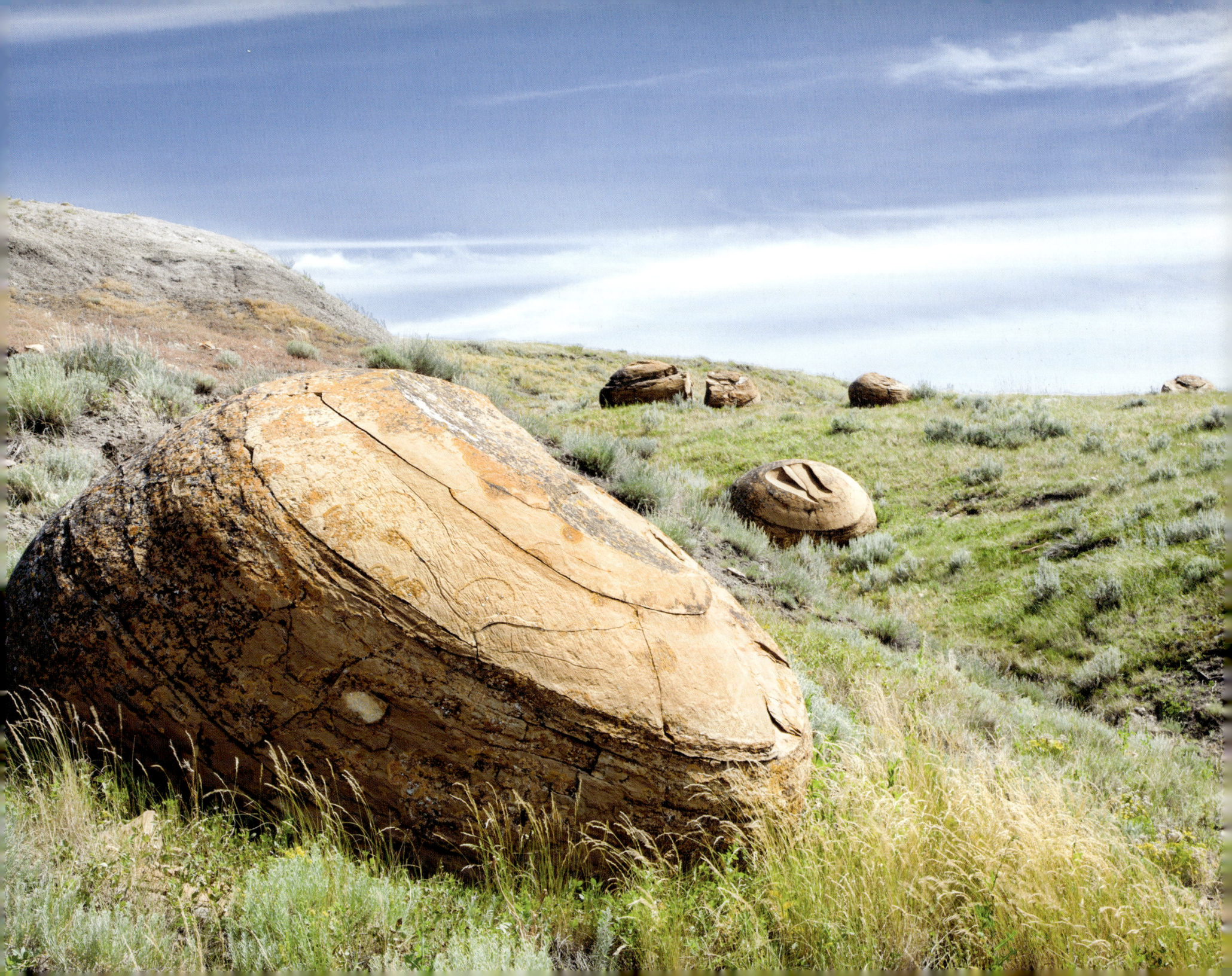

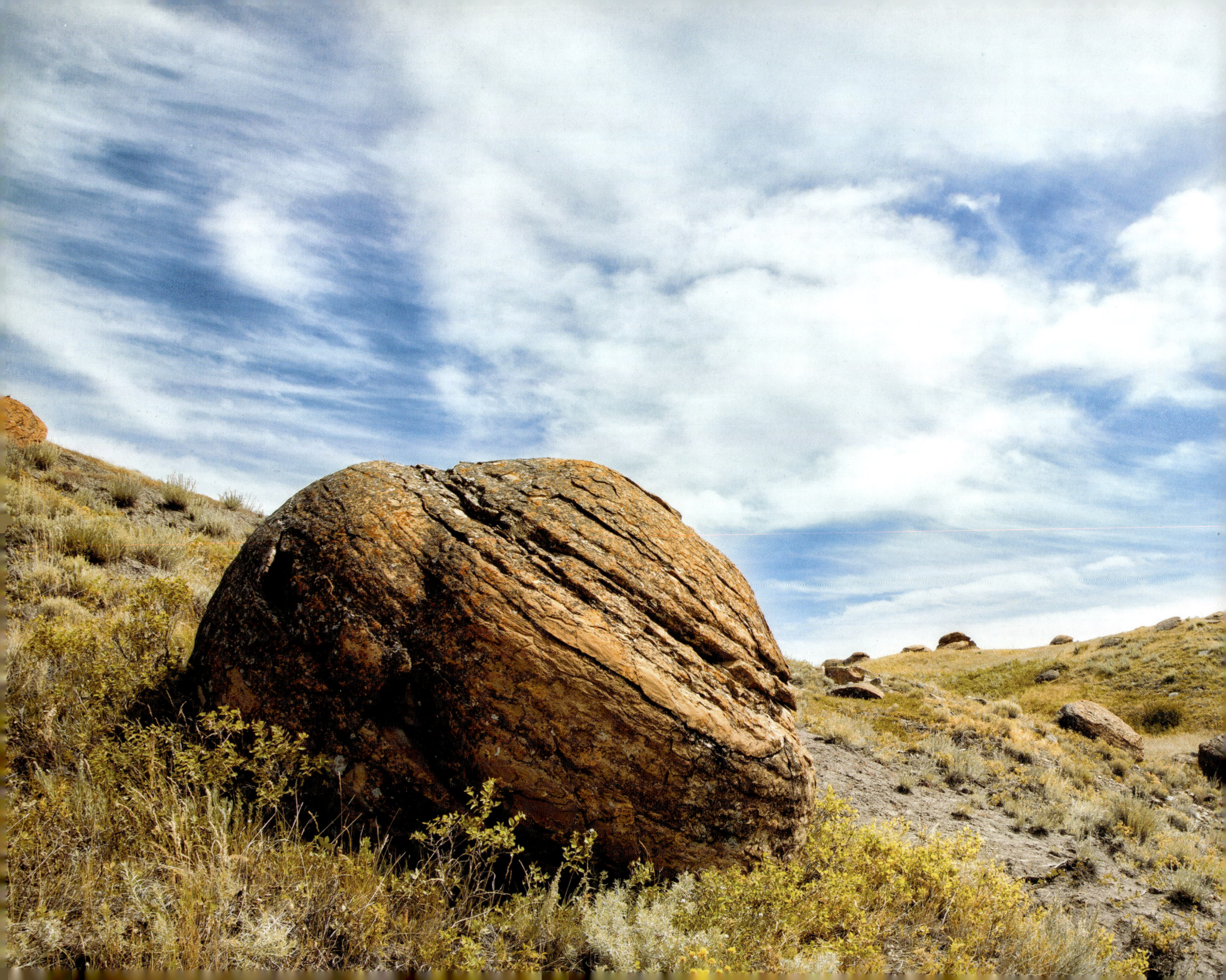

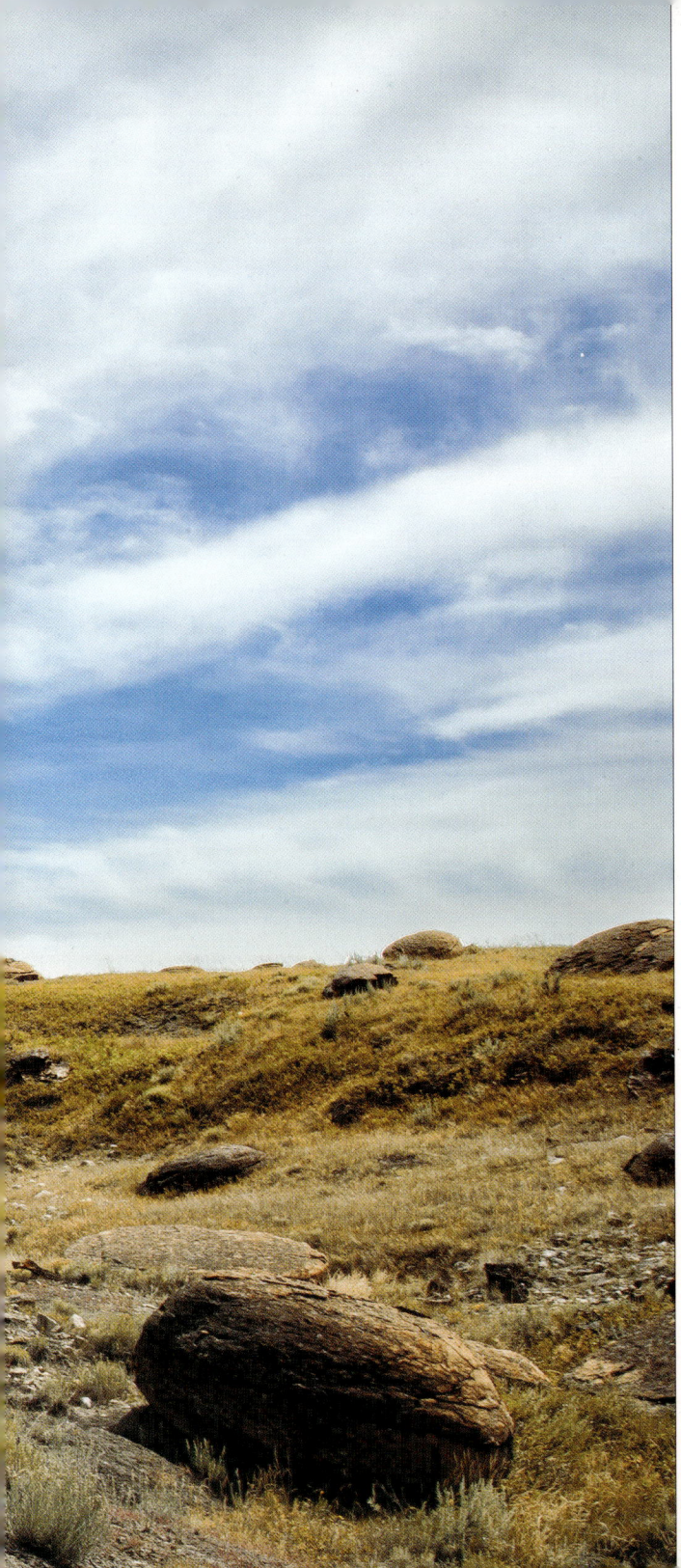

nearby Writing-on-Stone. The rest of this designated "natural area" lying to the south consists largely of grass-covered hills. Curiously, the concretions aren't demanding. They lie randomly on both the exposed and the grass-covered hills. Regardless of where they are situated, these huge red masses take centre stage (see page 156, the photograph "Atlas Shrugged" to compare concretions).

Red Rock coulee is located only 50 kilometres south of Medicine Hat, Alberta, and a short detour off Highway 887. If you are travelling south along Highway 3 toward Lethbridge, (the Crowsnest Highway), turn south at the village of Seven Persons.

The Red Rock Coulee Natural Area is a bit off the beaten path, but there is a sign indicating the turnoff. Still, with only the raw prairie and the Sweetgrass Hills on the horizon, if you blink you might miss it. Try not to. Sometimes, when you walk through this extraordinary landscape, the perception of your own individuality begins to fade and you are left to contemplate the miraculous wonders of nature. It is at this moment an overpowering peace flows through your body and you understand that you and the universe are one.

"The most beautiful thing we can experience is the mysterious. It is the source of all true art and all science."—*Albert Einstein*.

LEFT: A DISCOVERY OF STRANGERS
Red Rock Coulee terrain looks like scorched earth. It is a small badlands ecosystem with exposed strata, bentonite clay, and scree. To the south, however, the hills sustain a vibrant grasslands ecosystem.

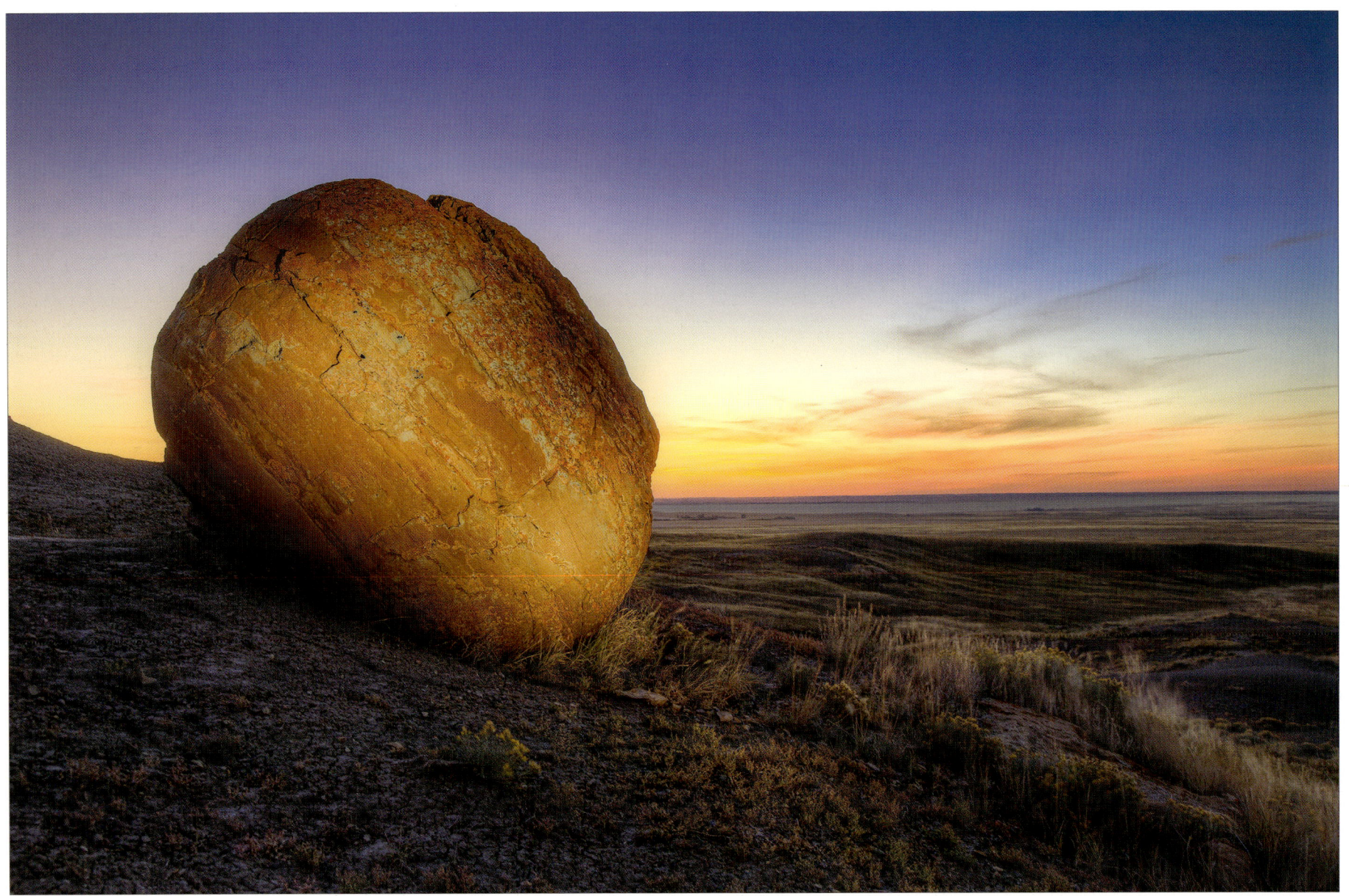

EVERYTHING ARRIVES AT THE LIGHT
The afterglow of the setting sun in two images fused together. I first exposed for the sky
and then reset my camera to light-paint the red rock.

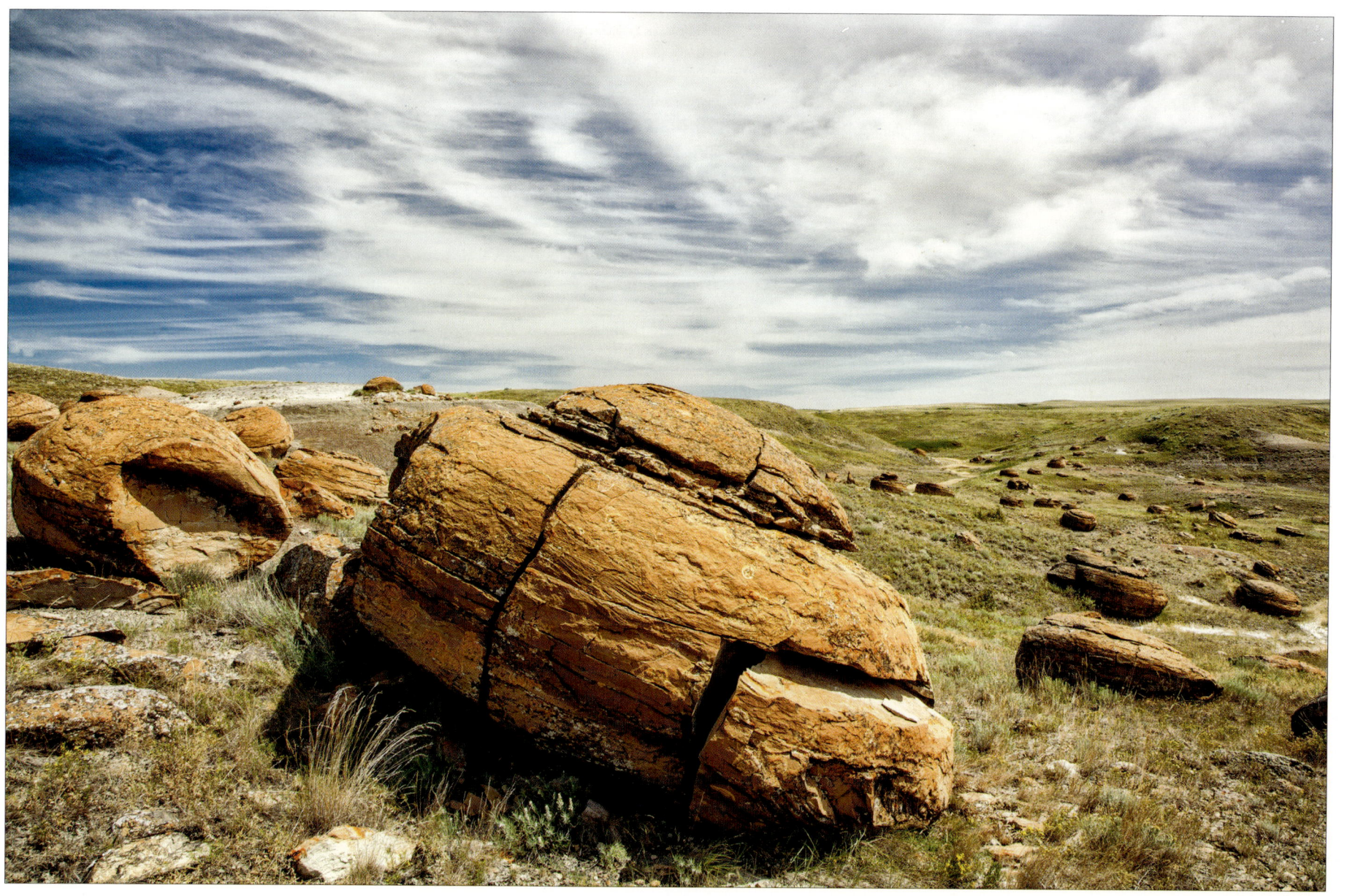

FUGITIVE PIECES
Each concretion has its own distinct character. This one looks as if it is about to hatch.

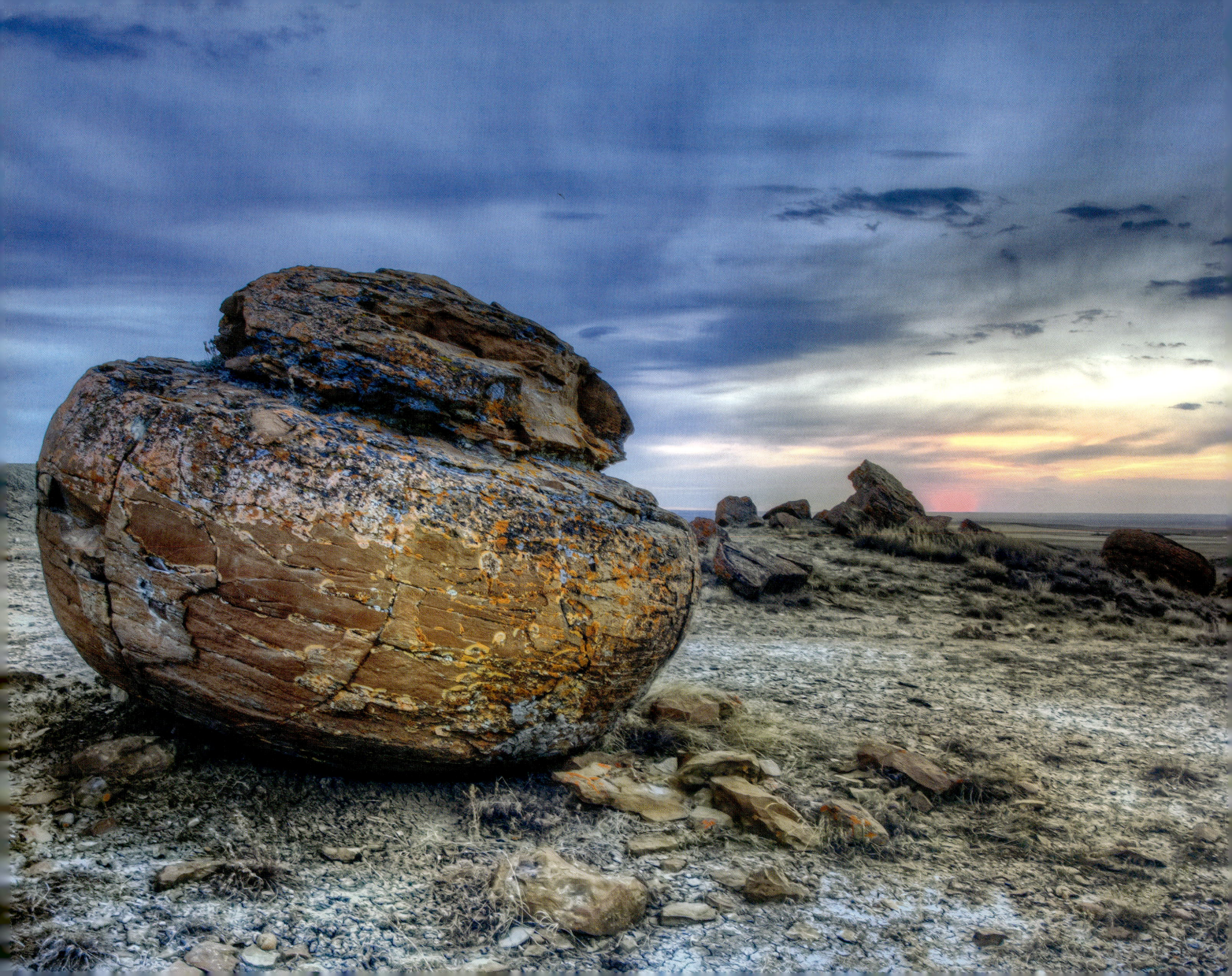

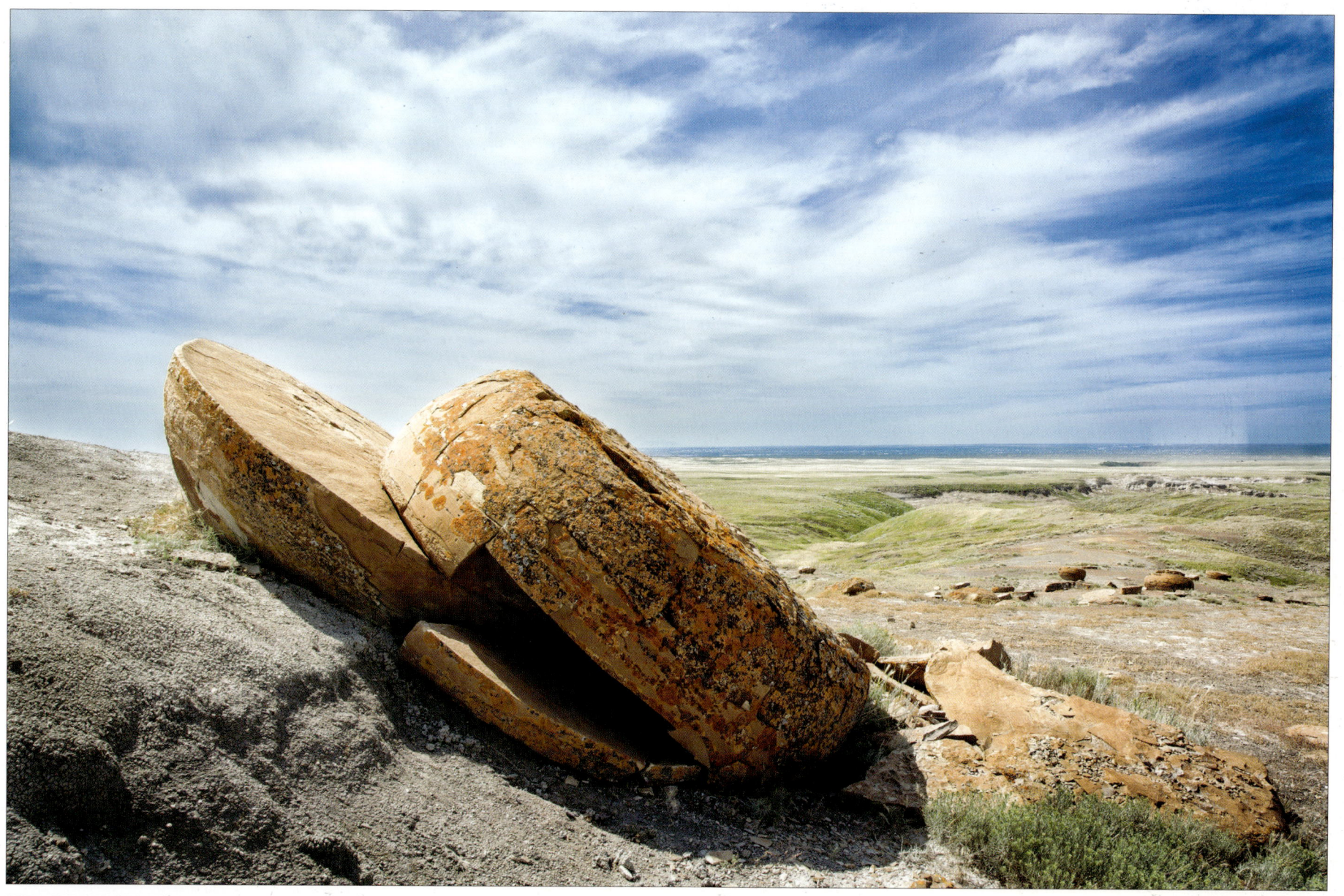

LEFT: THE AGE OF LONGING

ABOVE: FAULT LINES

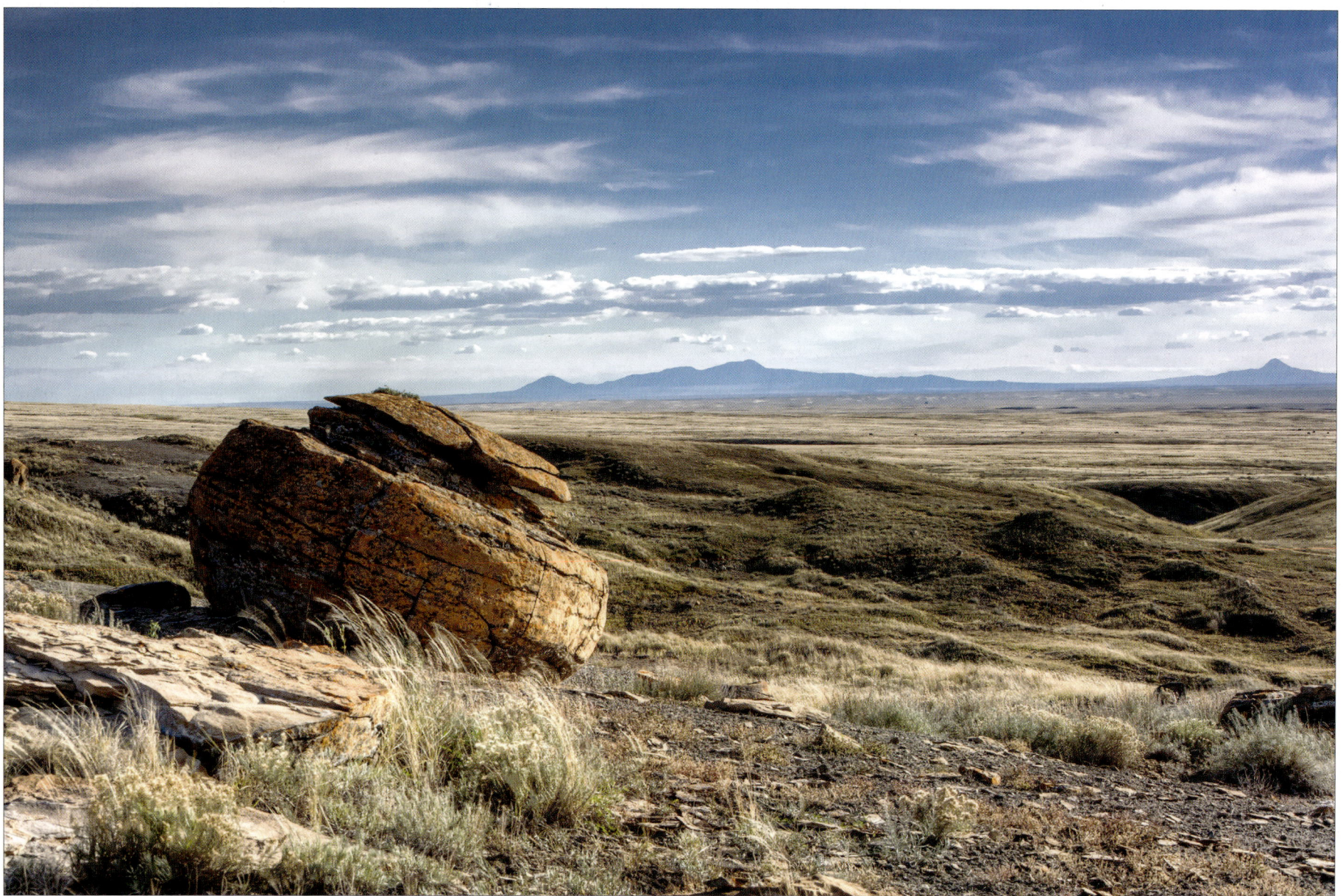

ABOVE: THE GUARDIAN
The Sweetgrass Hills in Montana, seen in the distance approximately 80 kilometres away, still dominate the horizon from Red Rock Goulee.

RIGHT: WANDERING SOULS (NIGHT)

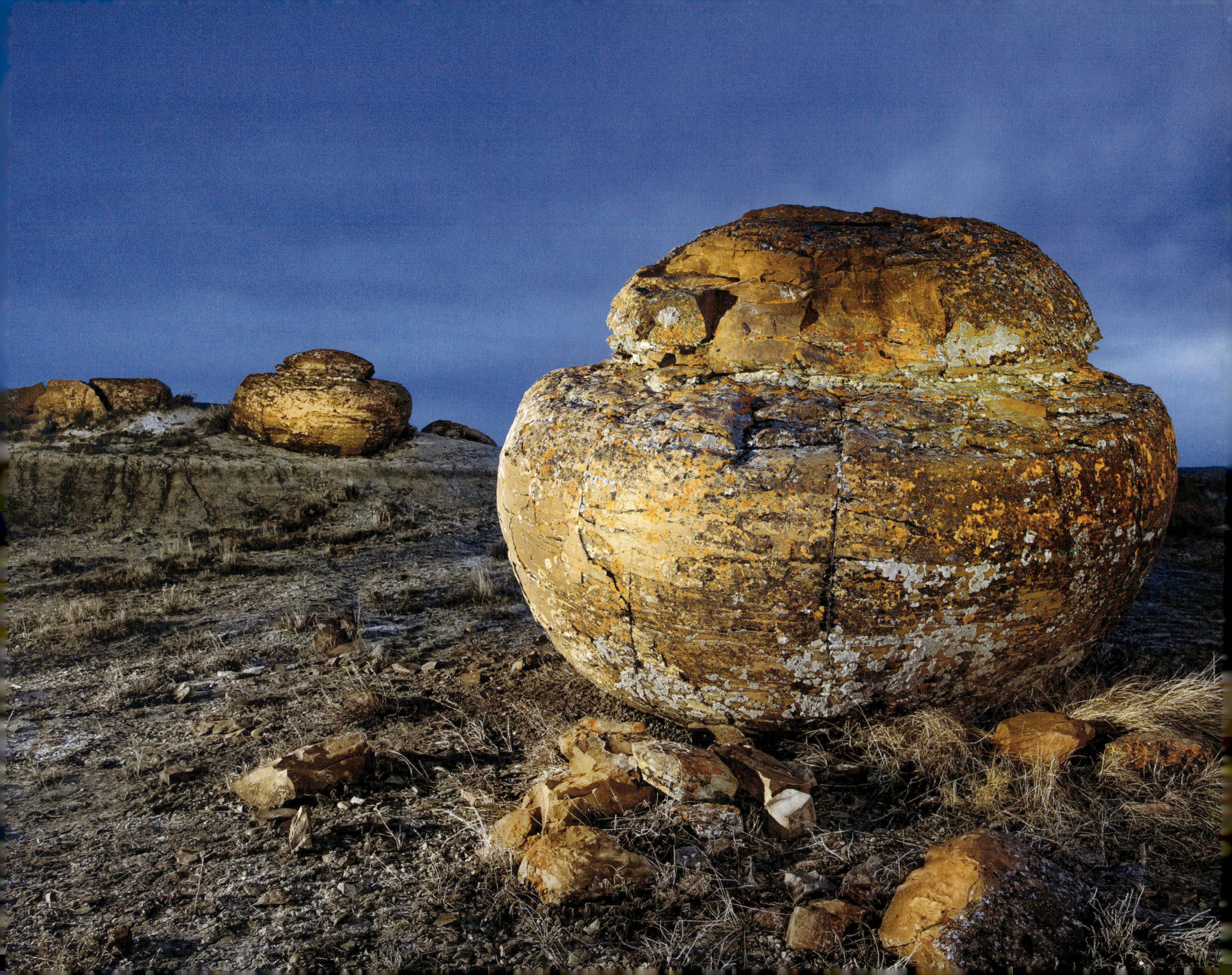

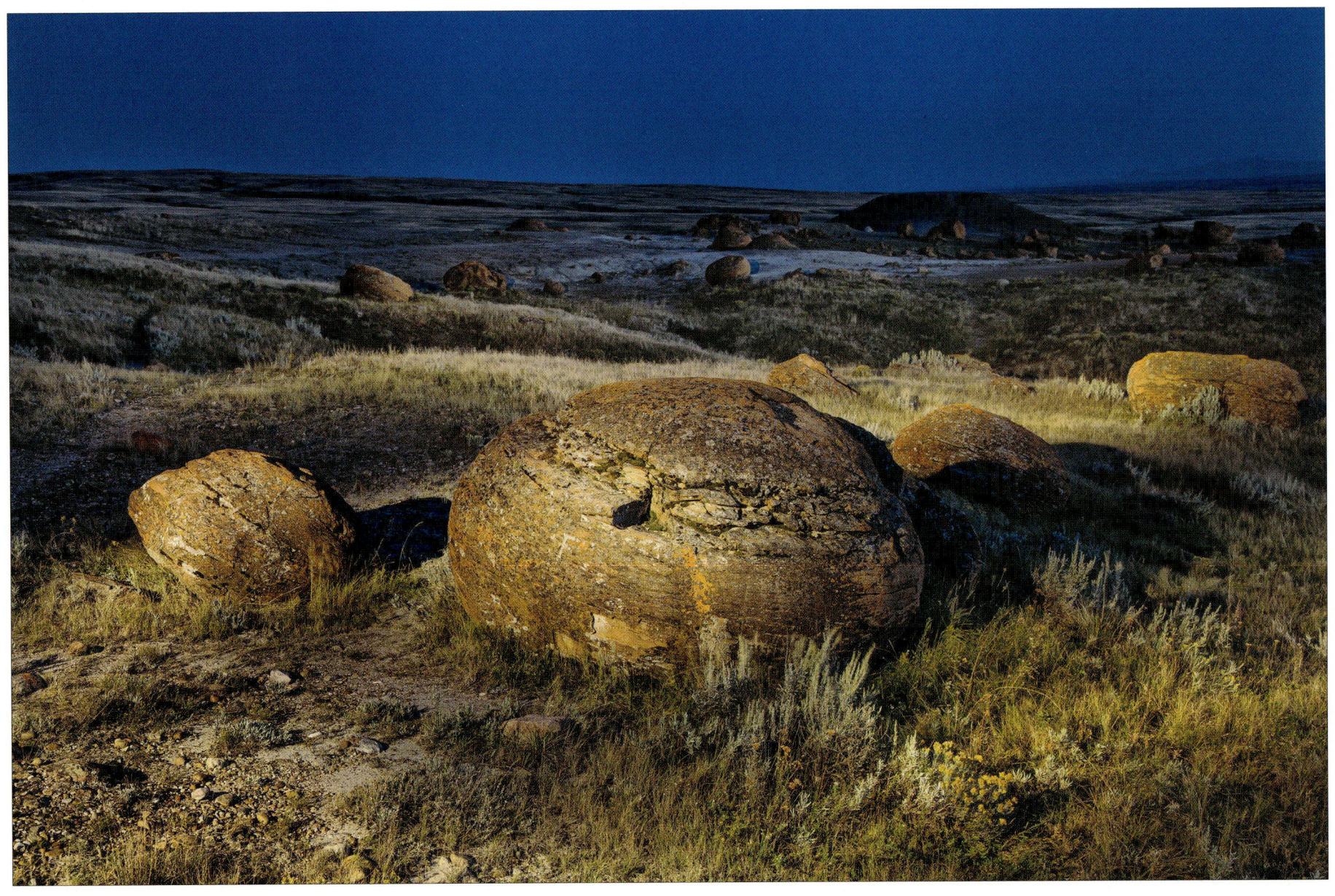

LAST SEEN

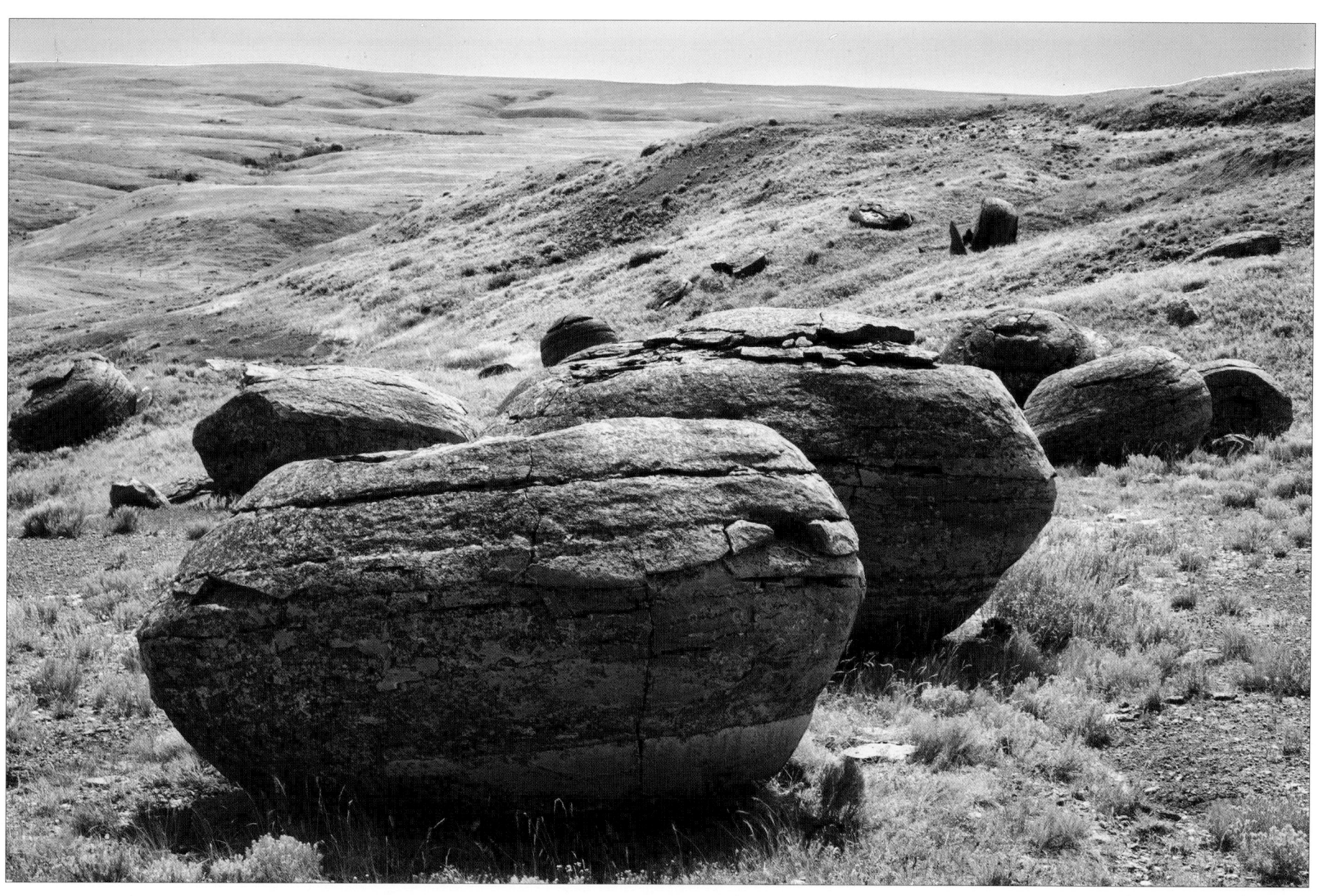

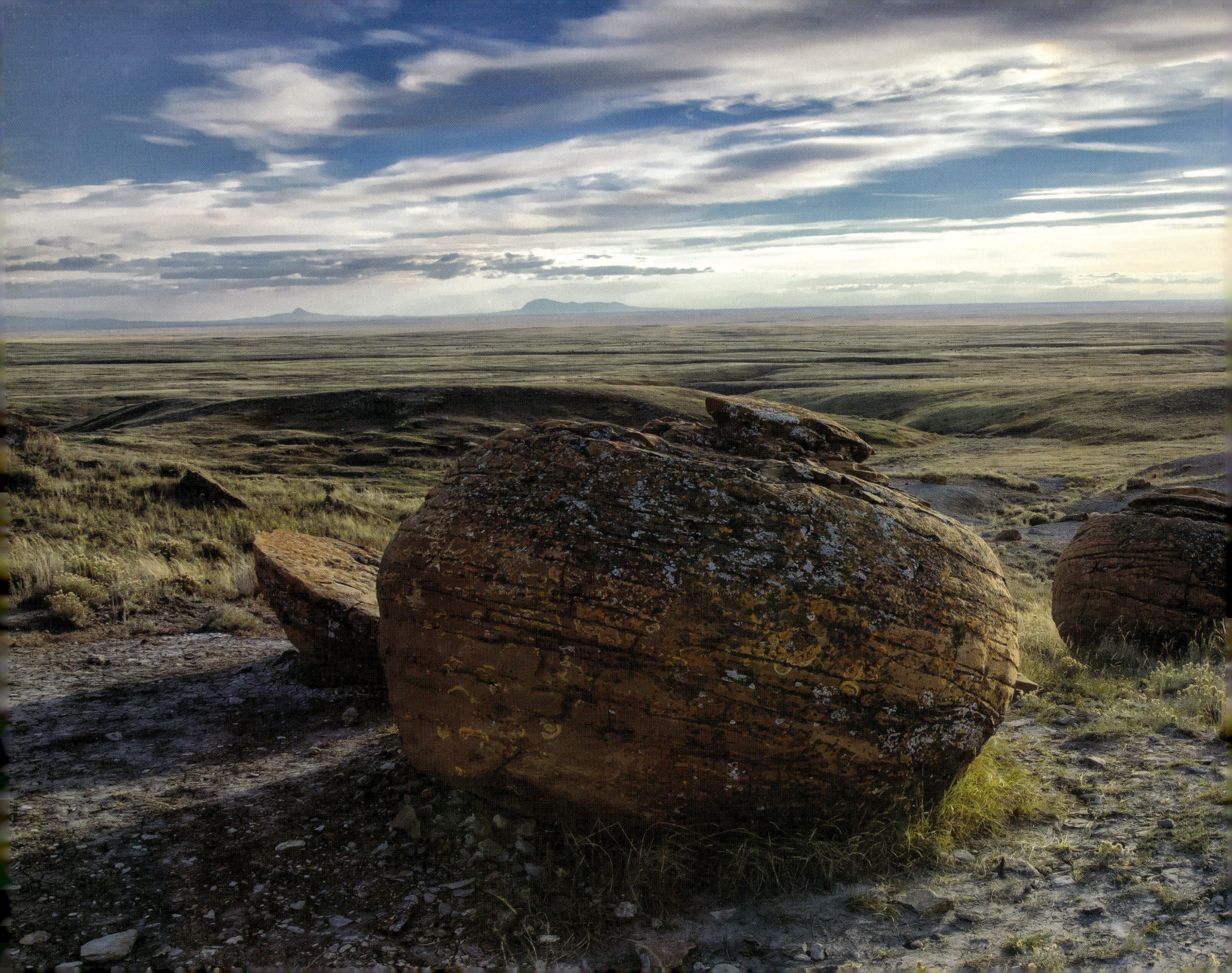

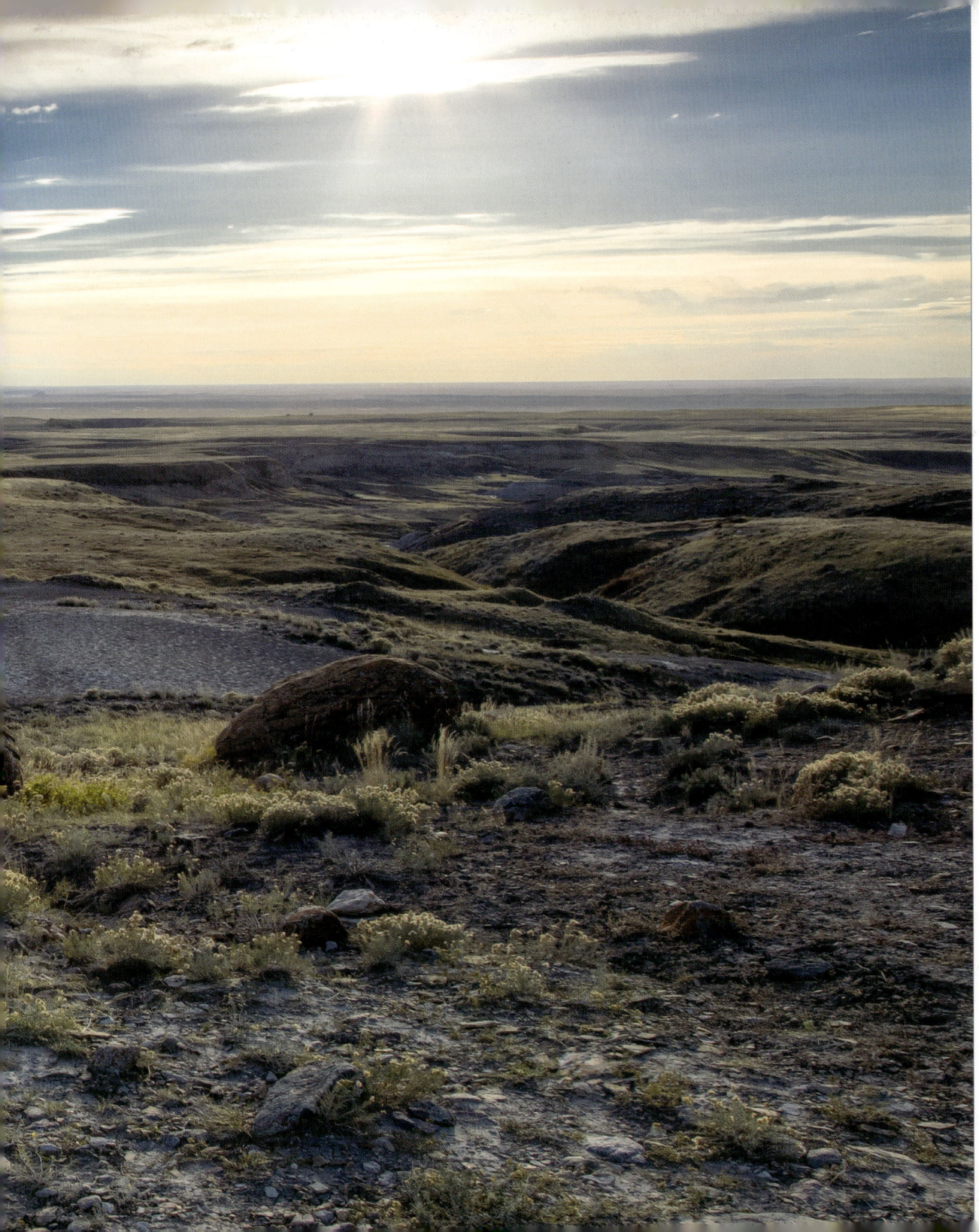

FIELD OF DREAMS

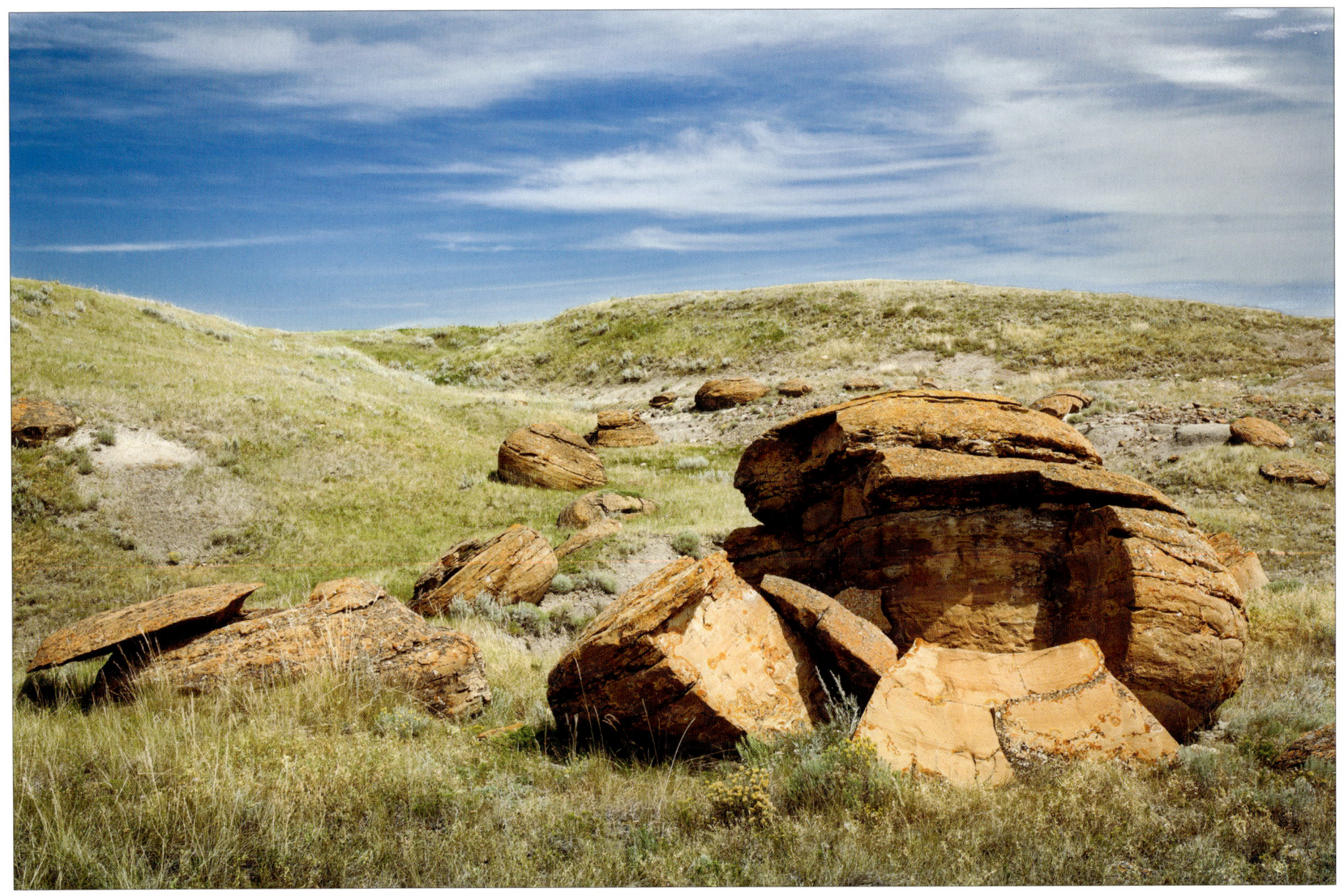

THE WRECKAGE

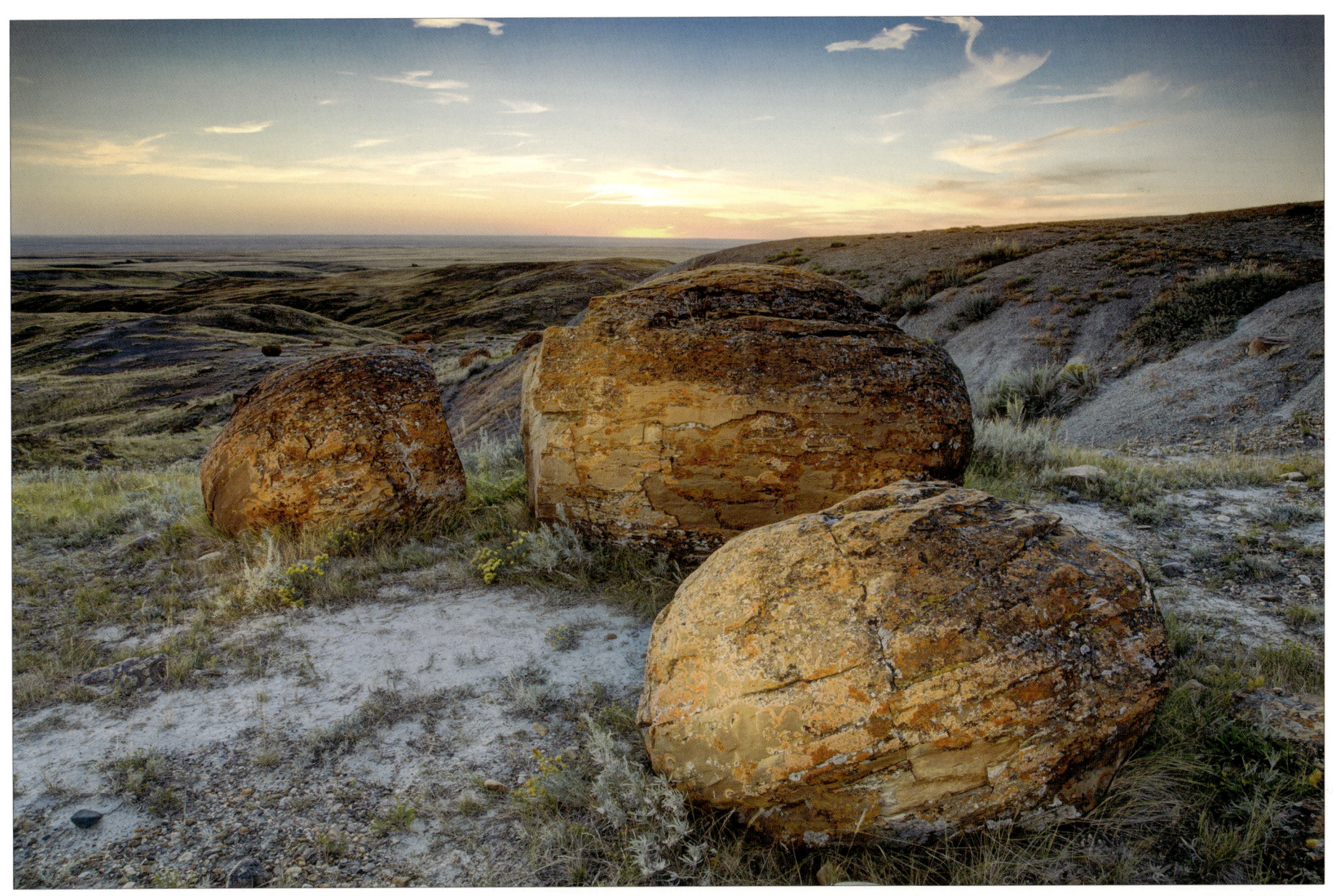

THE LUMINARIES

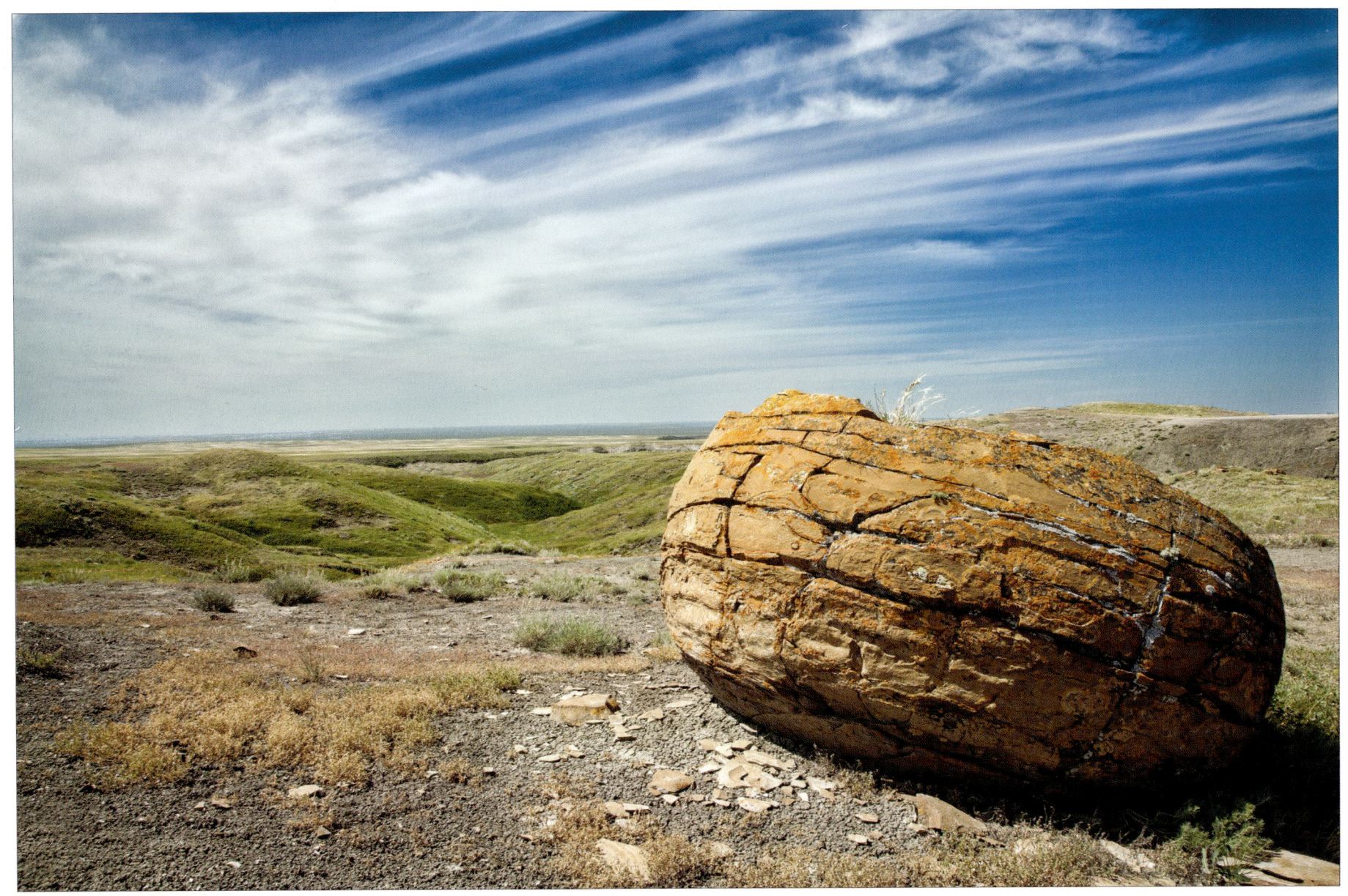

ABOVE: THE ANTAGONIST

RIGHT: WANDERING SOULS

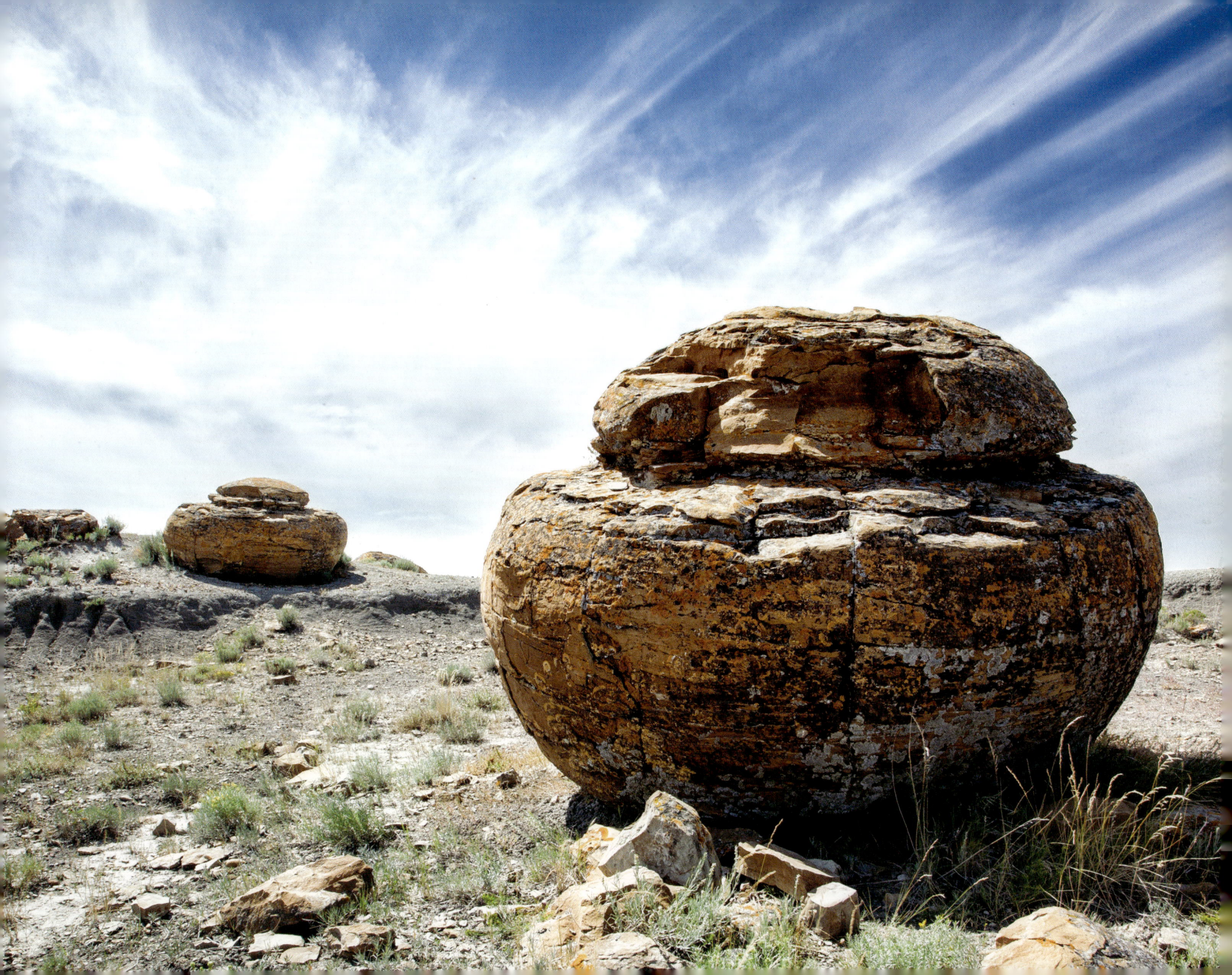

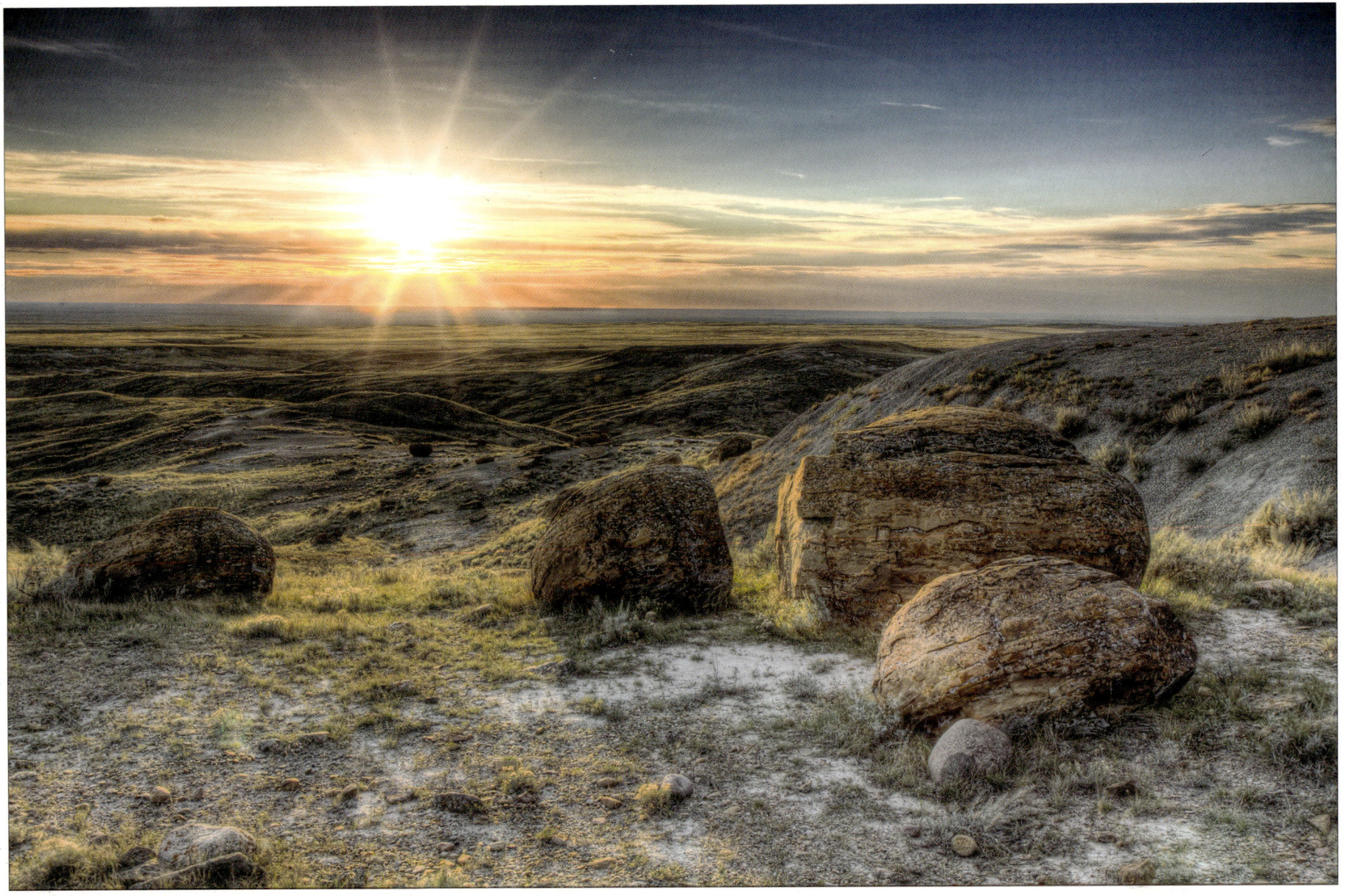

ABOVE: A TALE FOR THE TIME BEING

RIGHT: THE WATCH THAT ENDS THE NIGHT

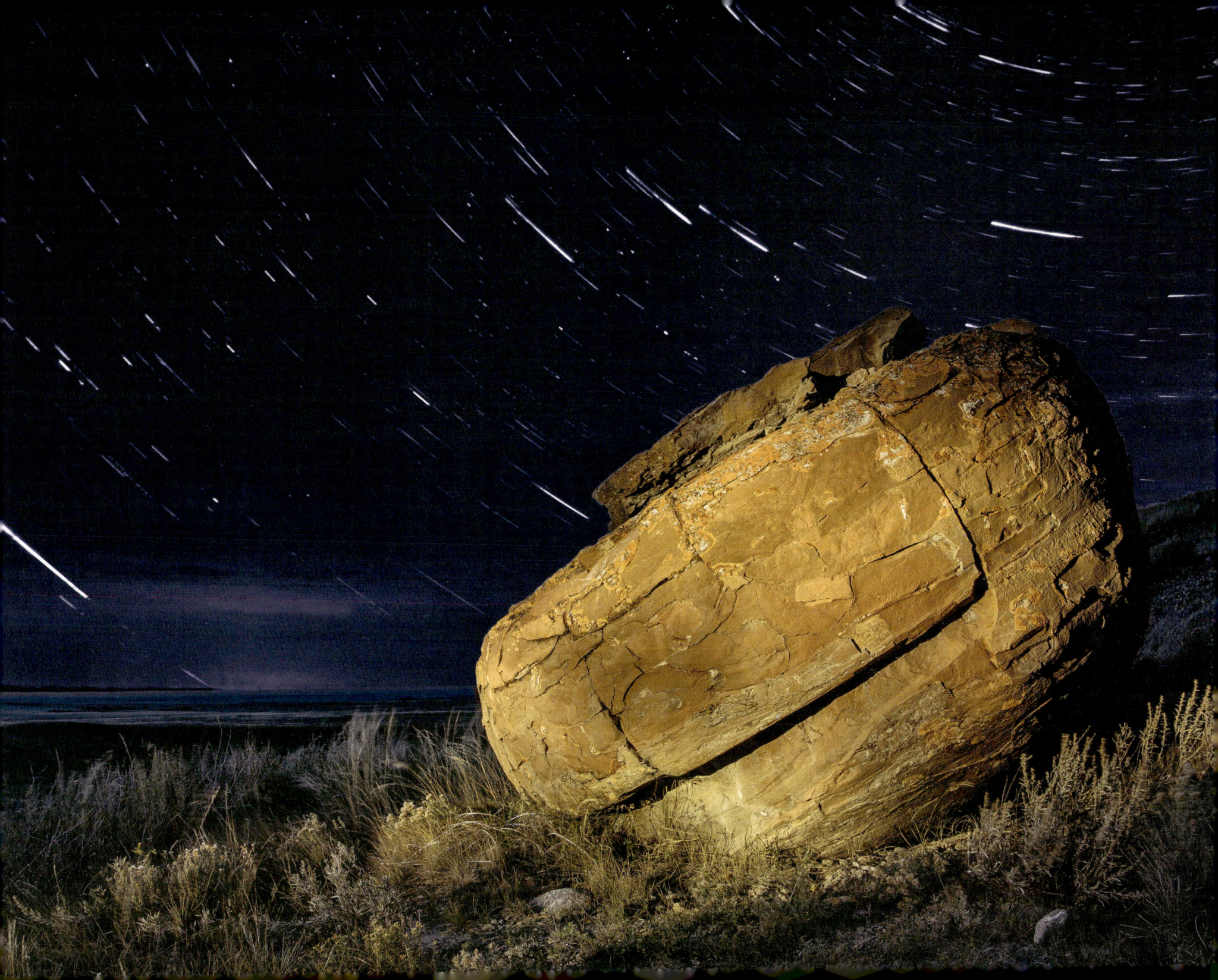

WRITING-ON-STONE PROVINCIAL PARK

Even before you reach Writing-on-Stone Provincial Park, located just north of the 49th parallel in southern Alberta, an isolated range of low-lying mountains called the Sweetgrass Hills rises up from the Montana grasslands and dominates the horizon. Soaring over 900 metres above the surrounding plains, these hills are massive enough to alter weather patterns. On a clear day, they still appear prominent from Red Rock Coulee approximately 80 kilometres away (see photograph "The Guardian" page 120). The Sweetgrass Hills even has its own unique ecosystem—separate from the prairies—with alpine fir trees and long, scented grasses for which it is named. First Nations people would make pilgrimages to these hills to gather and use the sweet-scented grass in their medicine bags and for ceremonial purposes.

The hills were also a hunter's dream, providing an abundance of game. The Sweetgrass Hills are entrenched as a sacred place in the culture of the Blackfoot people and, no doubt, to the many tribes who lived there before them.

While the Sweetgrass Hills dominate the landscape by their sheer physical presence, one glance at the magical rock formations at Writing-on-Stone Provincial Park, located along the banks of the Milk River, and the Sweetgrass Hills are completely forgotten. The views from the winding descent into Writing-on-Stone are simply breathtaking. As you drive down the twisting road into the valley, first impressions reveal what could easily be an ancient stone city in ruins. Stretching along both sides of the Milk River, and for as far as the eye can see, jagged rocks, decaying sandstone mounds, and stunning hoodoos populate the valley. The sheer density of these peculiar rock formations is astounding. Then, before you even have time to process the landscape, the road turns a hard left. Without warning, only several feet from the pavement, you encounter two of the most captivating hoodoos you will find anywhere (see "A Jest of God" page 133). Large, with impressive caprocks, these solitary twin hoodoos stand detached from the surrounding rocks and appear to be growing out of the grass-covered earth. If you didn't already know it, you will discover that Writing-on-Stone is a special place—even before you've parked your vehicle!

Aside from its dazzling badlands terrain, Writing-on-Stone Provincial Park, as its name suggests, is best known for having one of the most extensive collections of First Nations rock art in North America.

RIGHT: A JEST OF GOD
The slow twisting descent into Writing-on-Stone Provincial Park is so visually stunning you have to force yourself to blink. Then just when you think the view can't get any better…

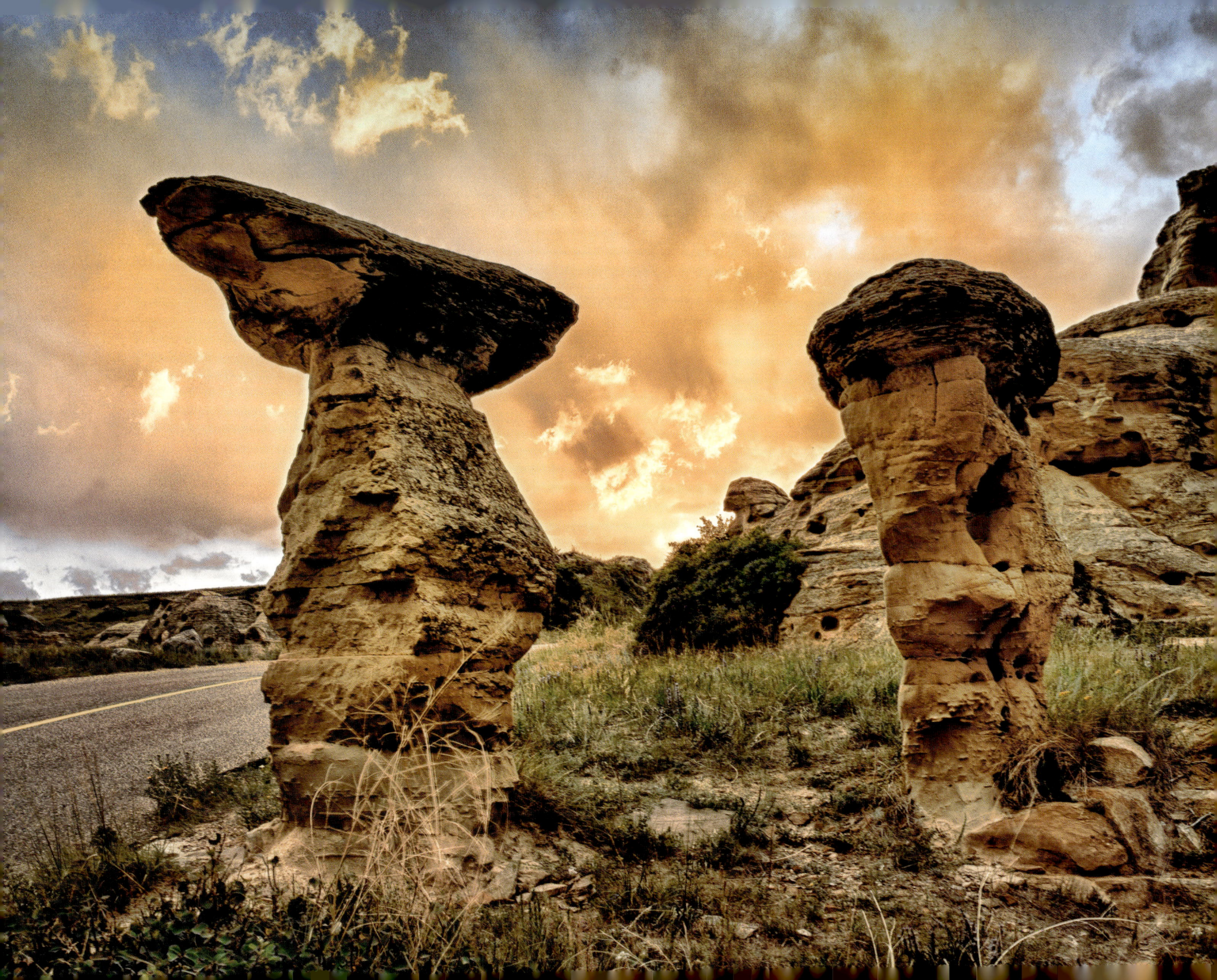

This consists of hundreds of petroglyphs and, to a lesser extent, pictographs. Petroglyphs are pictures or symbols etched into stone by using a tool of some sort. Pictographs are images that have been painted on rock, usually with a red ochre pigment.

Archaeological evidence suggests that Writing-on-Stone was inhabited 3,500 years ago, but it's difficult to ascertain who created this rock art and when. It's fairly certain that when the Blackfoot arrived and gained control of this area during the 18th century, some of this rock art was already present. Still, a good number of the petroglyphs appear to be Blackfoot in origin. In fact, the most elaborate artwork in the park is the "Battle Scene Petroglyph." Blackfoot elder, Bird Rattle, during his visit to Writing-on-Stone in 1924, attributed this artwork to the "Retreat Up the Hill" battle fought along the Milk River around 1866. This was a decisive victory for the Blackfoot over a combined war party of Gros Ventre, Crow, and Plains Cree, who lost over 300 warriors.

The large stick-figure scene, approximately 90 centimetres by 180 centimetres, depicts a group of warriors, some of whom are armed with guns and riding on horseback, attacking a circular encampment with tipis. The encampment is defended by a frontline of warriors, also with guns. Bullets, represented as repeating dots discharged from the guns, are fired by both sides. In the centre of the scene, you can clearly make out a warrior with a tomahawk attacking his foe in one-on-one combat. It's a powerful image. As the "Battle Scene Petroglyph" clearly shows, some of the rock art in Writing-on-Stone is biographical and depicts actual events. But many First Nations people believe the rock art was created by Shamans, or medicine men, as part of rituals to communicate with the spirit world.

While much of the rock art is located inside a restricted archaeological preserve and thus accessible only by guided tours (which are highly recommended), there are still quite a few petroglyphs and pictographs to see along the self-guided Hoodoo Interpretive Trail. The impressive but weathered "Battle Scene Petroglyph," protected behind a chain-link fence, is one of them.

The Hoodoo Interpretive Trail is one of two marked trails located in the park. The first half of this trail is fairly strenuous since it involves quite a bit of scrambling up and down the congested rocks. You need to keep your eyes open and watch where you put your hands because rattlesnakes like to sun themselves on the rocks. The second half of the trail is more spread out, so it's an easier hike. The trailhead begins at the west side of the campground and runs parallel to the Milk River. The trail length is listed as 4.4 kilometres, return. Since there is a paved scenic road overlooking the valley from the north, with lookout points and information signs, a good plan would be to hike half the trail, return to the trailhead, rehydrate, and then drive to the end of the trail near the Police Post Viewpoint, and hike the other half from there. The badlands heat can be scorching. Plus, the views from the scenic road are wonderful, especially with the Sweetgrass Hills in the background.

The other trail begins 50 metres west of the campground office in the valley. It winds its way through a crowded maze of tall and imposing rock structures climbing to the top of the hill, where the landscape opens up and the sandstone outcroppings become less crowded. The trail ends at the Writing-on-Stone Interpretive Centre, but I recommend that you continue eastward. Some of the best mystical hoodoos are located along this deer path.

If you have travelled to and explored other badlands, you will realize that Writing-on-Stone is uniquely special. While most badlands are semi-arid regions with abundant amounts of bentonite clay, interspersed with desert-like patches of exposed and barren strata, Writing-on-Stone is classified as a dry mixed-grass

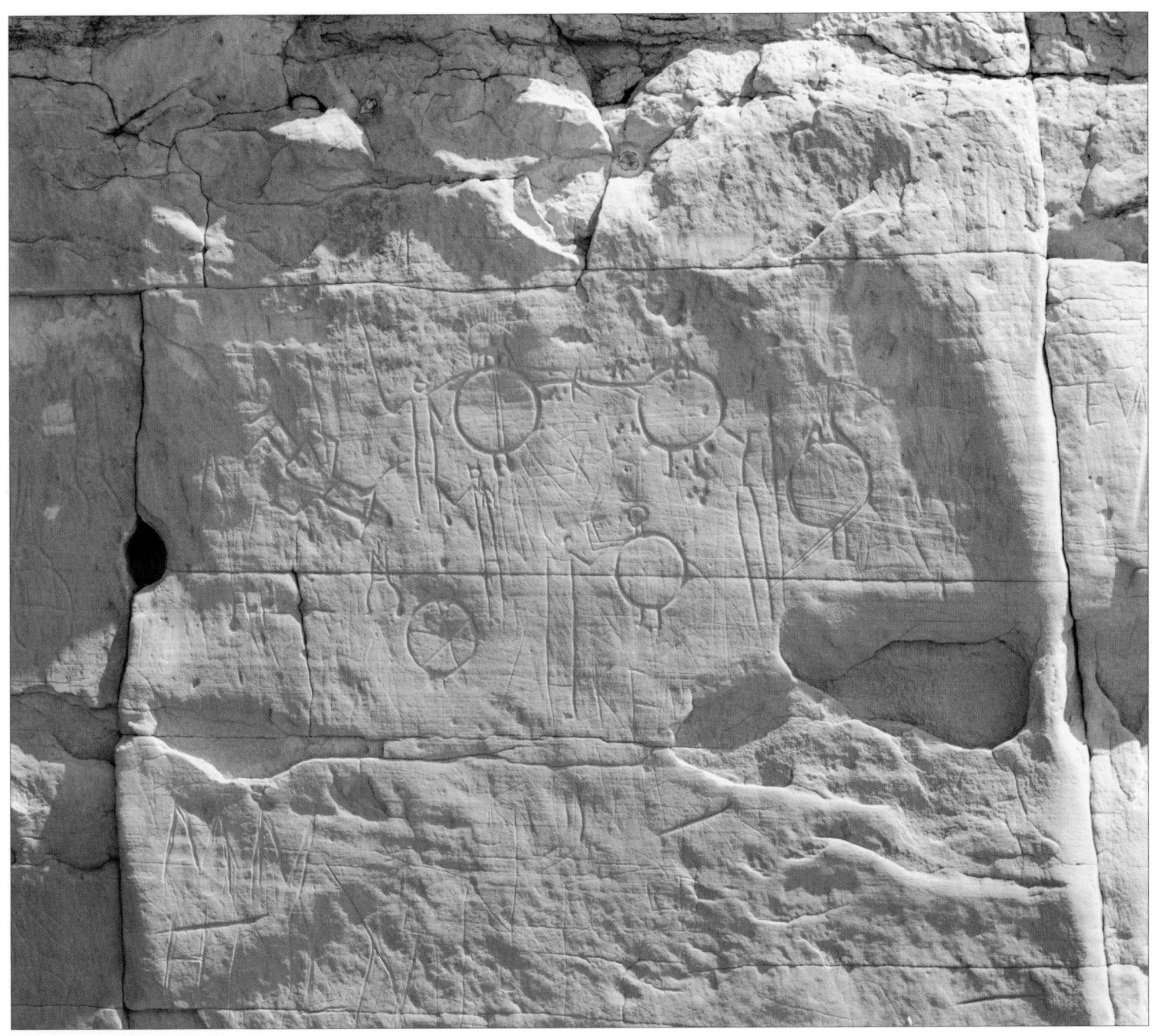

PETROGLYPHS

135

sub-region. So while the organic-looking rock mounds, pitted cliffs, and mystical hoodoos found here are definitely characteristics of the badlands, the terrain intermingled through and around this ravaged sandstone city is primarily a variety of grasses alongside large cottonwood and aspen trees. In other words, Writing-on-Stone is a prairie oasis. American Alexander Culbertson wrote: "... along the Milk River the buffalo didn't arrive here by herds but as far as the eye could see they were one solid mass of living animals..."

The buffalo no longer congregate here, but today the attraction to Writing-on-Stone remains just as compelling as it did to those who lived in this valley long ago. The wondrous rock outcropping and hoodoos become entrenched in your imagination. The First Nations rock art archived like history lessons among the cliffs consecrates this area as a sacred place. There is a strong spiritual presence emanating from this seductively enchanting landscape.

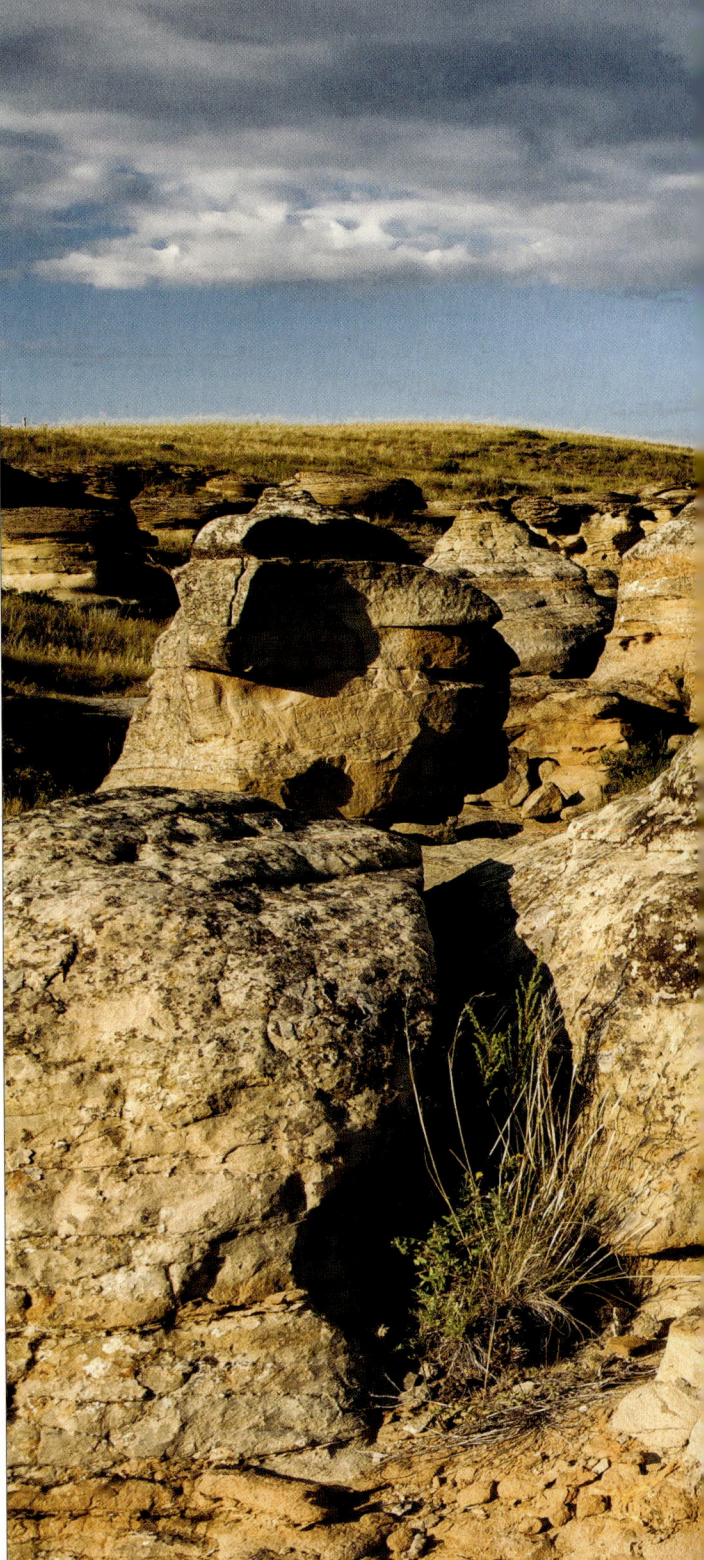

BURIAL GROUND
Flattened mounds resemble an ancient city in ruin.

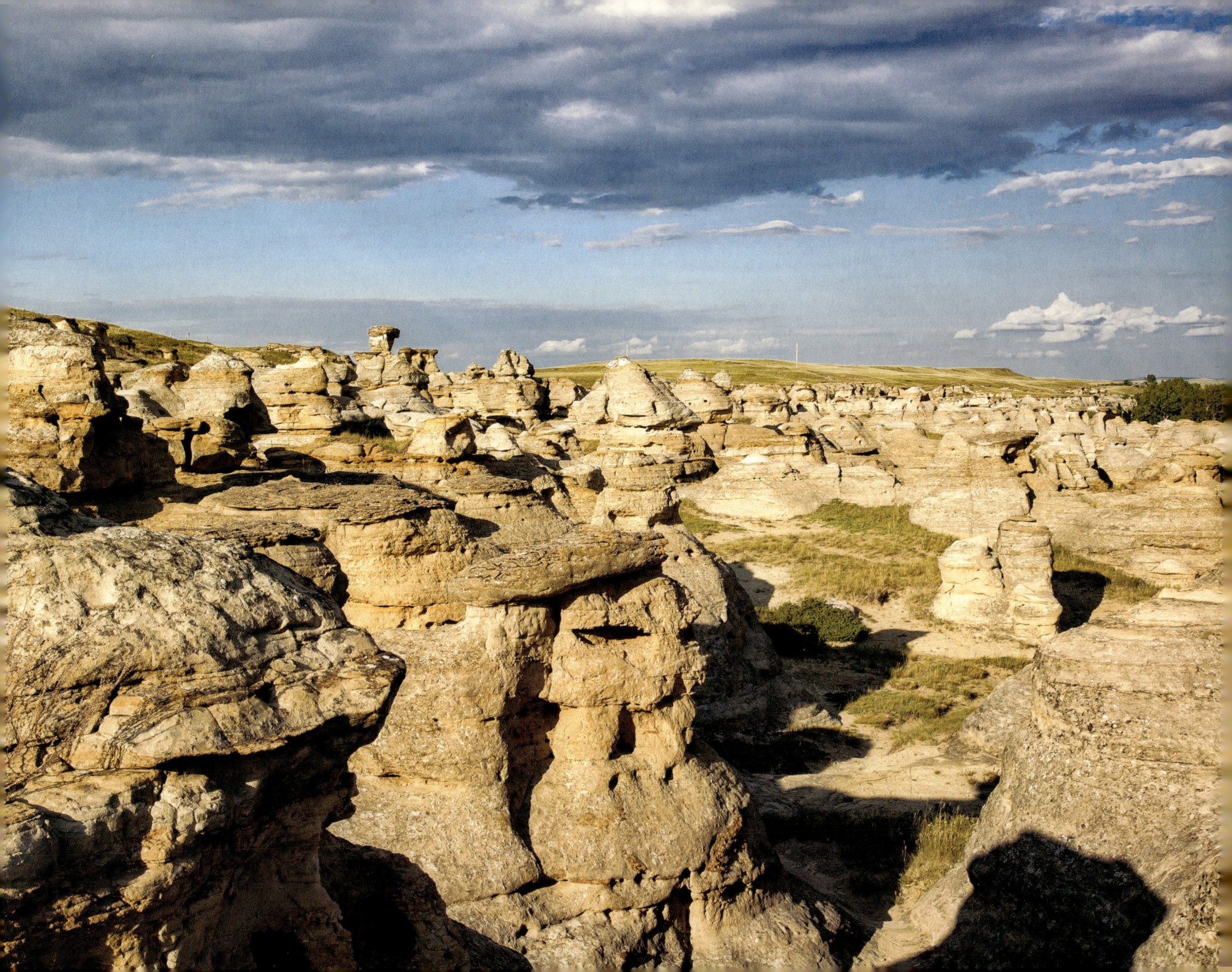

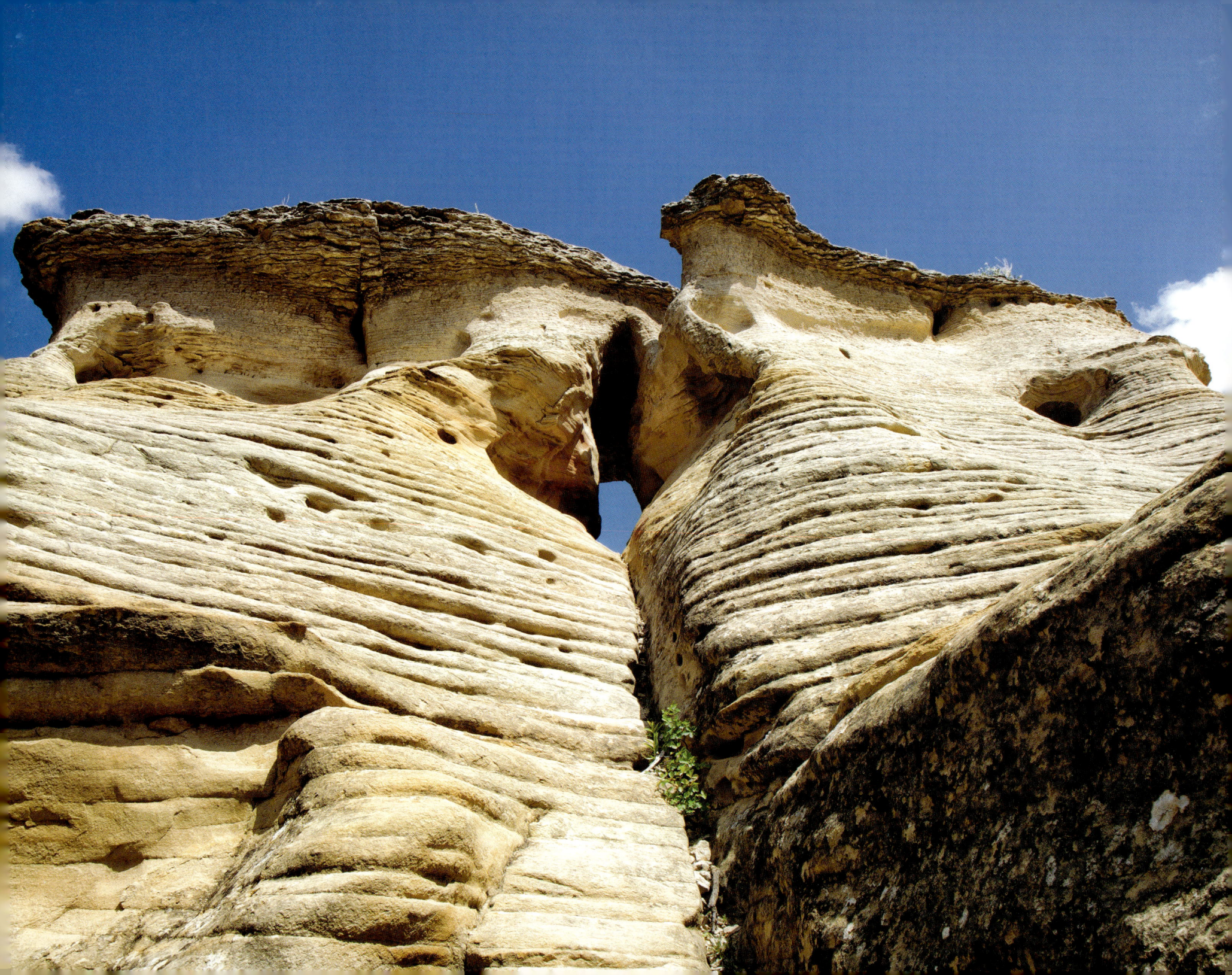

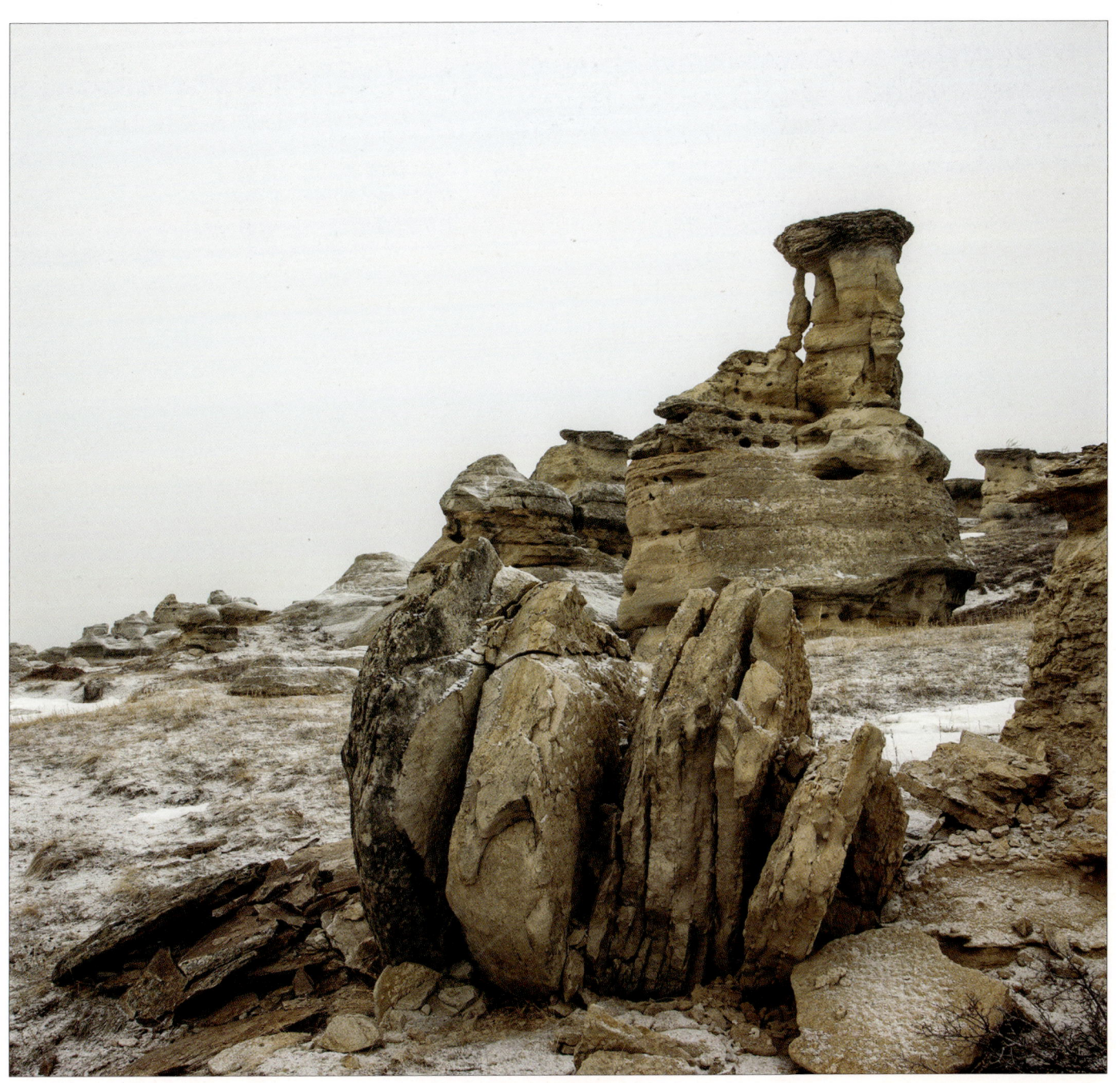

FAR LEFT: HISTORY LESSONS

ONCE A WARRIOR
I love the juxtaposition of the fallen hoodoo
with the larger, precariously balanced one
still standing.

139

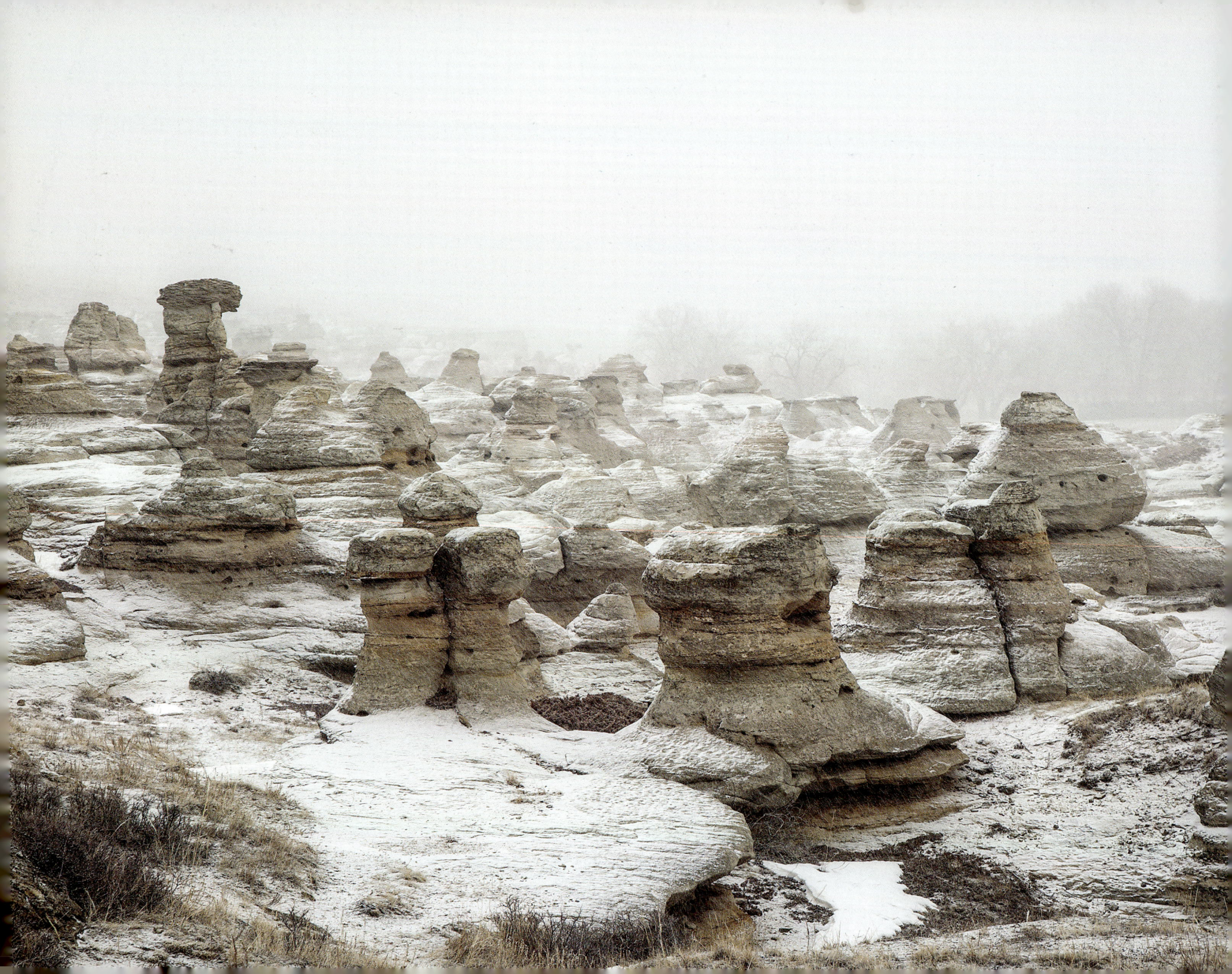

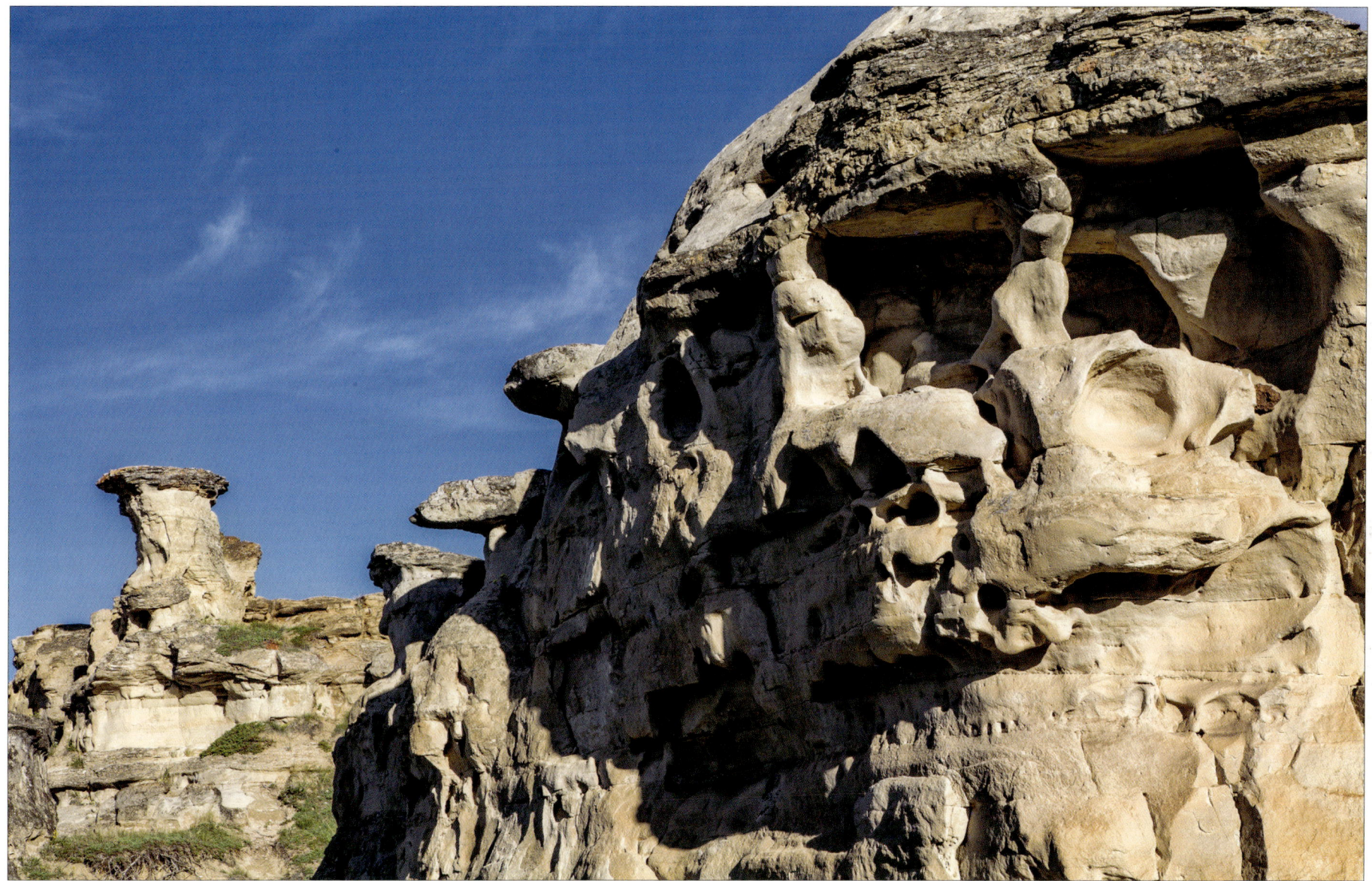

LEFT: THE WINTER PALACE

When I arrived here in February there was no snow whatsoever, despite forecasts for flurries. I patiently waited in Hotel Santa Fe (my SUV) and sure enough, the weather system rolled in like a blanket.

ABOVE: CALL OF THE WILD

Cavities and depressions eroded into rock.

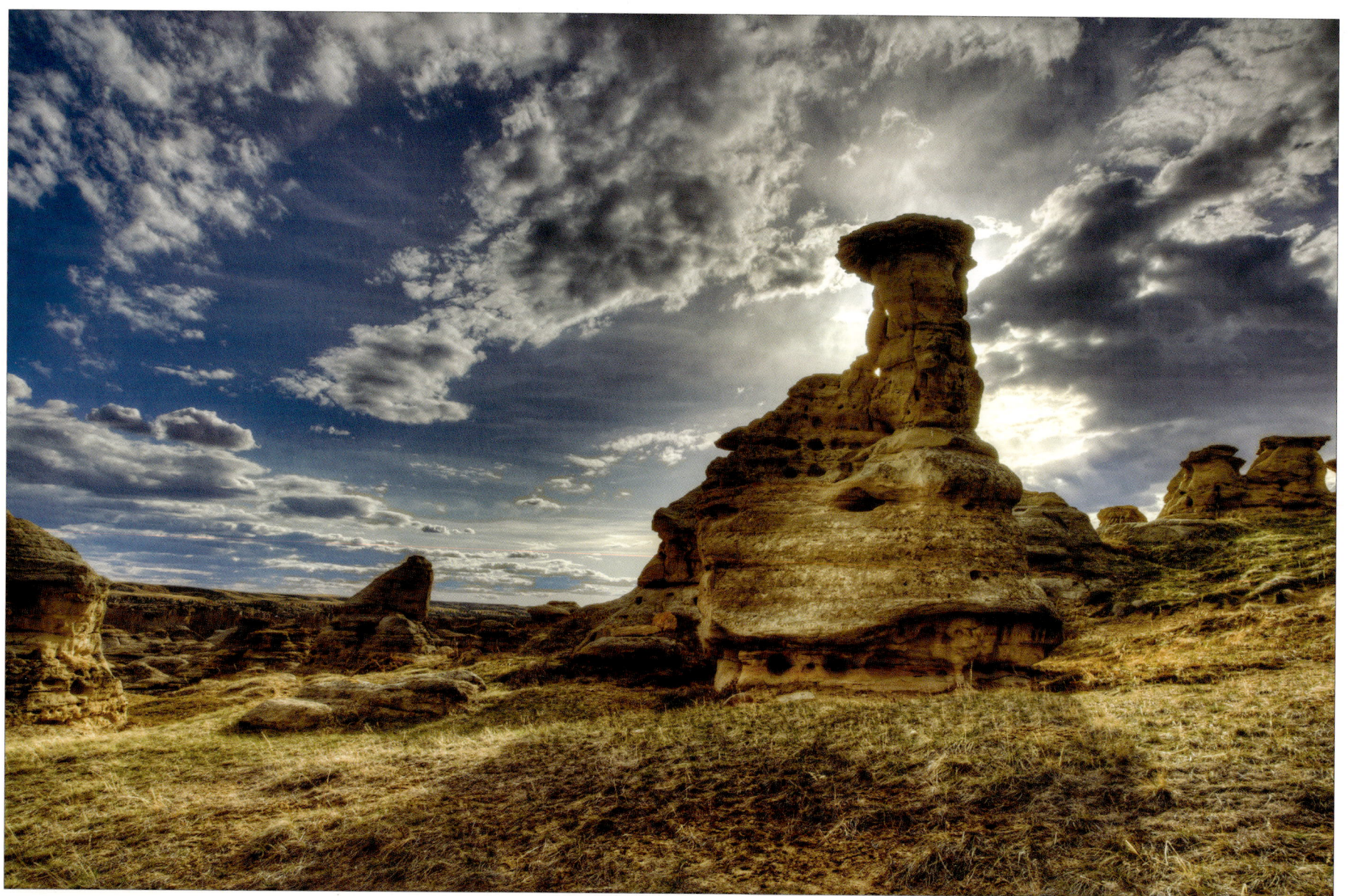

BAROMETER RISING

There is something about this solitary hoodoo's sculpted artistry that kept me coming back to it time and again. For this HDR photograph I made sure the sun was directly behind the hoodoo to accentuate the already dramatic sky.

142

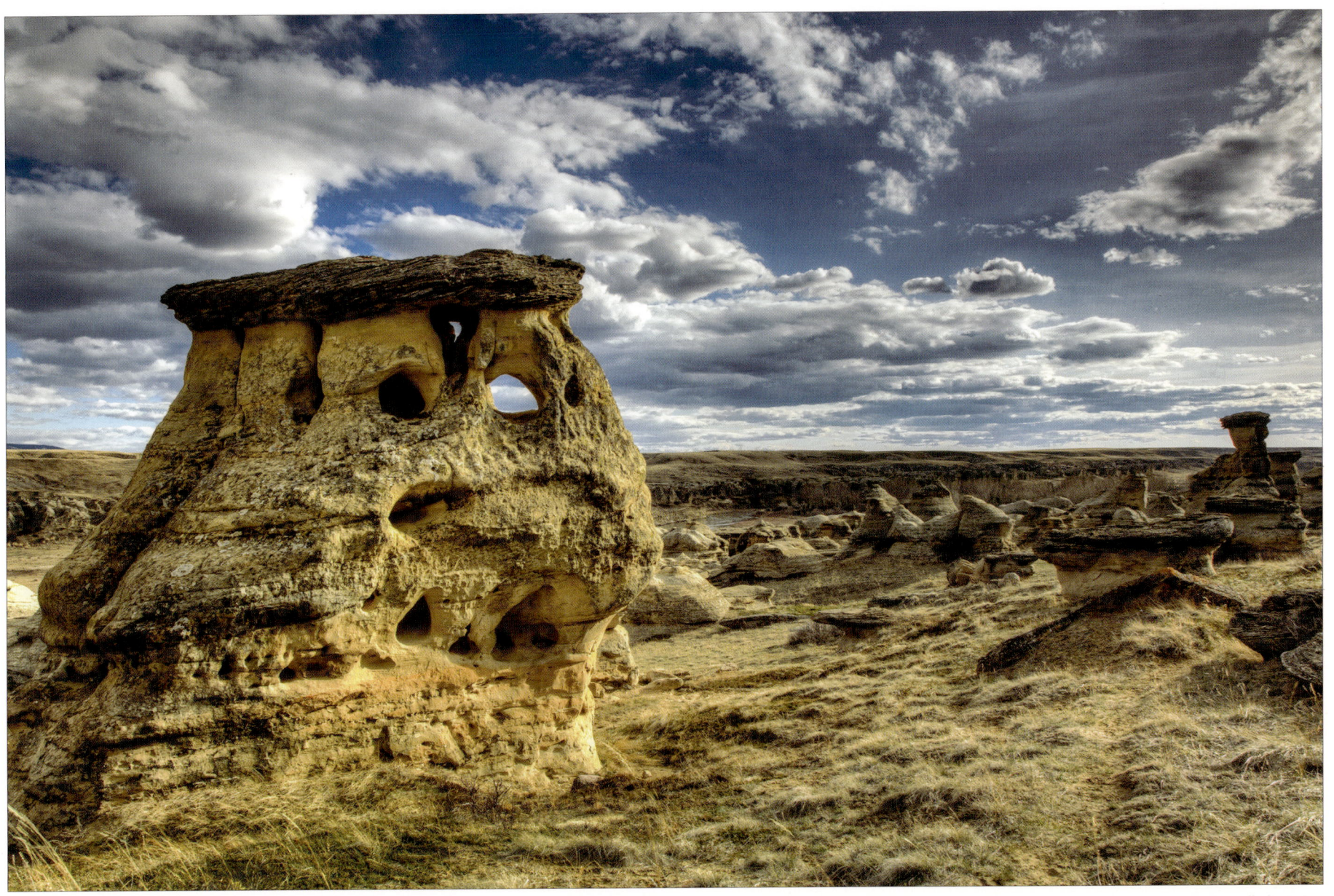

THE MEMORY ARTIST
The variety of textures and the vibrant light transform this grey
mask-like rock a golden yellow.

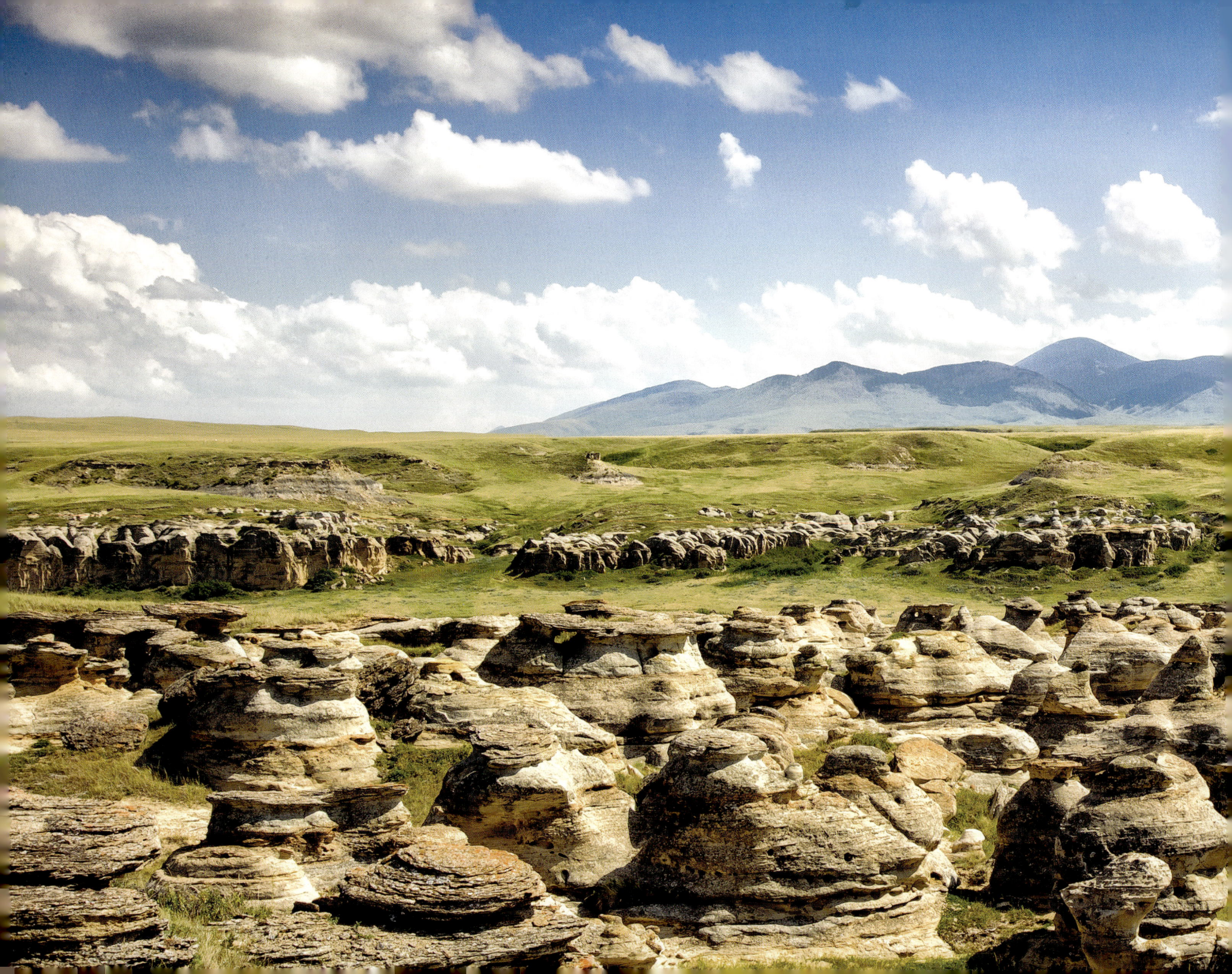

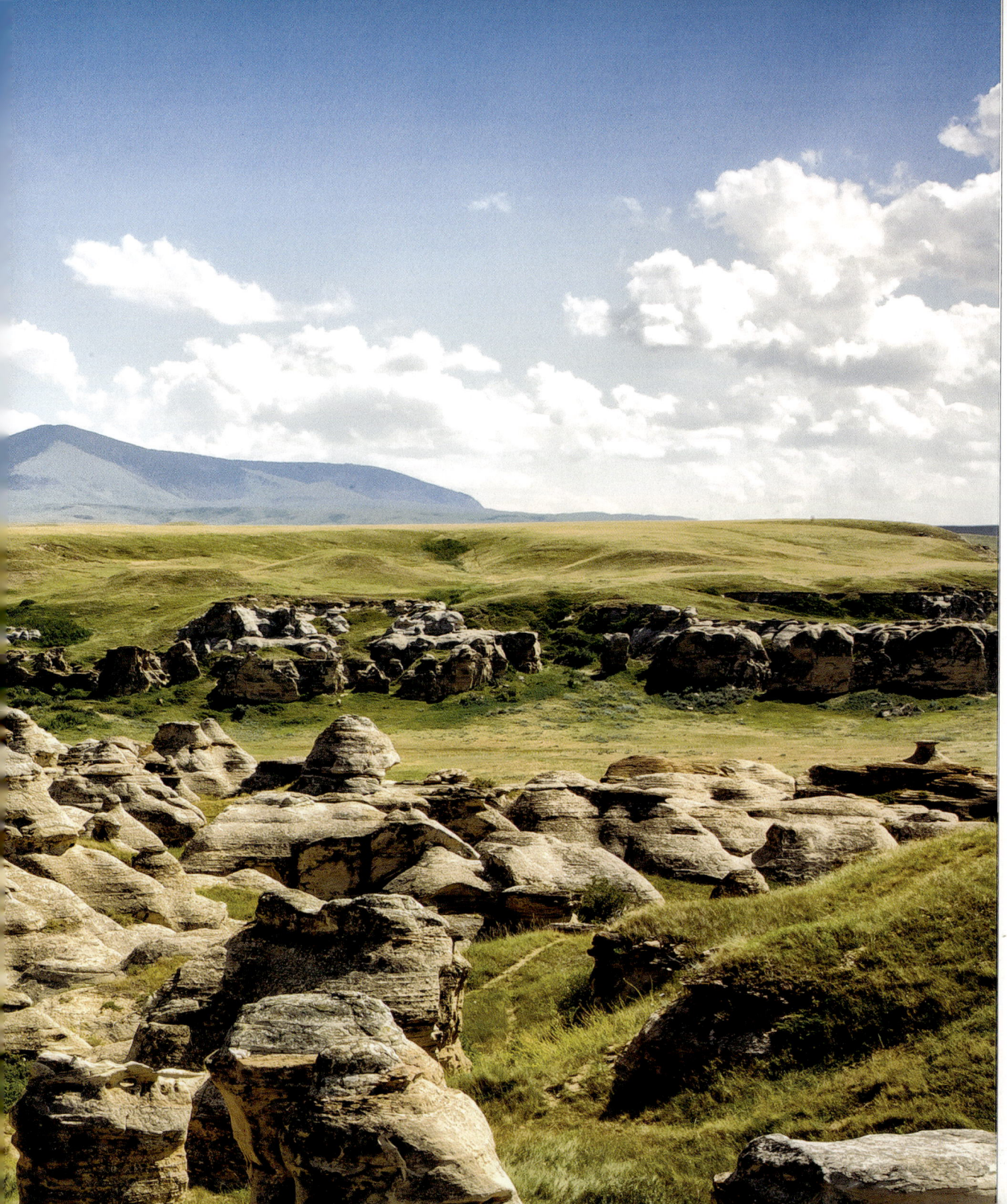

SWEETGRASS HILLS
A pilgrimage of stone as the Sweetgrass Hills weighs its rocky subjects. These low-lying mountains just across the Montana border are sacred to the Blackfoot. The unseen Milk River divides the foreground and the middle ground in this photograph.

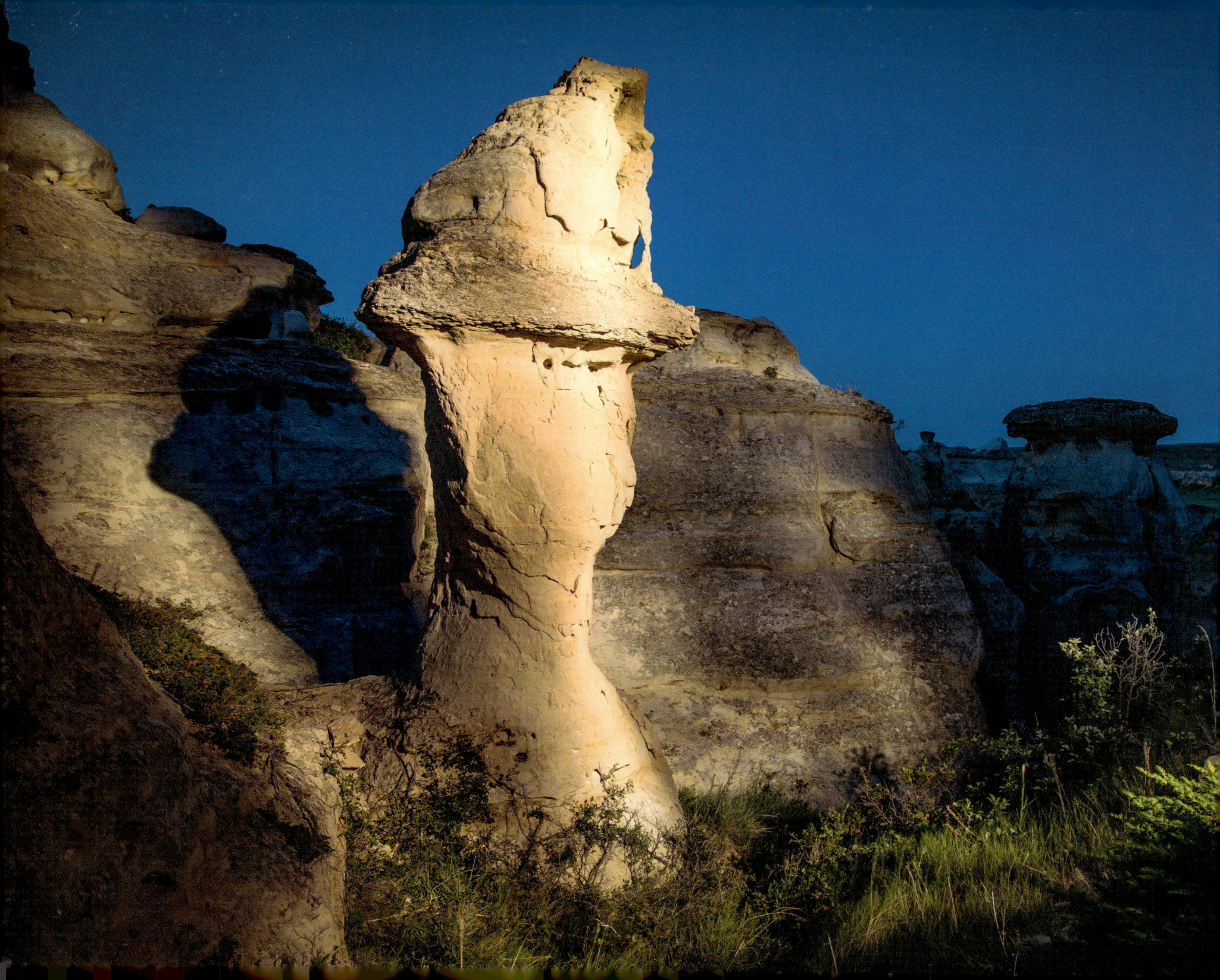

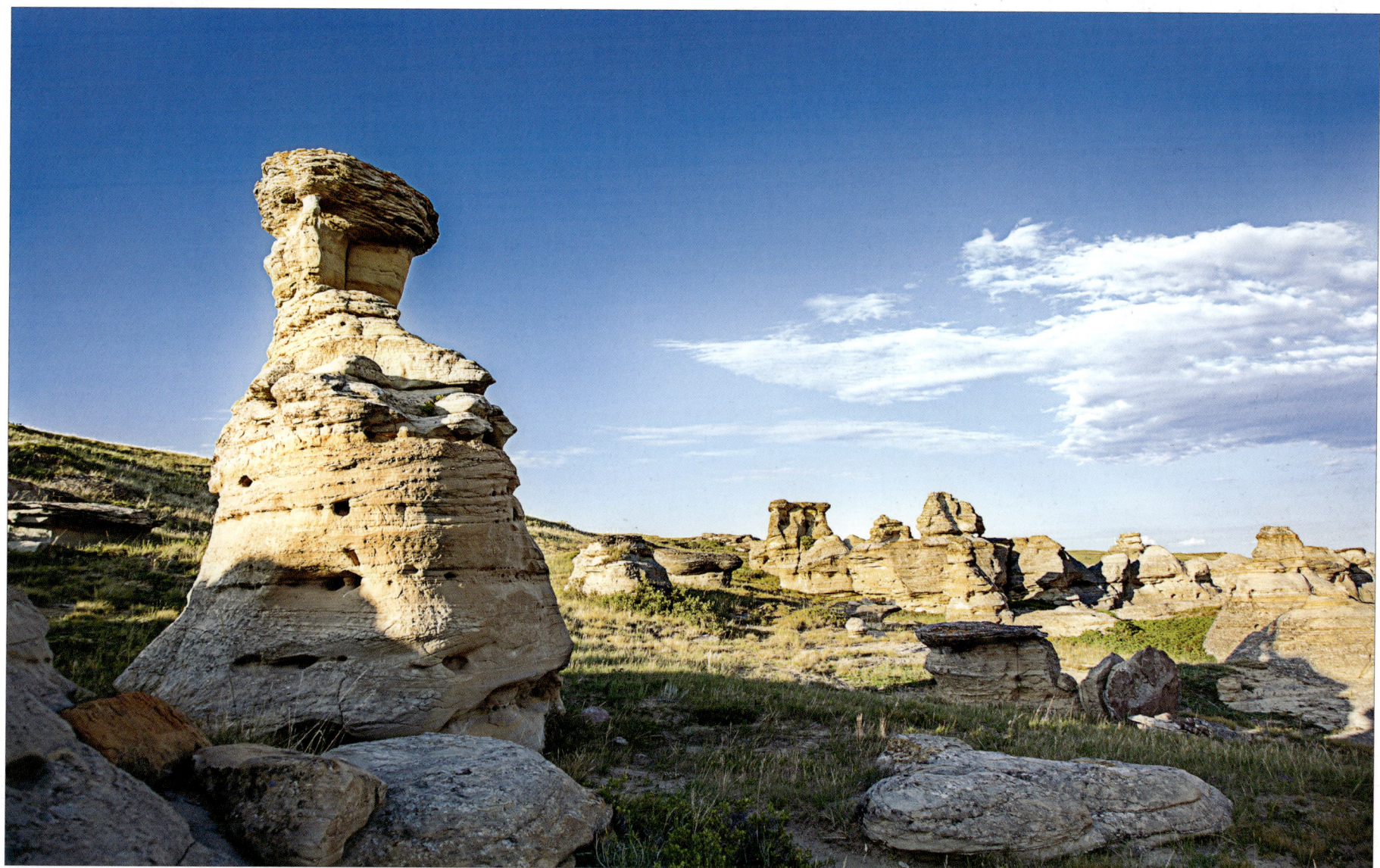

LEFT: BURIED IN STONE
This 6-metre "Cat in the Hat" hoodoo overlooks where Police Creek empties into the Milk River. It is situated in a bowl-like depression and surrounded by a natural rock amphitheatre.

ABOVE: WHISPERS AND LIES
The Blackfoot call Writing-on-Stone *Aisinai'pi*, meaning where the drawings are. For thousands of years this sacred landscape has drawn First Nations to its mystical rocks, eroded sandstone cliffs, and lush river valley.

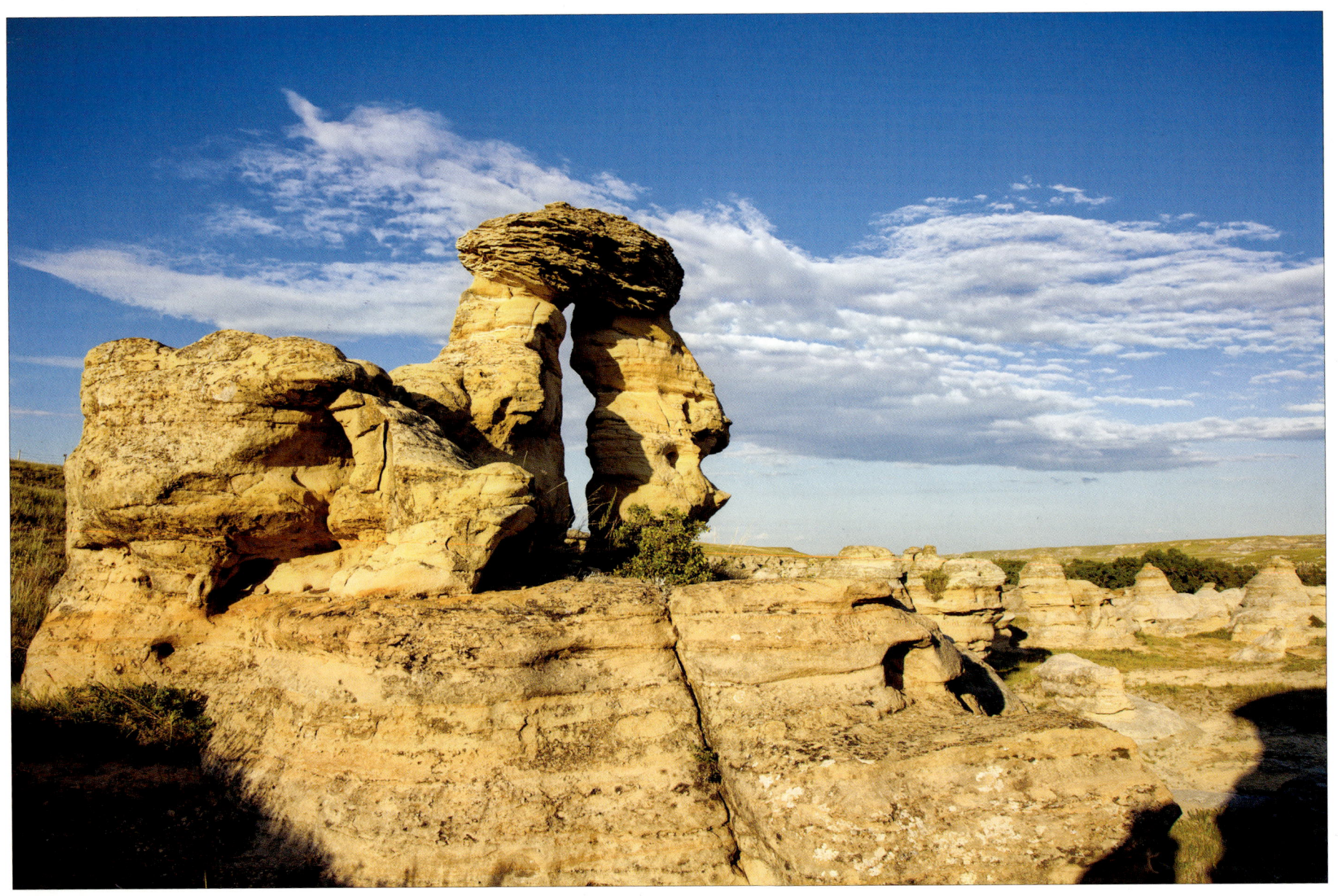

A FINE AND PRIVATE PLACE

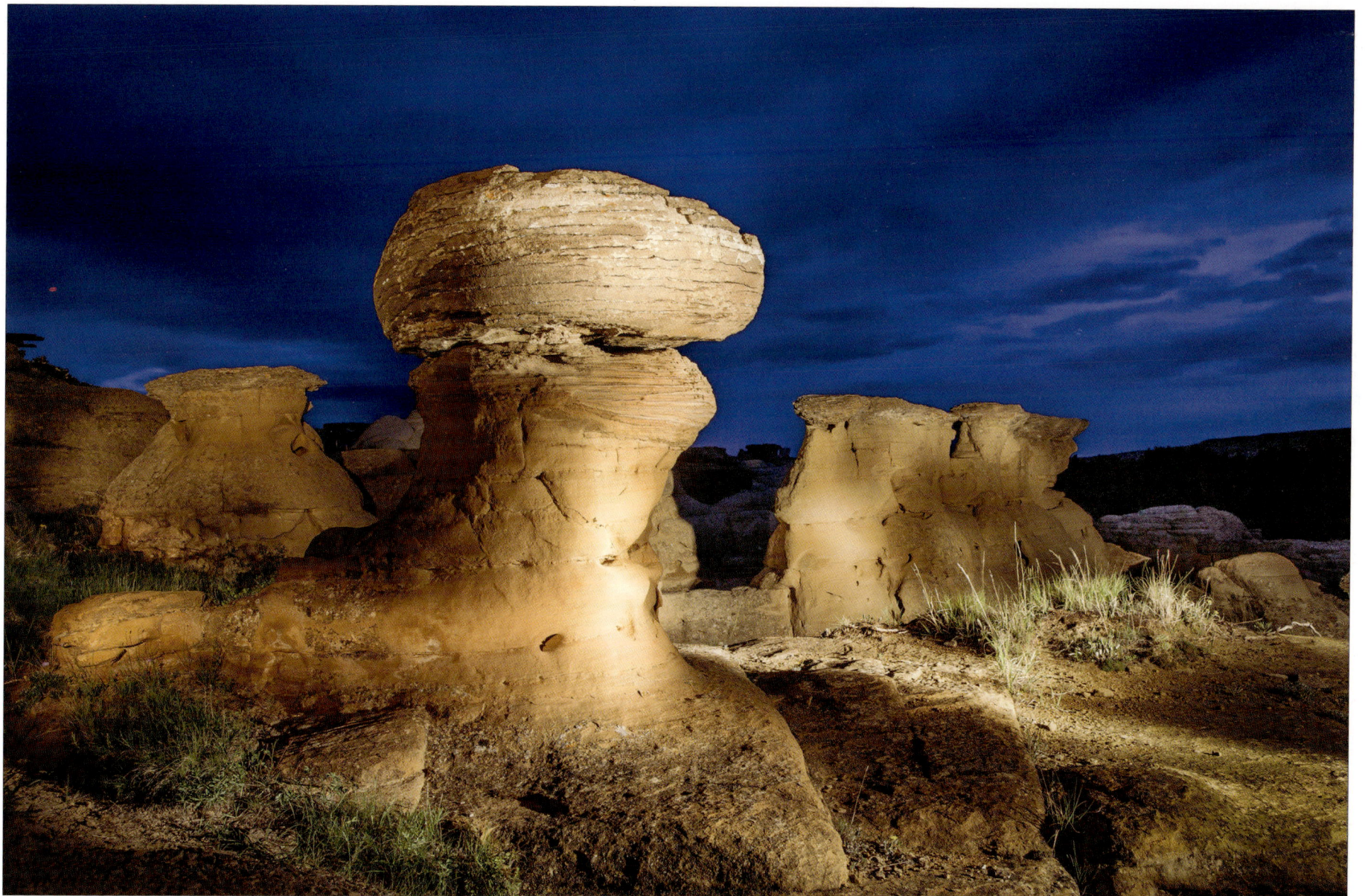

EARTH MAGIC

Being alone in the badlands at night can be unnerving, especially where there have been confirmed cougar sightings. During this light-painting shot I was panning my flashlight back and forth when I came across two eyes reflecting back at me, about 25 metres away. I panned away, then back, but the dots didn't flinch. I took a few steps in the opposite direction and whatever it was vanished.

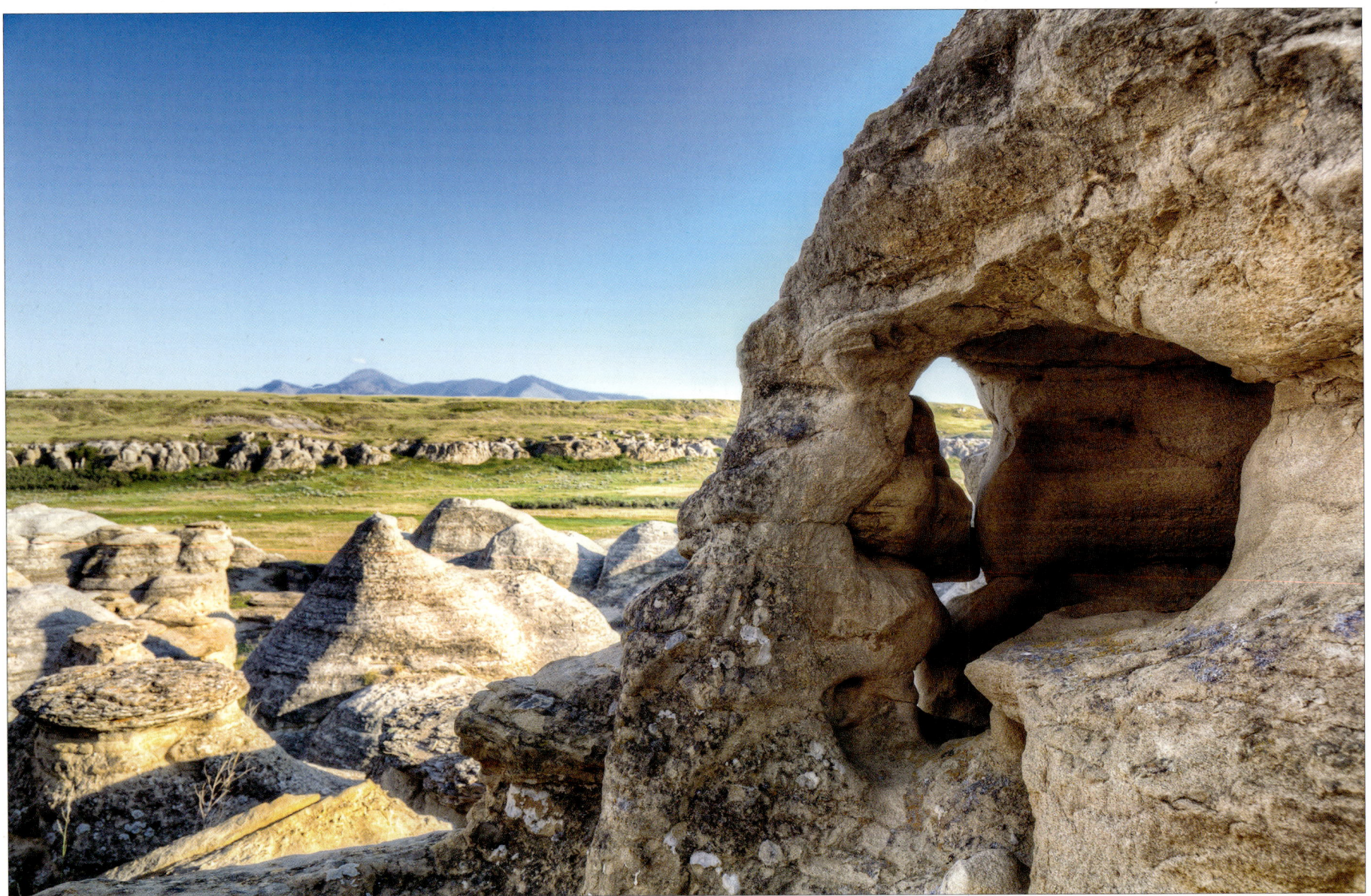

ABOVE: NO GREAT MISCHIEF

I think of erosion as a process beginning from the outside and working inward. But in this cavity it looks as if the rock is decaying from the inside out.

RIGHT: BEYOND THE GATHERING STORM

I had returned to my vehicle and was packing up my gear when, remarkably, a sliver of golden light ignited these rock outcroppings a bright golden yellow.

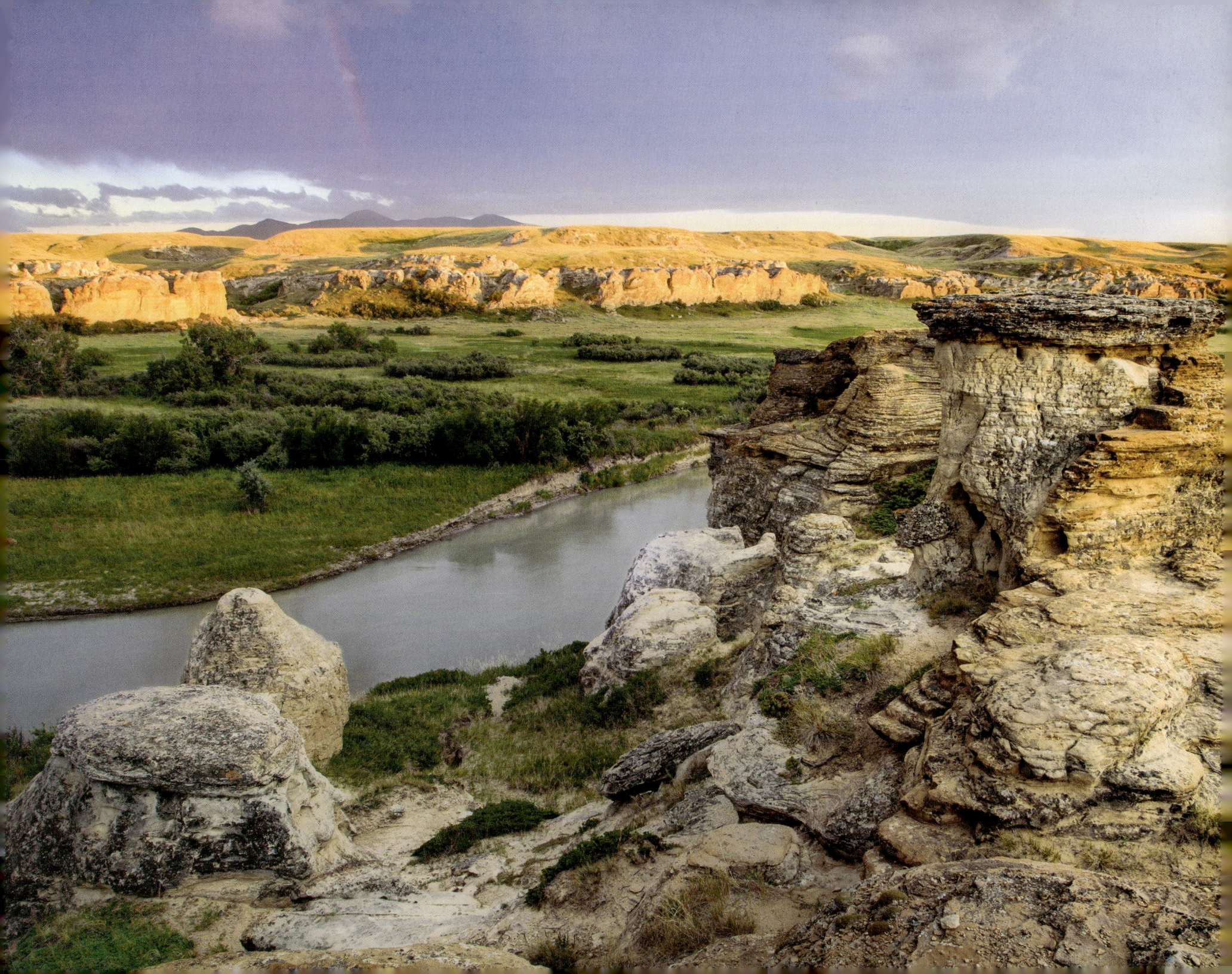

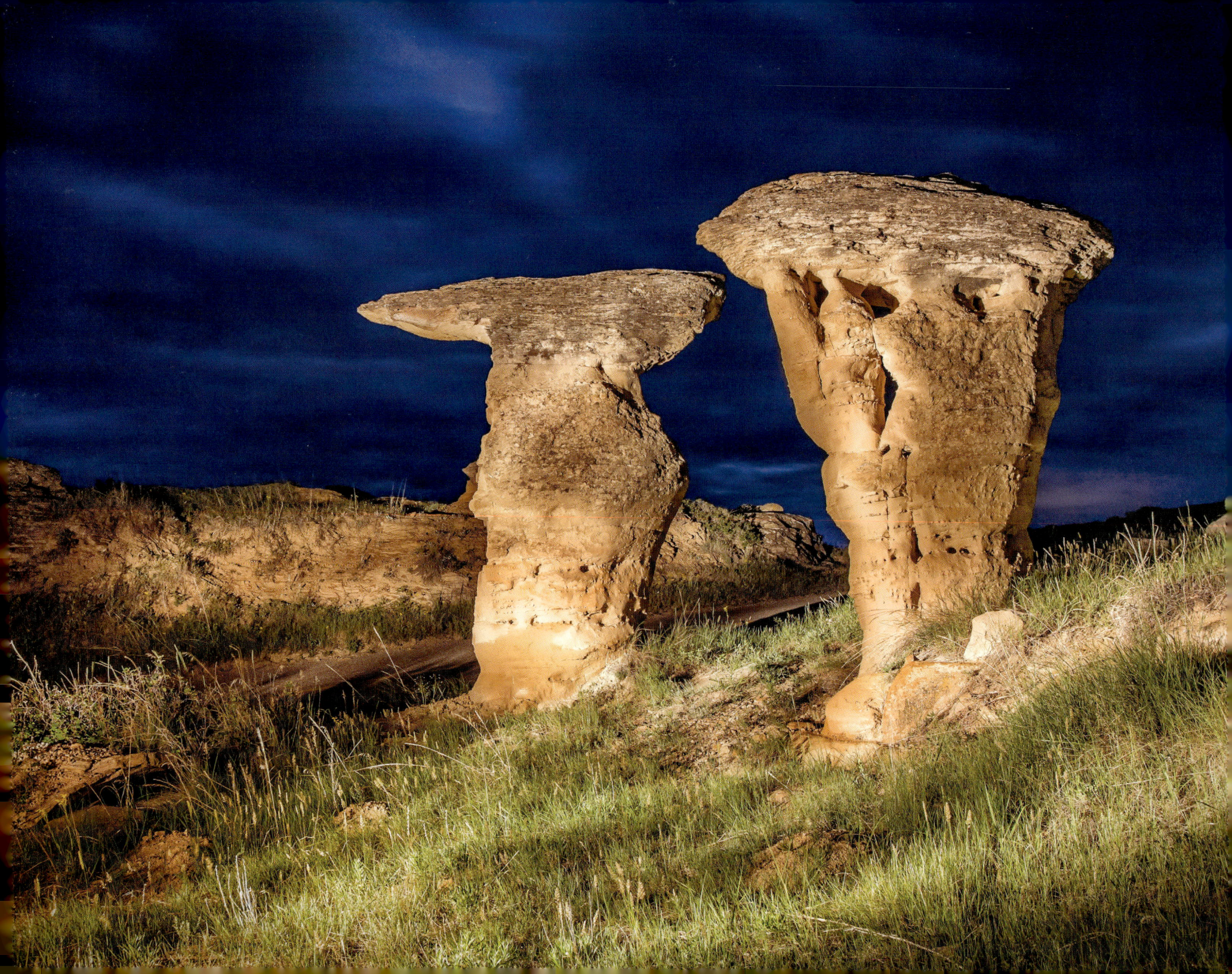

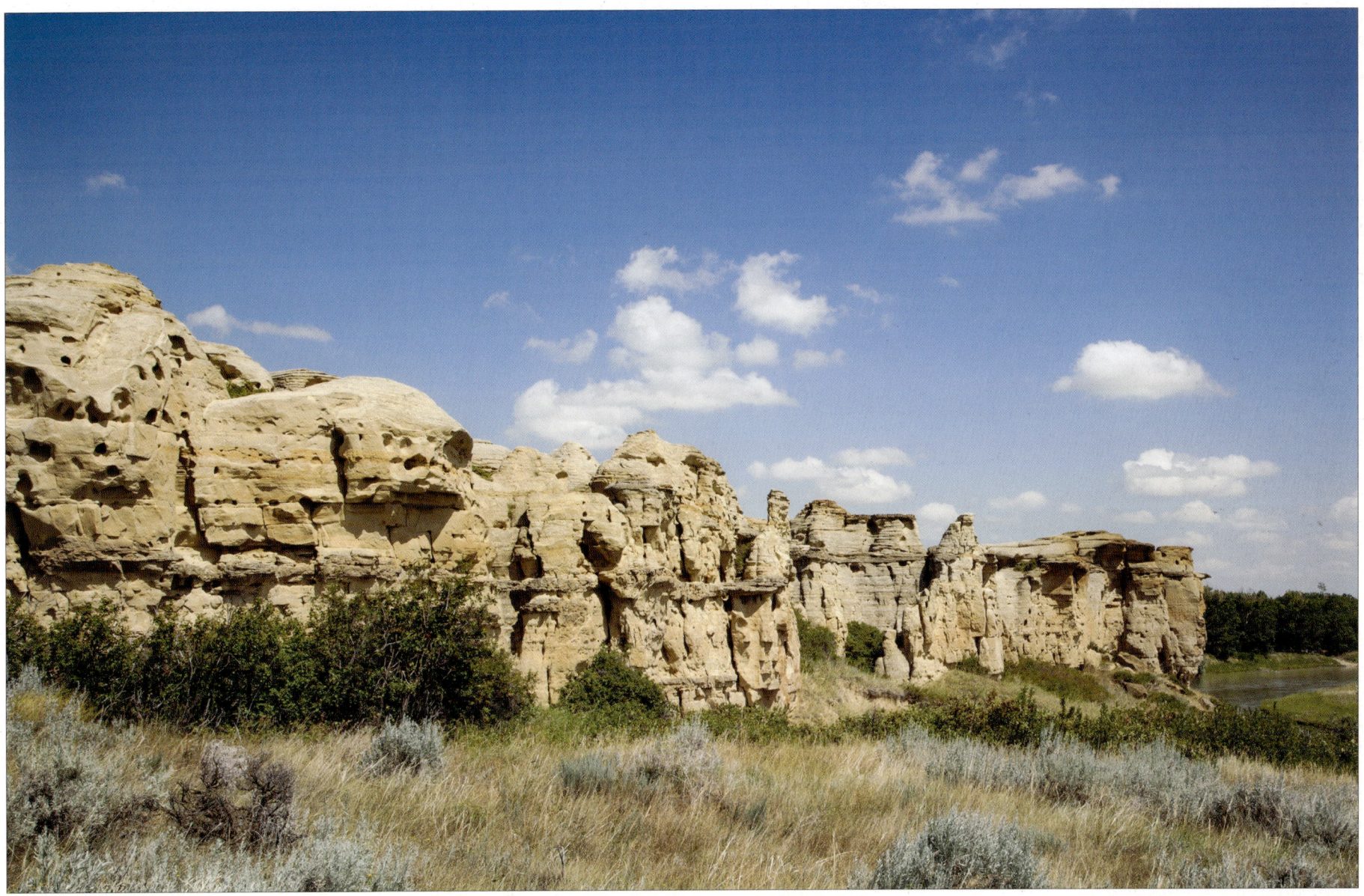

LEFT: A JEST OF GOD (NIGHT)

ABOVE: LIVES OF THE SAINTS

Cliffs taken near the banks of the Milk River with lush cottonwood trees in the background. While Writing-on-Stone is definitely a badlands, its topography is somewhat atypical in that there is very little, if any, exposed strata, fissured hills or bentonite clay. Here, rock formations and trees co-exist side by side.

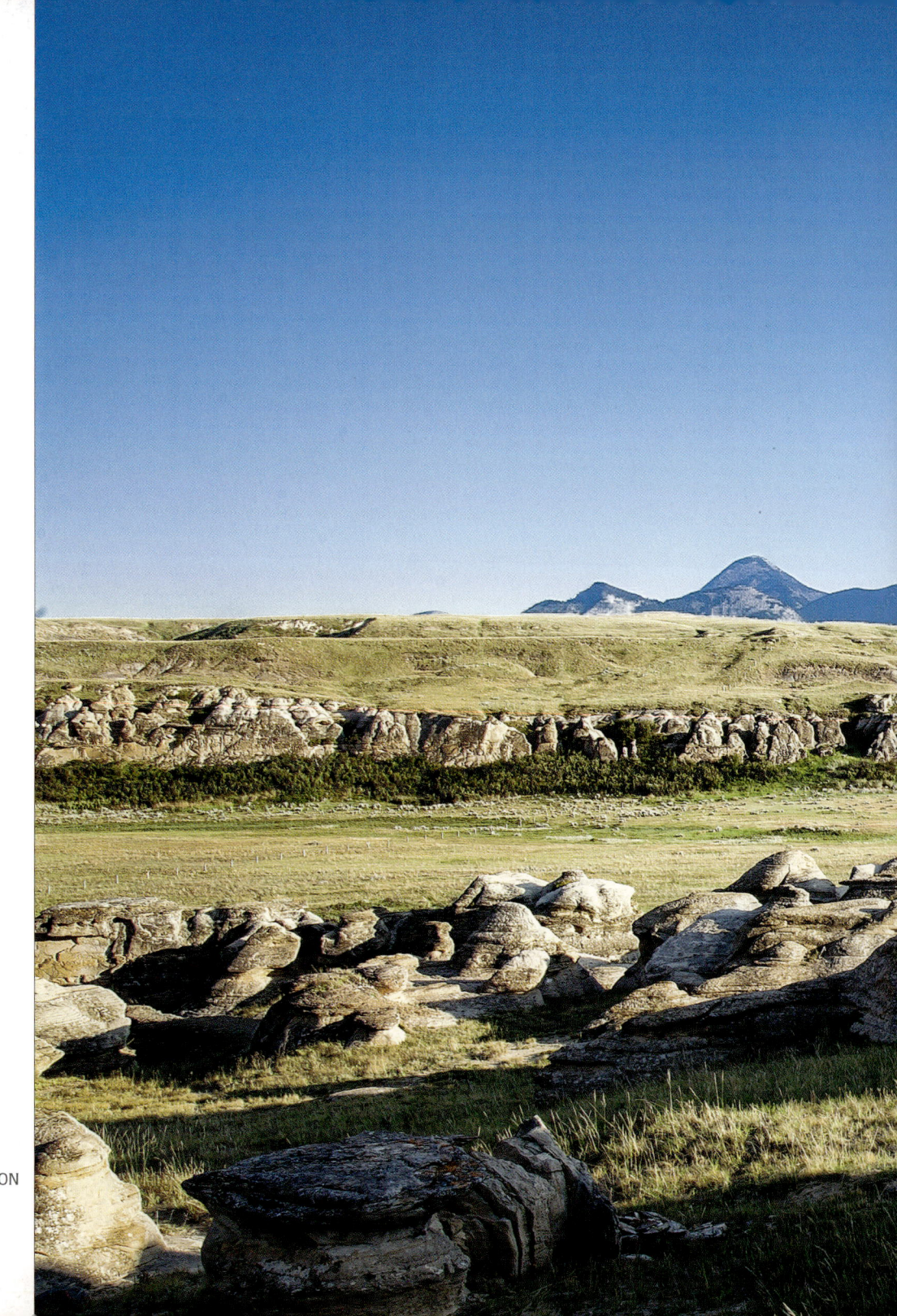

DISTANT BEACON

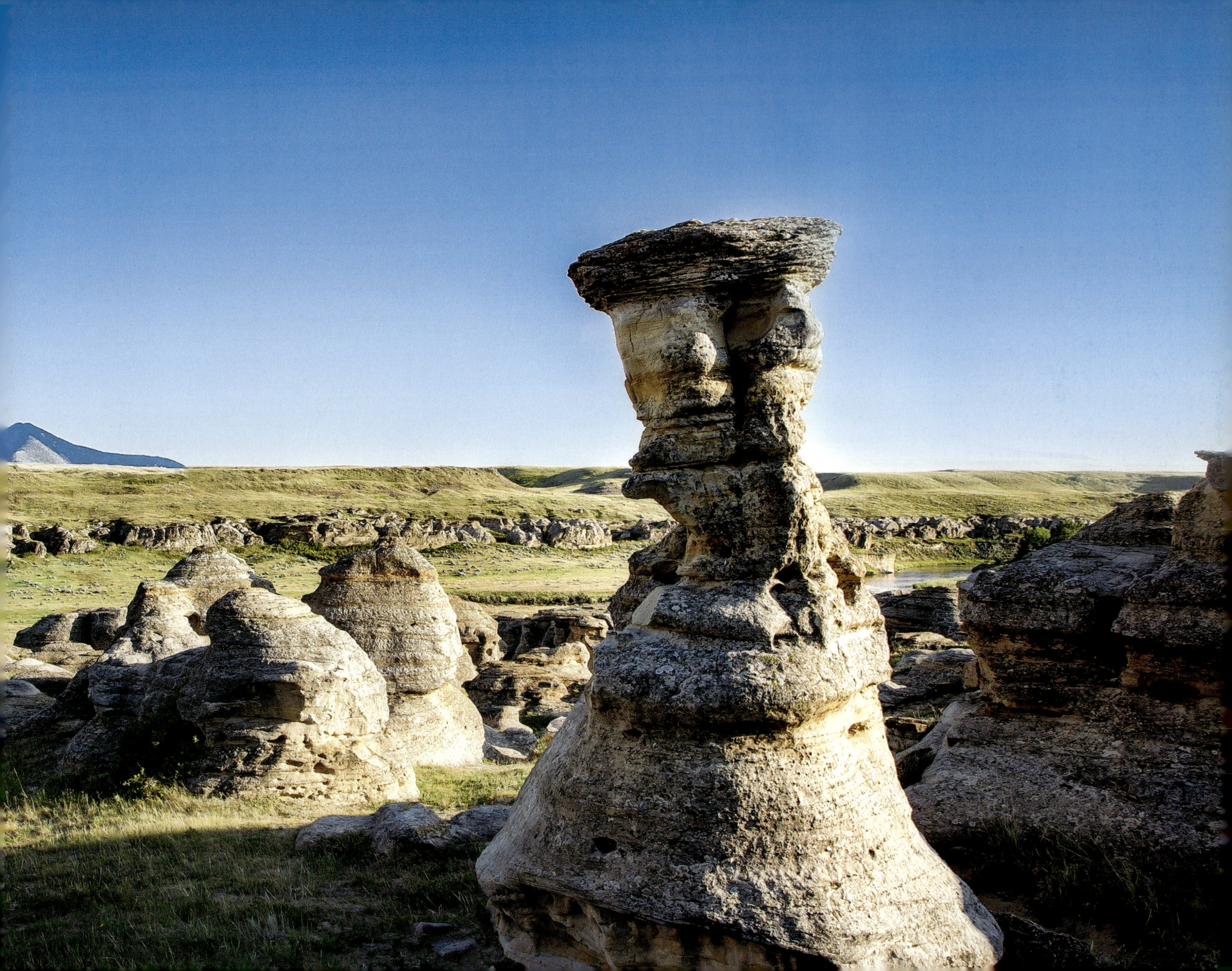

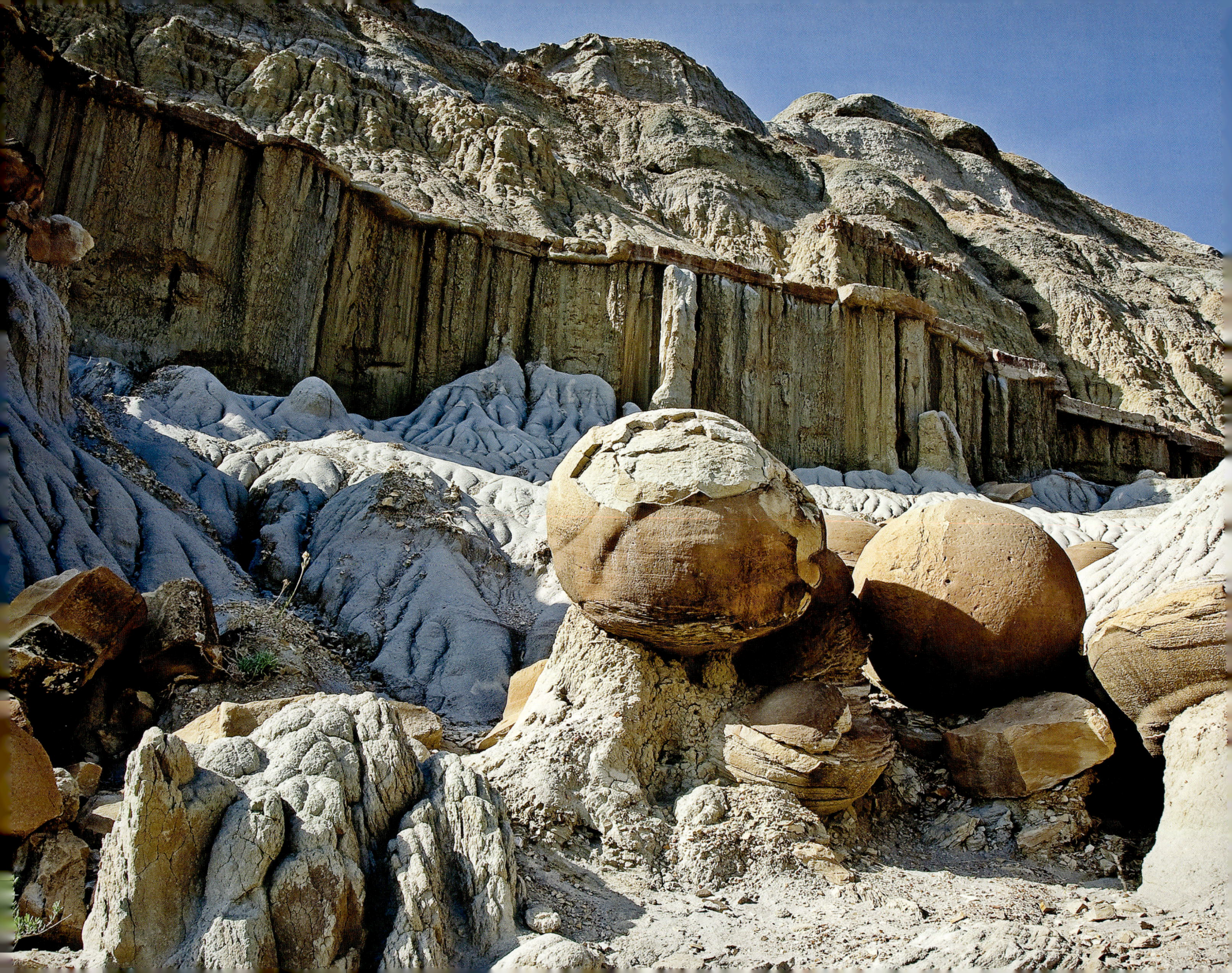

THEODORE ROOSEVELT NATIONAL PARK

There is an artfully calculated balance at Theodore Roosevelt National Park that, for me, brings to mind the title of the novel *The Beautiful and the Damned* by F. Scott Fitzgerald. There are no other badlands where pristine forests and dry broken badlands co-exist so eloquently (see the photograph "East of Eden" page 184). Theodore Roosevelt, the 26th President of the United States, was cognizant of this paradox, describing the landscape here as having a "desolate grim beauty." (Harmsen p. 34) And yet, the allure of the North Dakota badlands was undeniable to him. If one responds to the land in proportion to one's appreciation of its symbolism, then these badlands gripped Roosevelt like a spell.

Roosevelt was born and raised in a wealthy New York family and surrounded by society's elite. He was an active but often sick child, with asthma and stomach problems that persisted into early adulthood. He was highly educated and brilliant, having attended both Harvard and, later, Columbia Law School. At a young age, Roosevelt was enamoured by adventure stories, especially those of the American West. So, at the age of 24, when he was invited to hunt buffalo in North Dakota, TR jumped at the chance. It was 1883 when he first travelled to what was then known as the Little Missouri badlands—often called the "Little Misery." He was so enthralled by the country that he came back the following year and invested a sizable amount of money in land and cattle and built Elkhorn Ranch. Roosevelt resided, on and off, in North Dakota for the better part of three years before family pressures and his political career prompted him to return east.

The badlands were everything TR expected and more. It was a tough country for tough men, and Roosevelt was able to cast aside his privileged upbringing and thrive in the Wild West. His life was threatened on one occasion; on another, the boat he used on the Little Missouri River was stolen by three men, including a known criminal named Mike "Redhead" Finnegan, who had recently shot up the town of Medora. Roosevelt and two of his men went after the thieves, surprised them, and retook the boat at gunpoint. When TR finally sold his stake in his badlands ranch in 1897, the final tally on his investment was a $20,000 loss. Yet he might have called it the best investment he ever made. TR fully embraced this landscape and it, in turn, left a lasting

ATLAS SHRUGGED
Photographing the badlands can take considerable time, planning, and exhausting hikes. This photograph, though, couldn't have been easier: a paved road, a sign that read "Concretions," a 20-metre stroll, and picture perfect cannonball concretions positioned perfectly against a fantastic badlands background.

impression on him. So much so that he wrote three books about his experiences, romanticizing what he called "the perfect freedom" of the West. In fact, Roosevelt wrote, "I would never have been President had it not been for my experiences in North Dakota."

In stark contrast to Roosevelt's deep attachment to the North Dakota badlands, Lieutenant Colonel George Armstrong Custer had a very different impression of this devil's wilderness. In 1876, only seven years prior to Roosevelt's first trip to the Dakotas, multiple conflicts between the United States Army and Native American peoples had escalated to the point where the U.S. Army was actively pursuing Sitting Bull and his band of Sioux across North Dakota. Custer, commander in charge of the 7th Army Cavalry Regiment, was expecting a conflict to arise just outside the Little Missouri badlands. For whatever reason, Custer chose to march his men through the badlands. During this time, Custer wrote to his wife: "We found the Little Missouri River so crooked and the Bad Lands so impassable that in marching fifty miles today we forded the river thirty-four times. The bottom is quicksand. Many of the horses went down frequently tumbling their riders into the water..." One month later, in Montana, Custer and his troops

were ambushed and massacred at the Battle of the Little Bighorn by a contingent of tribes led by Sioux Chief Sitting Bull. The Badlands of North Dakota is a landscape that not only captivates and alarms one's imagination, but it also impacts people's lives in a very real way.

Theodore Roosevelt National Park, located in western North Dakota, consists of three separate units (a North Unit, a South Unit, and Elkhorn Ranch) that comprise over 70,000 acres. The North Unit is located just south of Watford City on Highway 85. The larger South Unit is located on the outskirts of the town of Medora along the I-95. The Elkhorn Ranch home is no longer there but the original site is still historically important. The Elkhorn Ranch is situated between the two units but is difficult to access, and a high-clearance vehicle is recommended.

The magnificently proportioned views in both units are awe-inspiring. Theodore Roosevelt National Park has perhaps the widest spectrum of terrain of any badlands region. There are large forested areas, scarred ravines, serrated hills, hogback ridges, a significant petrified forest, and the aforementioned "Little Misery" river. Hoodoos and rock formations with alternating layers of sandstone, siltstone, mudstone, and bentonite clay populate the broken

hills and make for an extraordinary visual experience. Complementing the landscape is an extensive array of flora and fauna. In fact, the wide diversity of plant life in the park is challenged only by that of a rainforest. There are over 400 plant species. To add to this, the park is perhaps best known for its wildlife. You are almost guaranteed to see buffalo anywhere there is a flat stretch of grass. There are close to 400 elk that roam the South Unit. Bighorn sheep inhabit the North Unit. Apparently, a small population of rarely seen mountain lions resides in the South Unit hills. Other animals include wild horses, deer, pronghorn antelope, badgers, coyotes, porcupines, and over 180 different types of birds. The best times to see wildlife are at dusk or dawn.

With this rich variety of wildlife, it's no wonder Theodore Roosevelt loved these North Dakota badlands. It is one of the reasons why, during his presidency, Roosevelt established the Forest Service and enacted legislation to preserve America's national forests, parks, and monuments. His administration placed close to 230 million acres under public protection, including these fabulous North Dakota badlands now named after him.

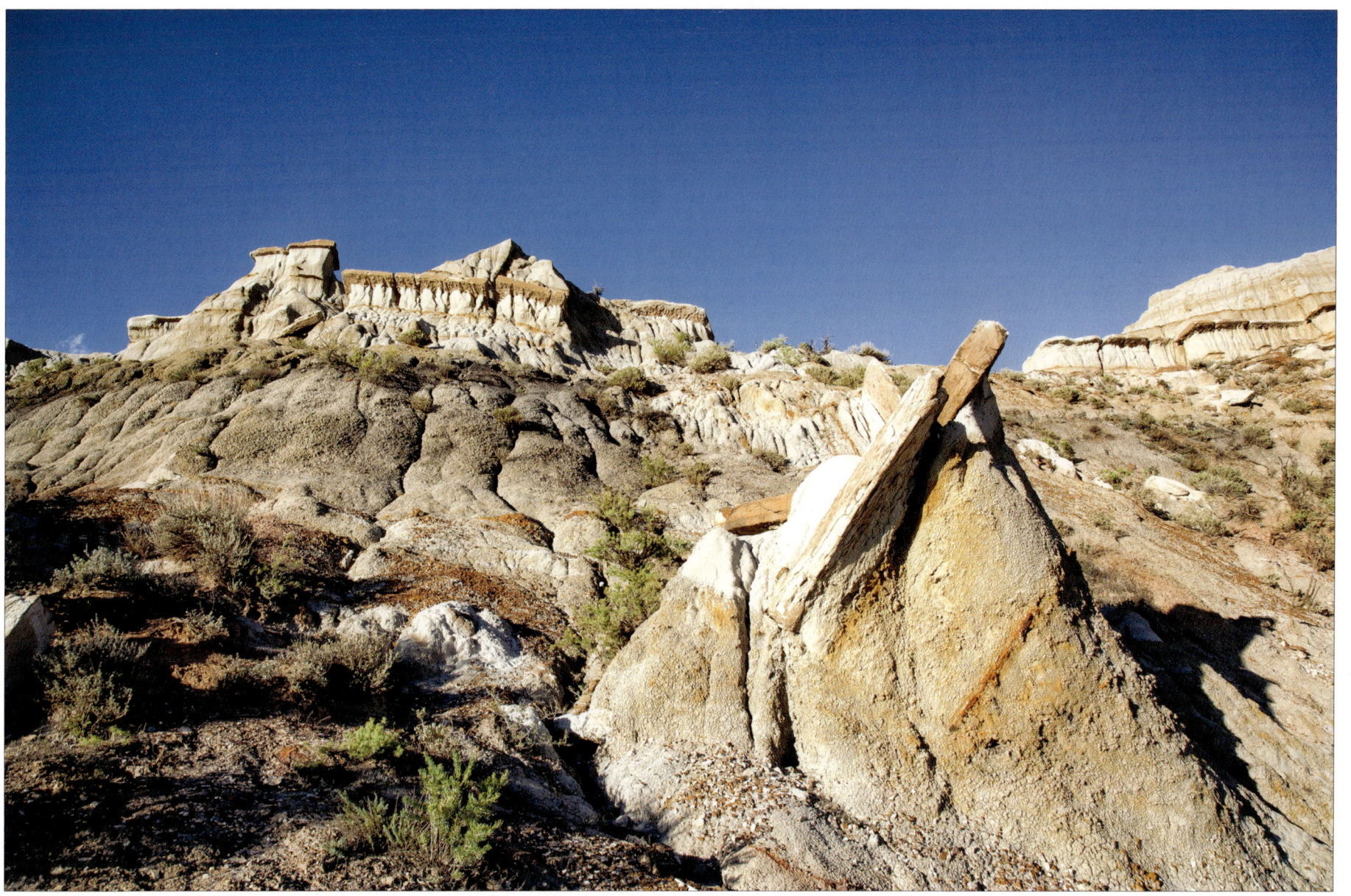

EXECUTIONER'S SONG
For me the badlands, rather paradoxically, convey a tranquil violence. Roosevelt himself described them as having a "grim desolate beauty." This angled stone slicing into the bentonite clay reminds me of an executioner's axe.

159

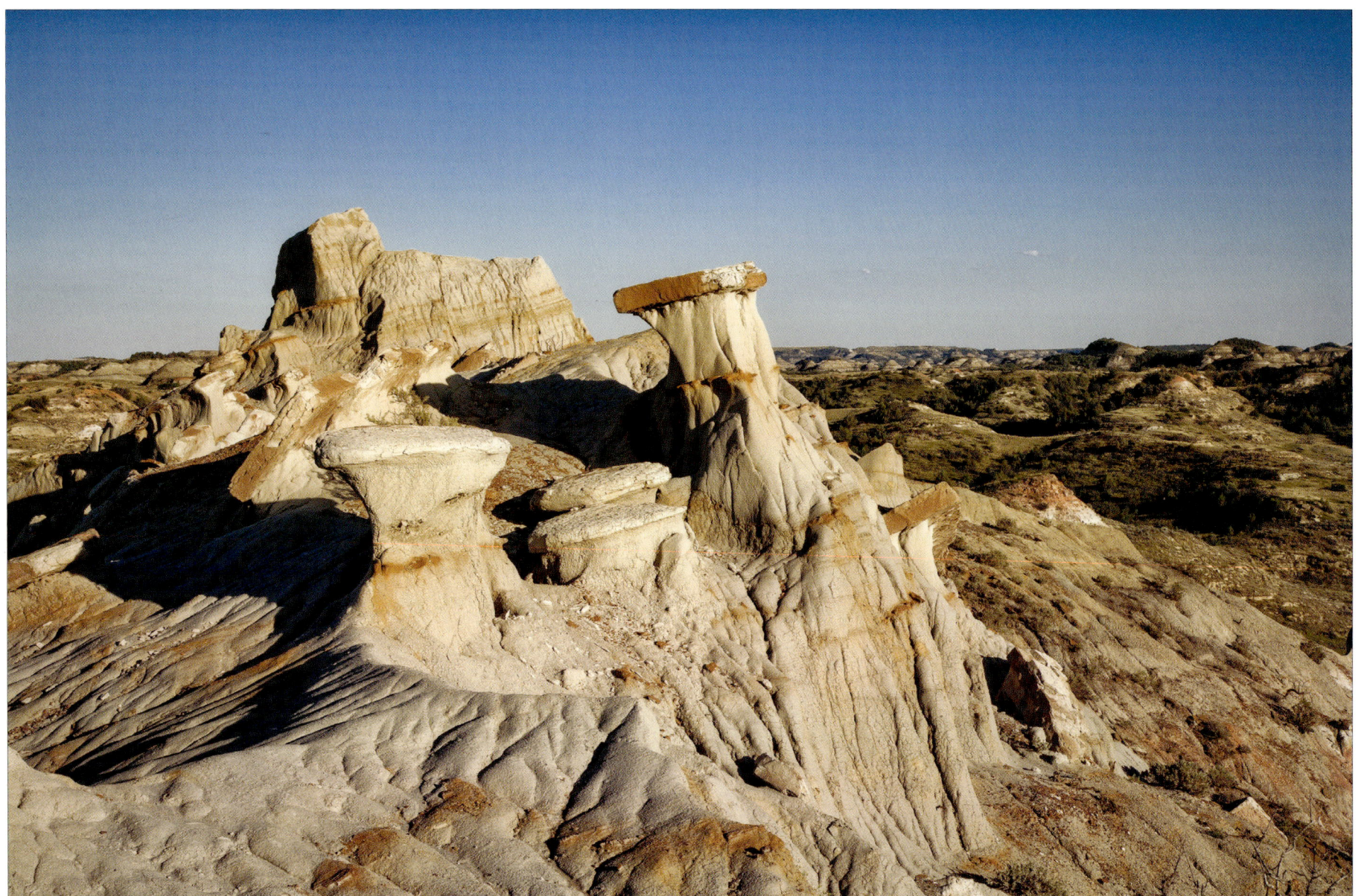

MOURNING BECOMES ELECTRA
In photography, dusk and dawn are called the golden hours. The shadows
are long and the light is warm and pristine.

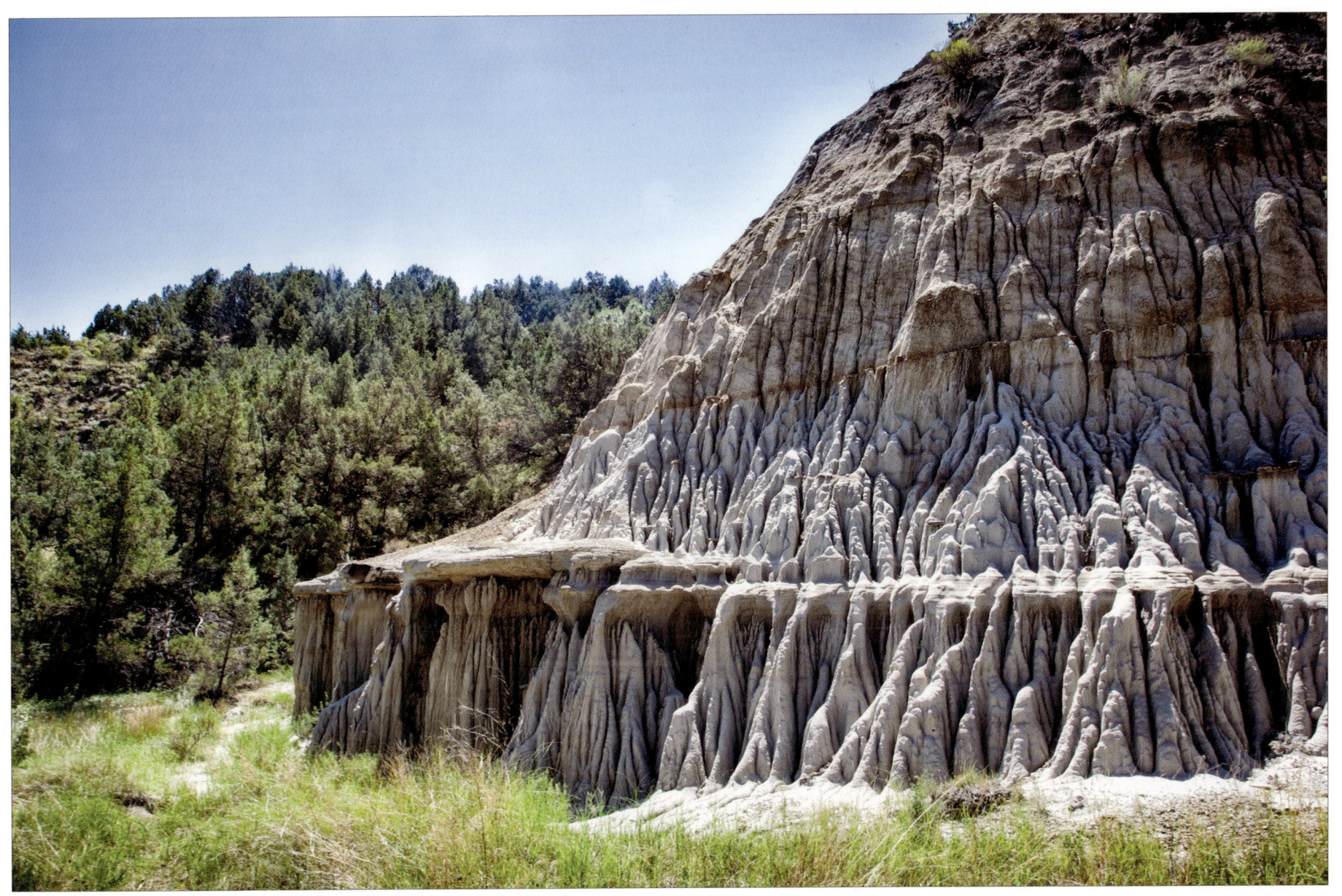

THE BEAUTIFUL AND THE DAMNED
This sterile hill with its deep-channeled fissures seemingly melts in the sunlight.
A grove of rich pine trees thrives in the background.

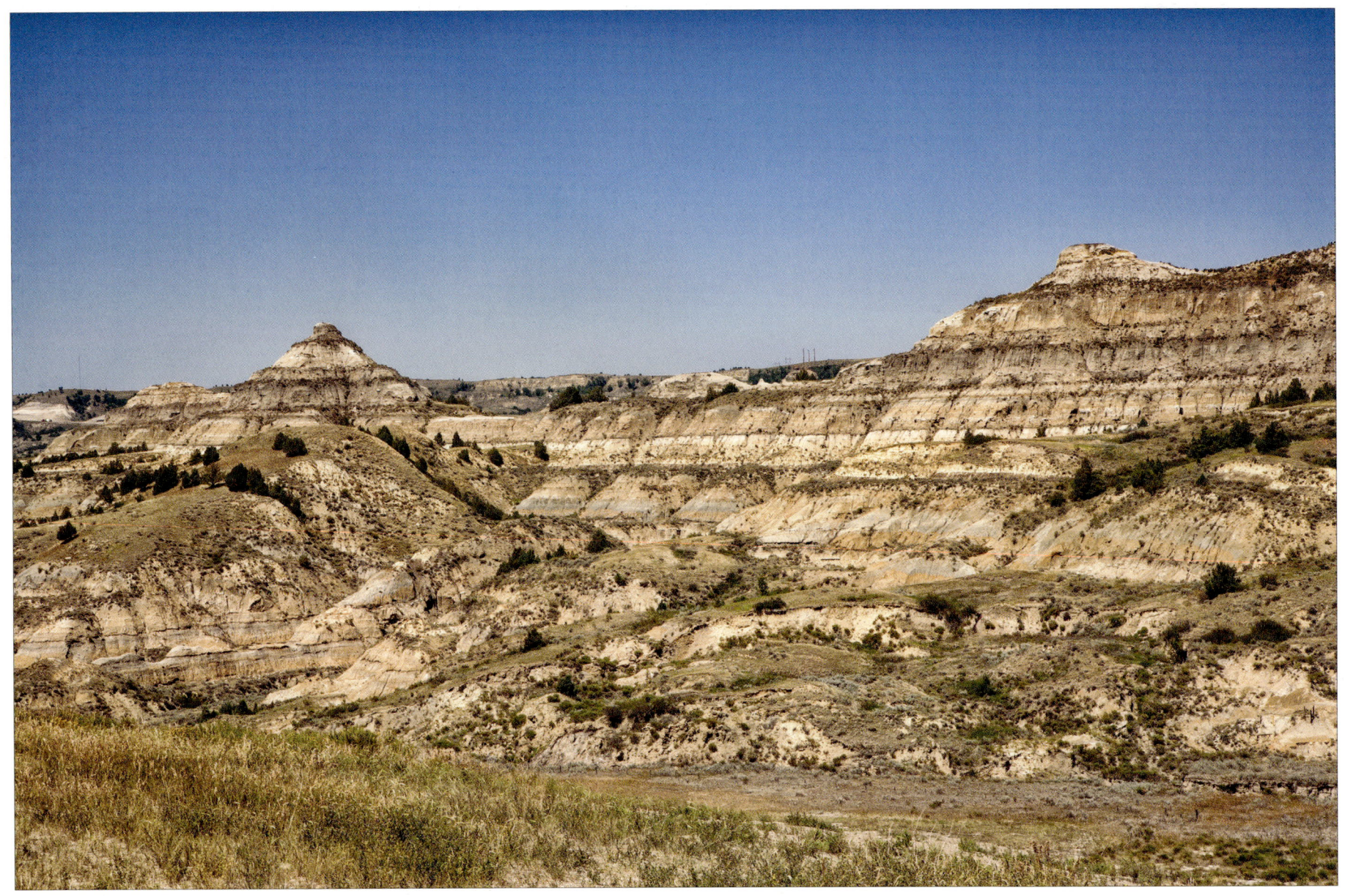

WONDERLAND

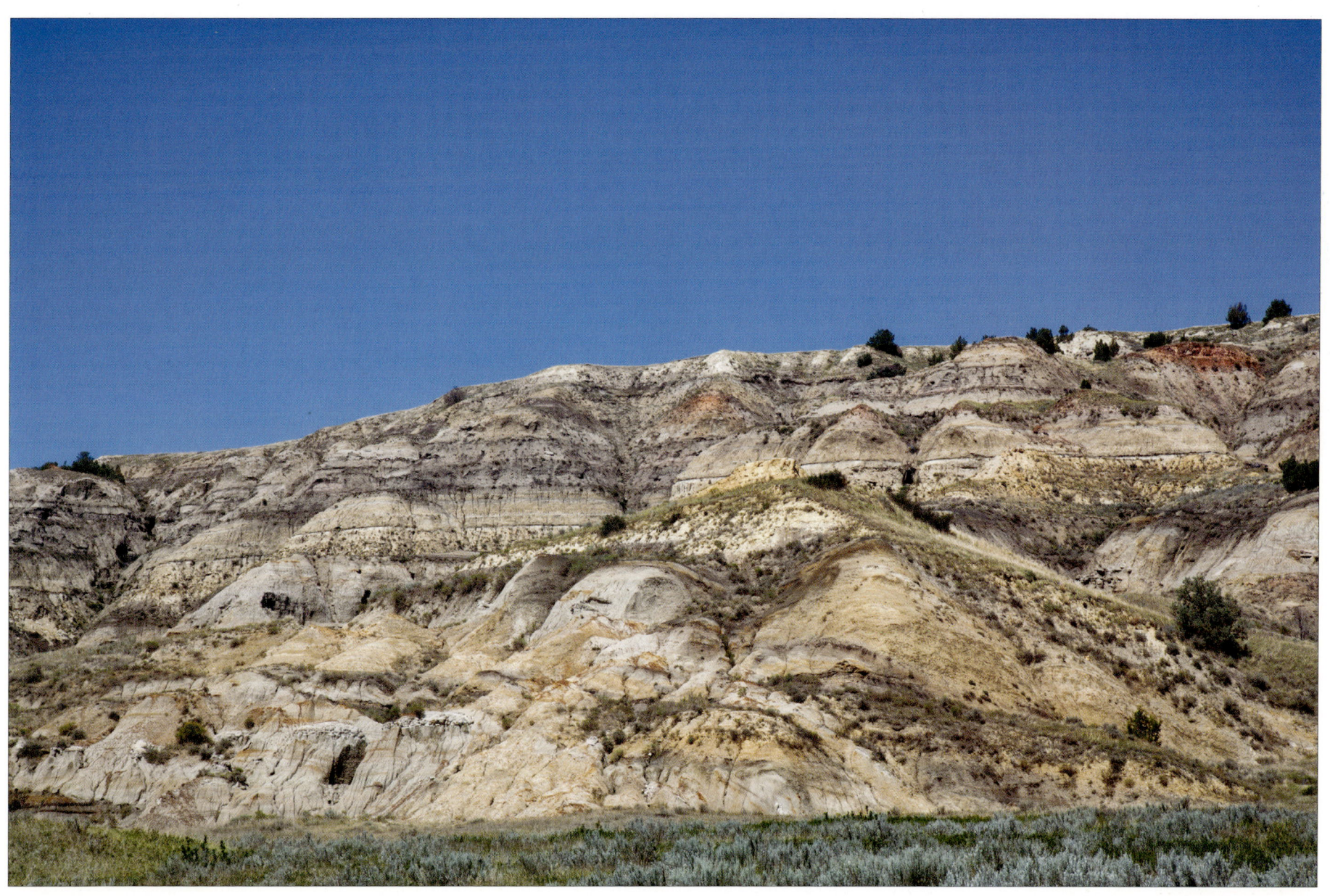

GRAVITY'S RAINBOW

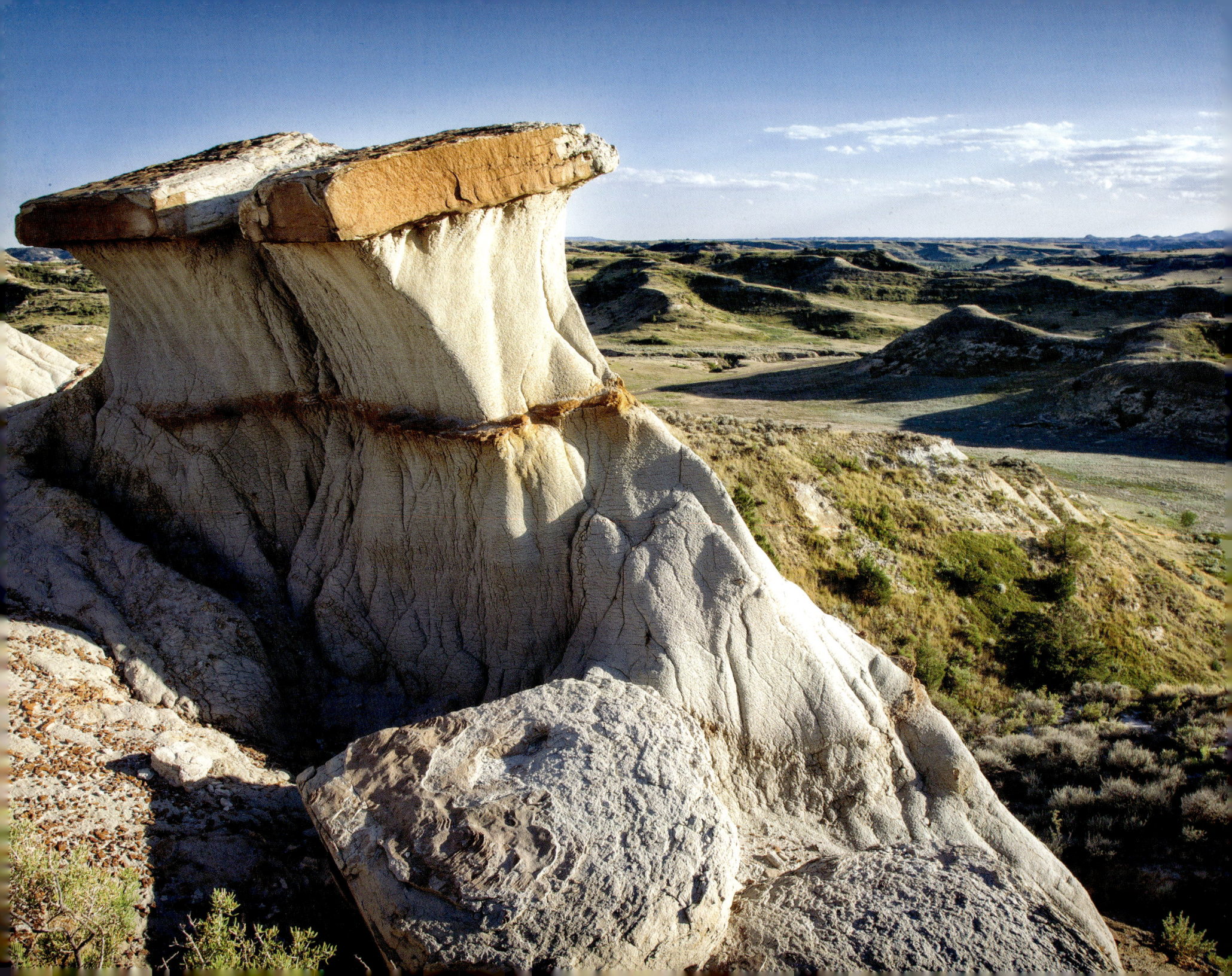

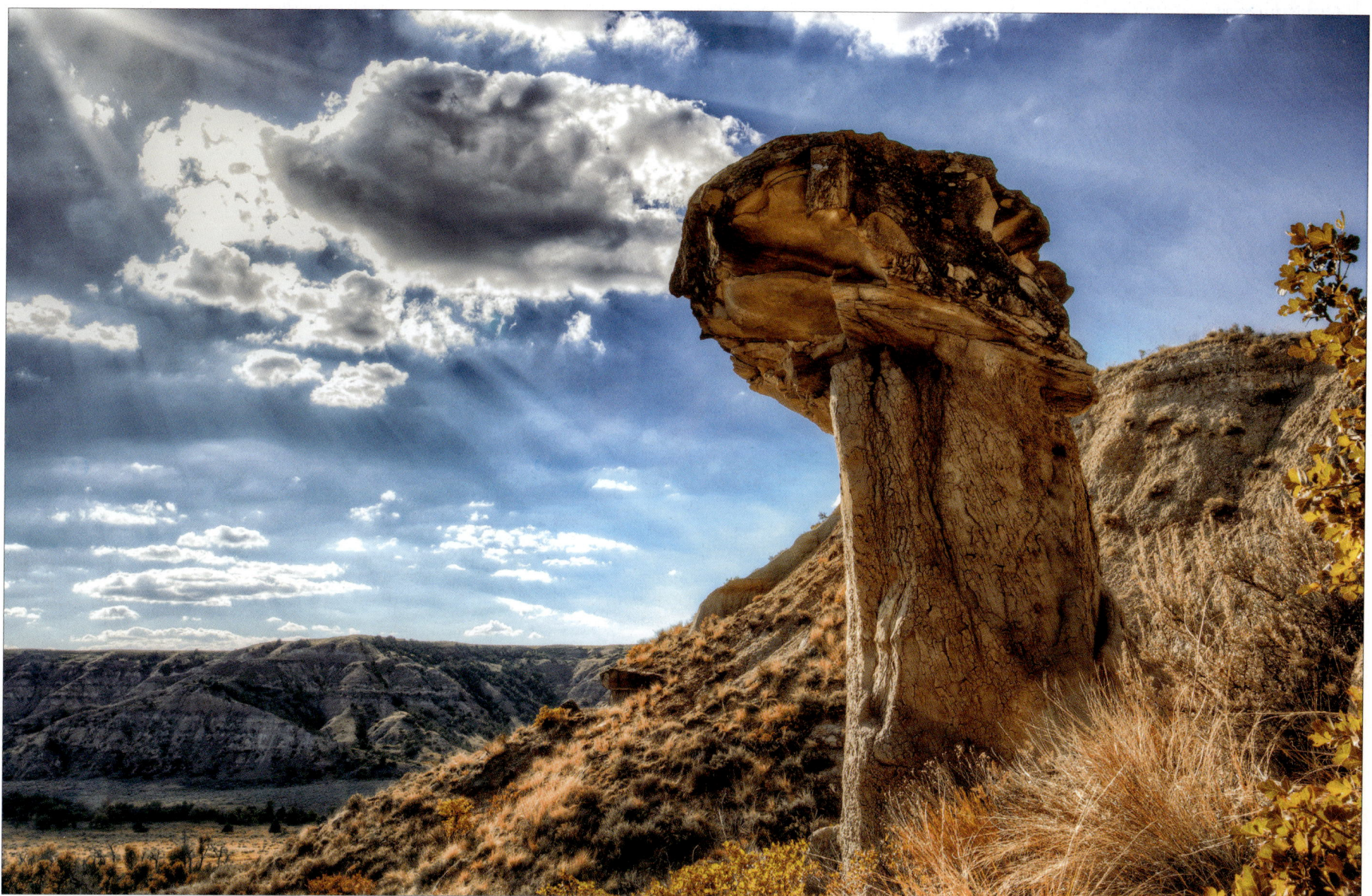

LEFT: THESE THOUSAND HILLS
Conflict and transformation sublimate this choreographed panorama, as these twin hoodoos overlook the undulated hills in the North Dakota badlands.

ABOVE: ANGEL PEAK
I scrambled high up this sharply inclined hill—using my tripod as support—to capture this dinosaur head hoodoo.

165

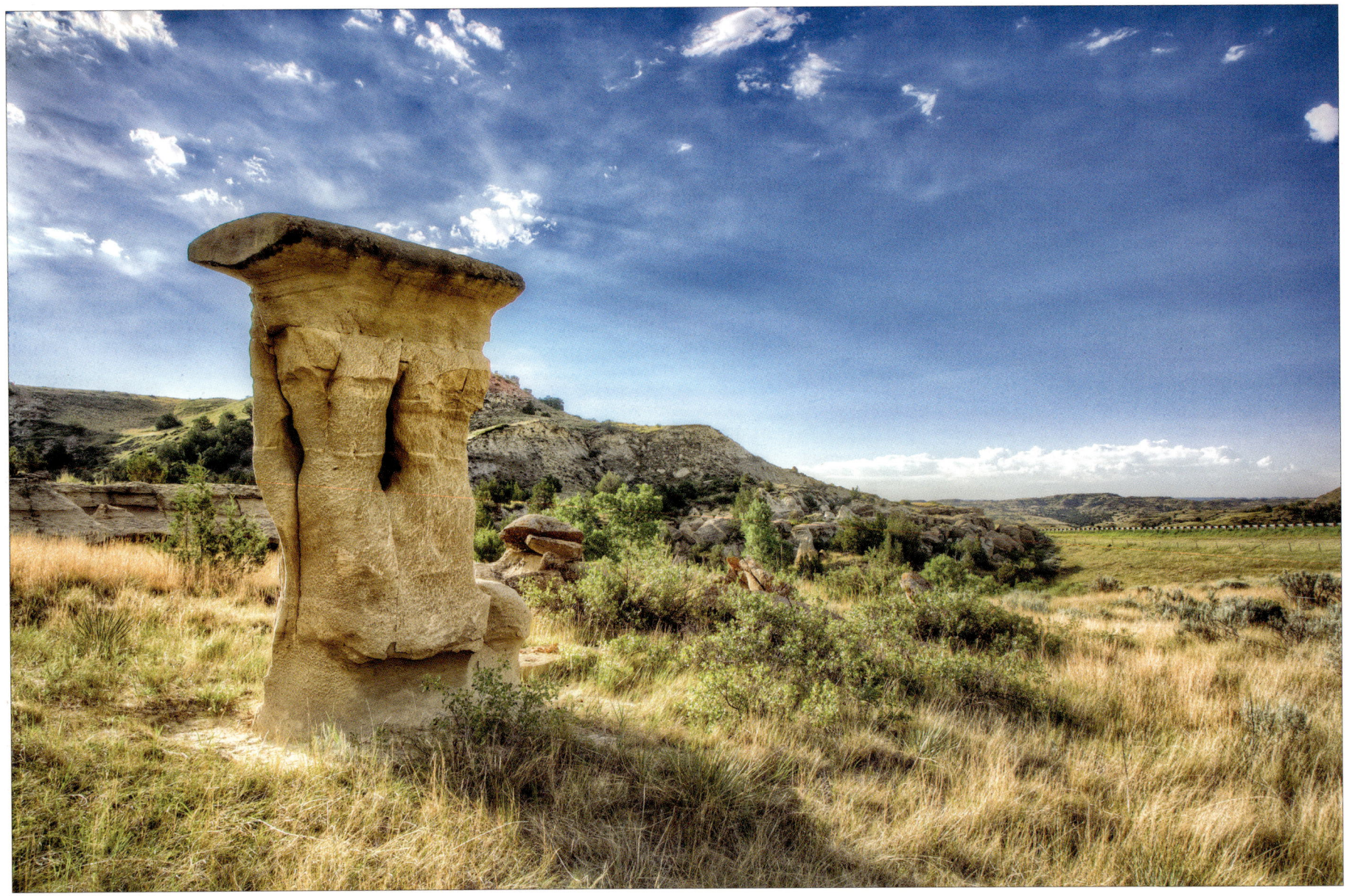

DESOLATION ANGEL

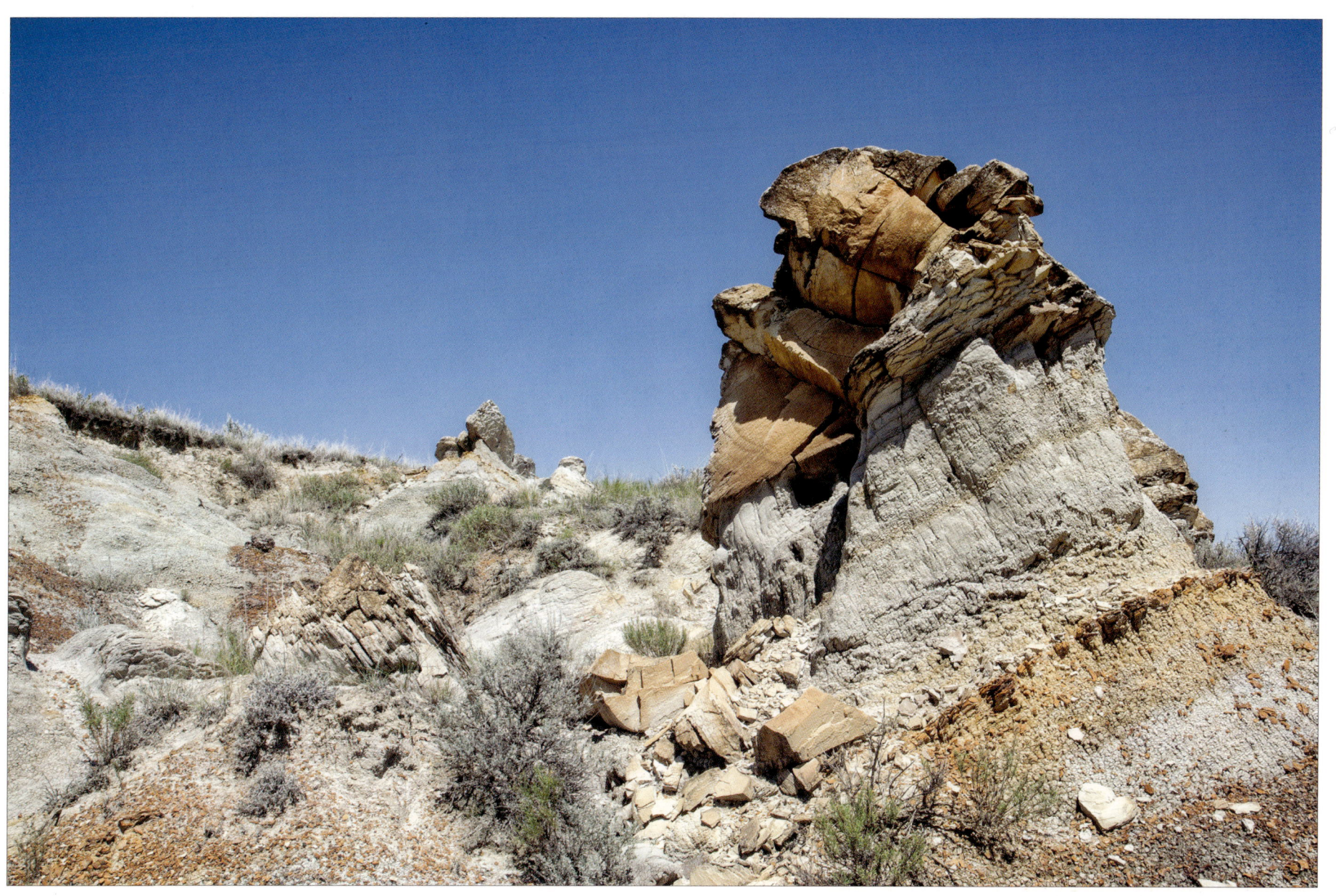

WHERE THE WILD THINGS ARE
"There must be more to life than having everything." Quote from Maurice Sendak's book of the same name.

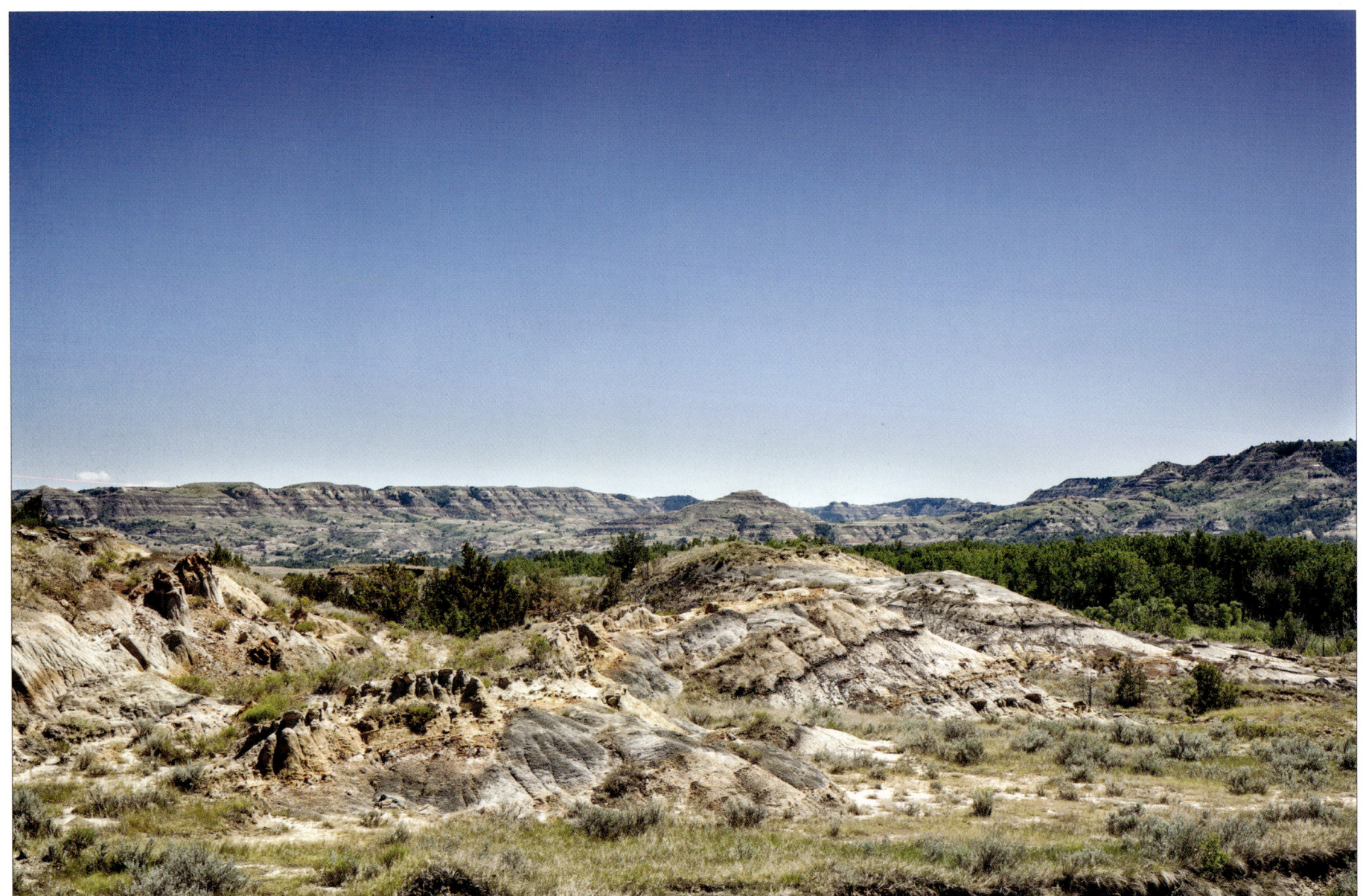

A MOVABLE FEAST

While the badlands exude an undeniable rough and rugged appearance, this belies a very tenuous infrastructure beneath the ground where "slumping" is a regular phenomenon. This is typically due to the underlying bentonite clay that expands when wet and contracts when dry. The result is a land that collapses and moves as if it were alive.

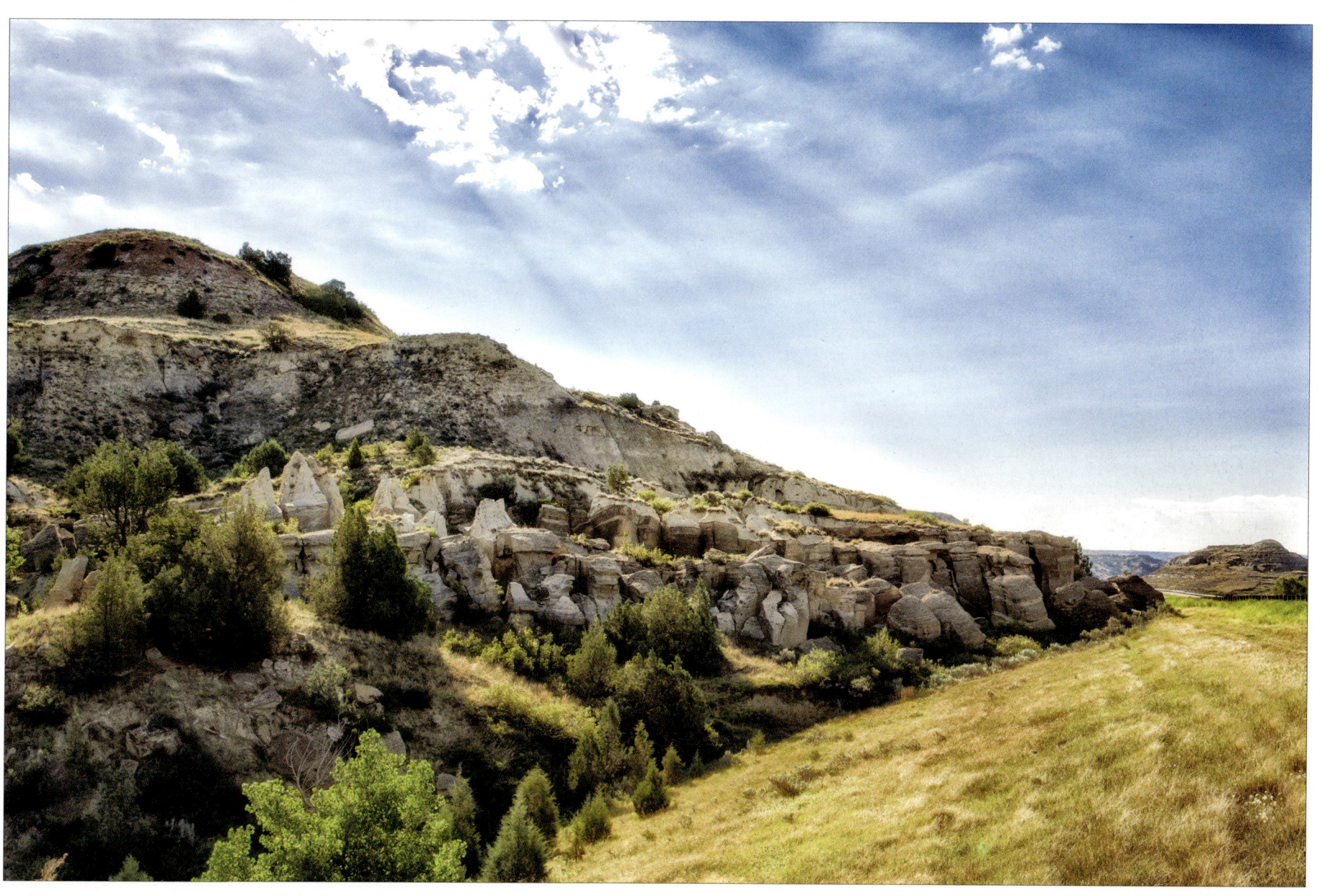

SOUND AND THE FURY

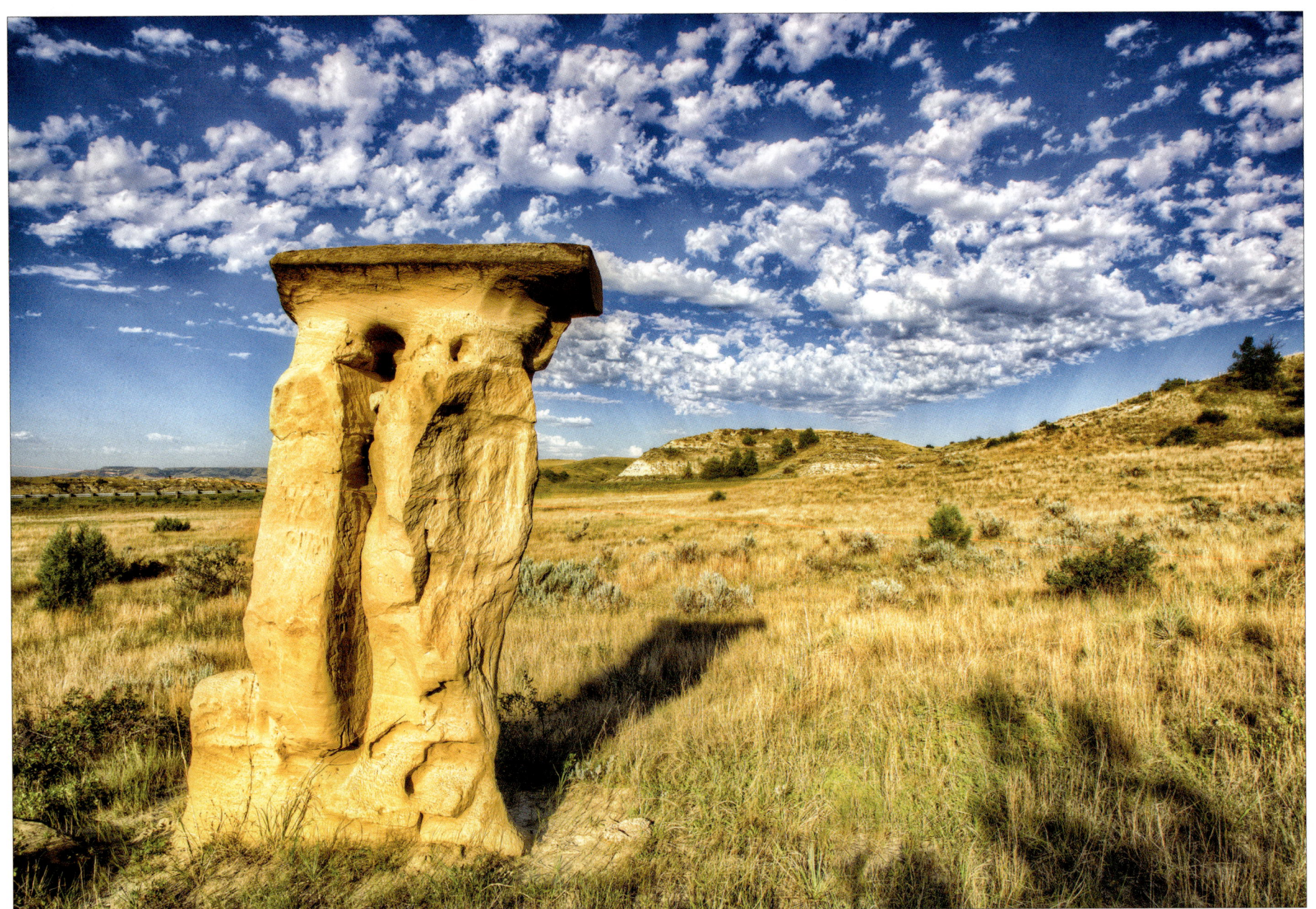

DESOLATION ANGEL
This 3-metre solitary hoodoo looks like it belongs on Easter Island.

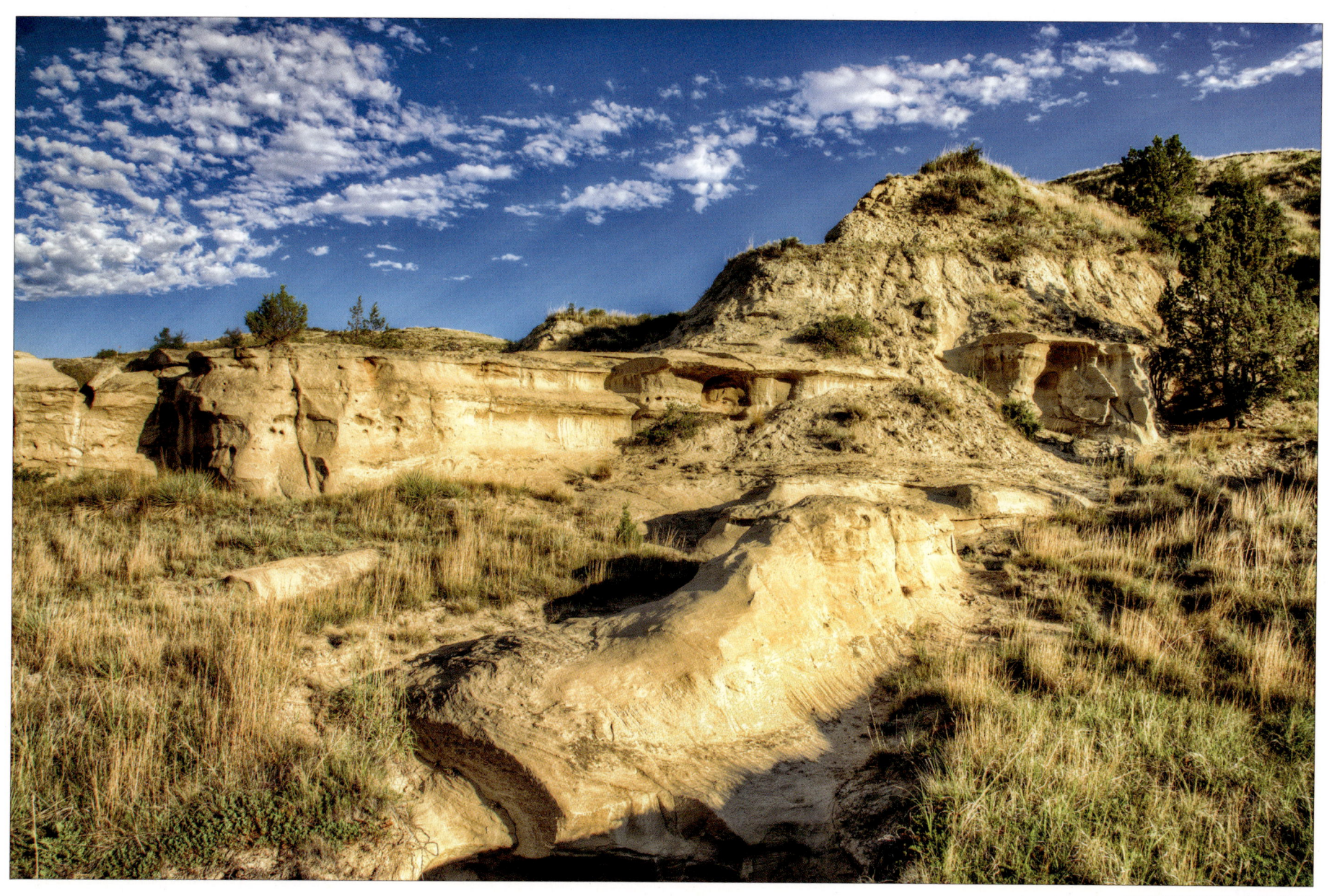

ON THE PULSE OF MORNING

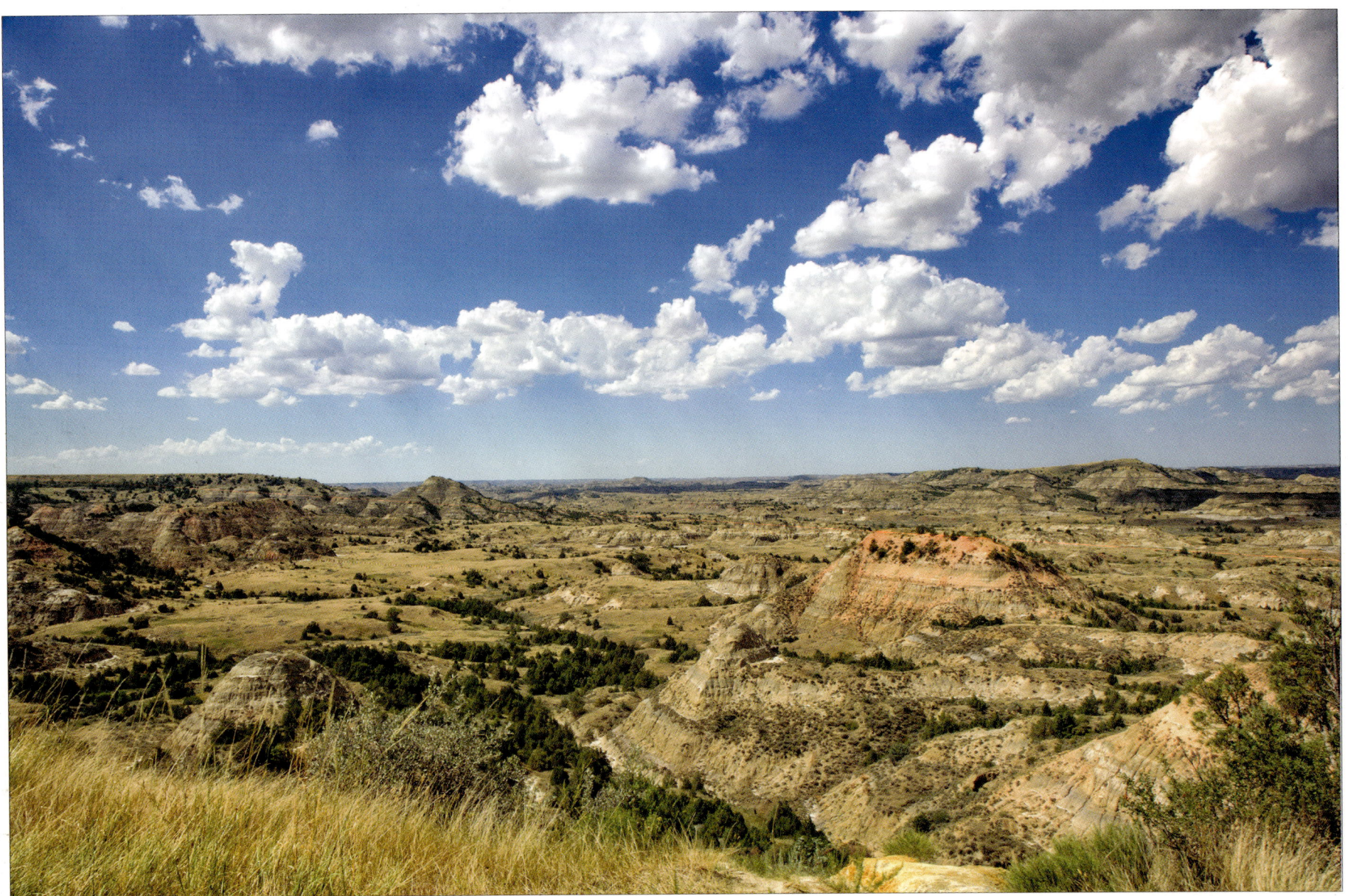

ABOVE: PAINTED CANYON

RIGHT: HEART OF DARKNESS
This bizarre mound with shallow caves and strangely eroded
windows is hauntingly visible from Interstate 94 just west of Medora.

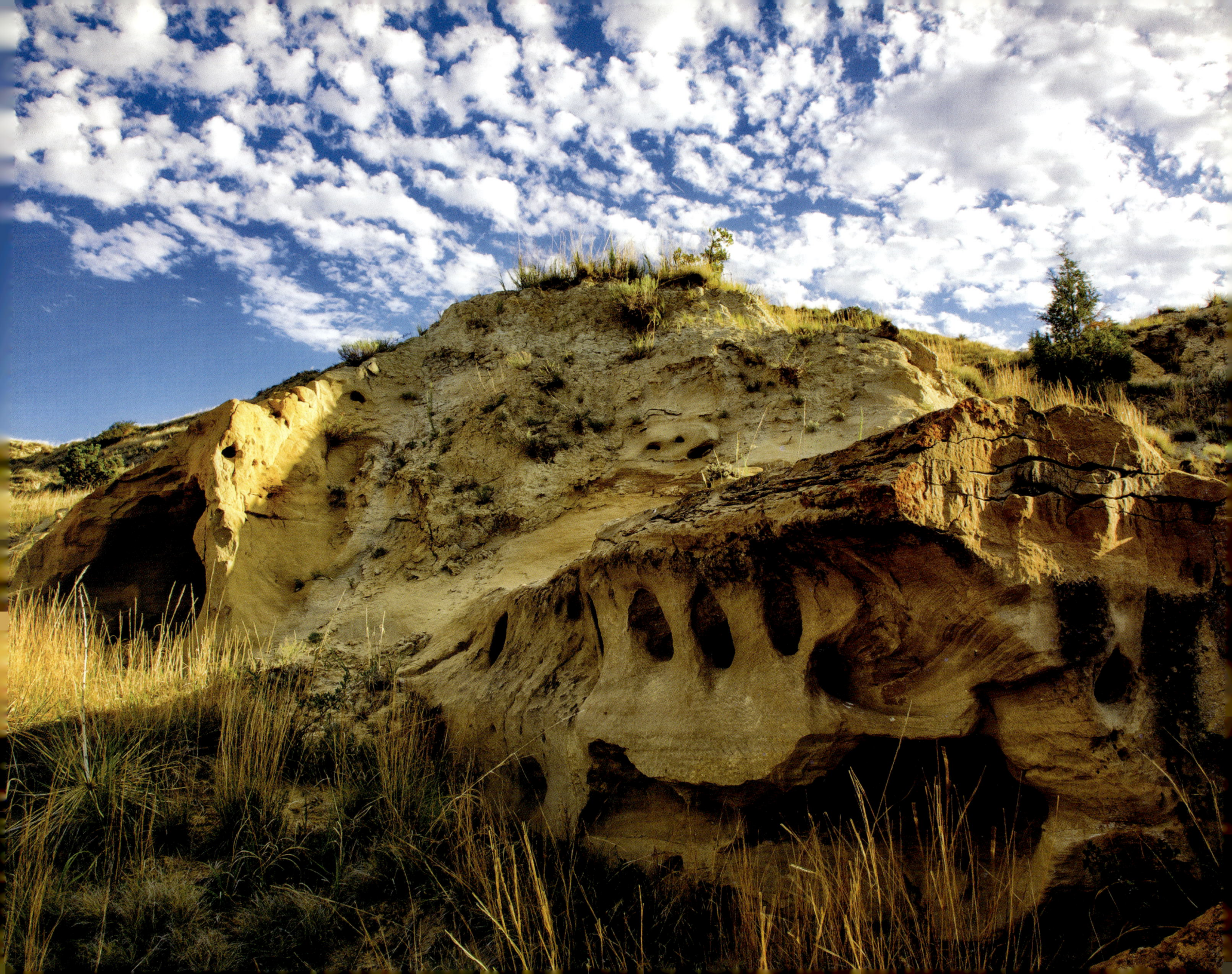

MAKOSHIKA STATE PARK

Of all the badlands in the Northern Great Plains, Makoshika State Park is, without question, the "baddest" of the bad. Pronounced Mah-KO-shi-kuh, which means "bad earth" or "bad spirit" in Lakota, the park is located just outside the town of Glendive in eastern Montana, and along Interstate Highway 94. Here, in this surreal landscape, the peaks are higher, the cliffs are sharper, and the hundreds of monument-like hoodoos that line the broken hills reverberate with mystery and isolation.

You might not make this assessment when first approaching this magnificently disfigured terrain. It doesn't feature the usual grand descent into a scenic badlands vista like Dinosaur Provincial Park in Alberta or the North Unit of nearby Theodore Roosevelt National Park in North Dakota. In fact, there really isn't much of an approach at all. The surrounding landscape begins to get more rugged as you approach the park, but there is only one way to access Makoshika and that is through the town of Glendive. Since the town sits at the doorstep of the park entrance, there is no growing sense of anticipation. This completely changes, however, when you pass through the park gates. You are instantly transported into this seemingly cataclysmic upheaval. It's like a war zone where the landscape has been ripped up and exposed from the inside out. Majestic walls of rock spring up like a fortress and surround you. It's a bit unnerving since, moments before, you were driving through the friendly confines of Glendive, only to be ambushed by Makoshika. You immediately realize that this mutilated land has its own secret history.

To begin with, unlike most badlands, Makoshika does not have a river dissecting it. For whatever geological reason, the mighty Yellowstone avoids these badlands; instead, it winds its way around the opposite side of Glendive about a mile away. Makoshika has numerous coulees but, with the absence of a river flowing through this terrain, the reddish brown earth has a distinct scorched and barren look to it. Having said that, however, Makoshika is the largest state park in Montana, comprising over 11,000 acres of wilderness, so there are significant areas of the park that retain enough moisture to allow junipers and gorgeous ponderosa pine to exist and thrive. A curious anomaly that's especially captivating is seeing a lush pine tree growing out of a rock or the middle of a seemingly barren wasteland where nothing else is allowed to grow.

Geologically speaking, the rock formations found here are older than most badlands. You can glean this by the depth and colour of the soil layers. The lower layers of the Hell Creek

RIGHT: INFINITE JEST
Makoshika is a visual stockpile of mystifying hoodoos and bizarre sandstone formations.

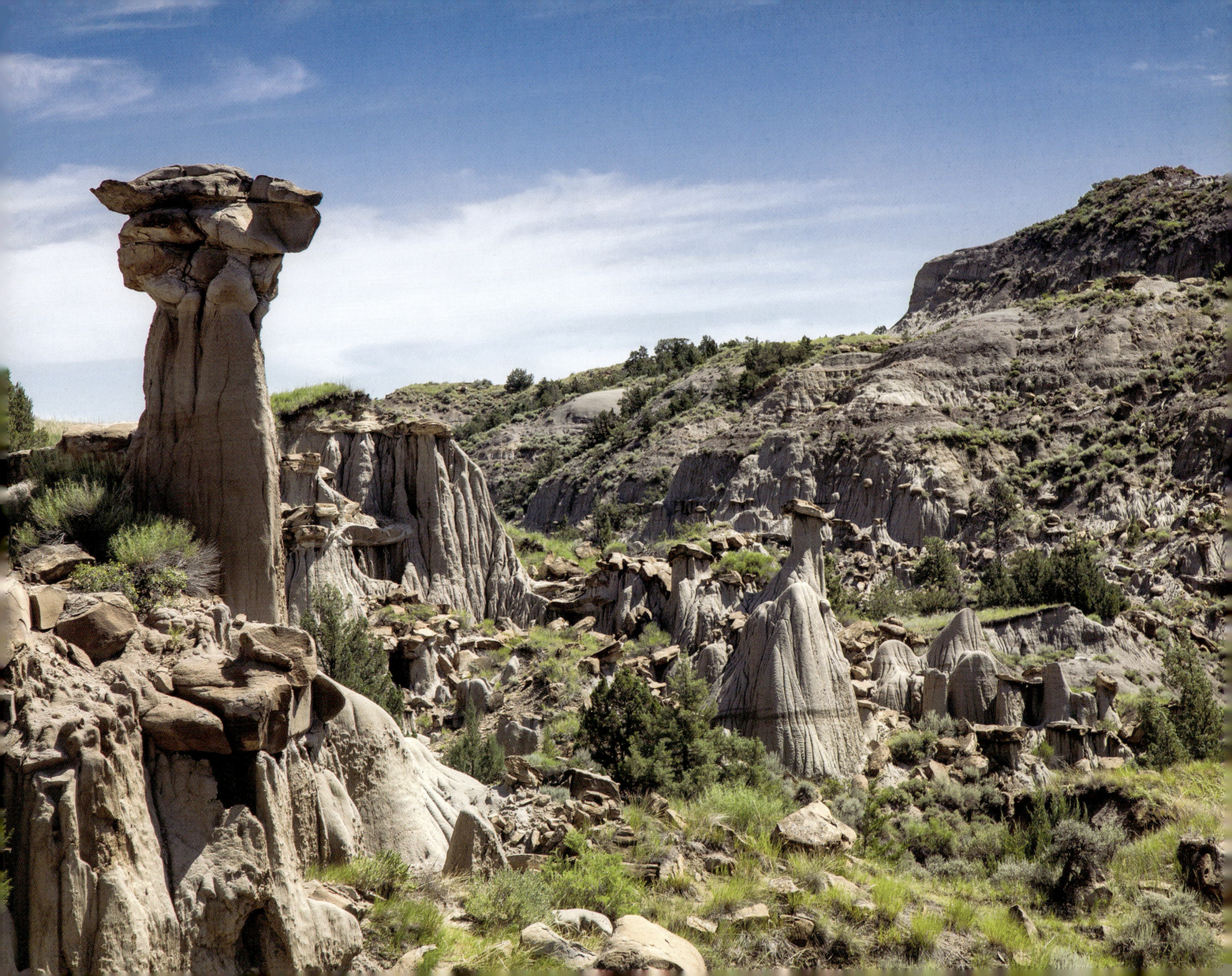

formation, dating back 65 million years, contain a number of world-class Cretaceous dinosaur fossils, including Tyrannosaurus rex, Triceratops, and Edmontosaurus. In 1991, a stunning example of a juvenile female Triceratops skull, measuring 1.7 metres long and weighing 270 kilograms, was excavated here. In 1997, the largest and most complete skeleton of the rare Thescelosaurus was discovered in an undisclosed area and airlifted out of the park. A cast representation of this dinosaur is on display in the Makoshika Visitor Center. If you want to see more dinosaurs, you can visit the 20,000 square foot Glendive Dinosaur and Fossil Museum. The museum displays 23 full-size dinosaur exhibits and sponsors digs in the badlands, giving people the opportunity to experience paleontology first-hand. A unique aspect of the museum is that it presents its exhibits in the context of Biblical creation.

While paleontology is a relatively recent pursuit, it is clear that early Native Americans were aware of these remarkable fossils. You don't have to look far in Makoshika. The highlight of the self-guided Diane Gabriel Trail loop is a hadrosaur, or duck-billed dinosaur, encased in a sandstone slab and clearly visible along the trail. Several Lakota legends make reference to mythical monsters, including the Unktehila and the Thunderbird that destroyed it. The Unktehila was an evil water serpent featured in the Lakota oral creation myth. According to Lakota Sioux storyteller, Kevin Locke: "There are many different kinds of Unktehila, but most were like huge reptiles with scaly skin and horns ... we know the giant Unktehila lived because our people found their bones in the Badlands." Native American peoples would carry away these giant fossils, believing they had special powers, and use them for healing and spiritual rituals. Locke goes on to say, "The bones were a physical manifestation of the evil forces the Unktehila represented," which is perhaps another reason why the Sioux referred to this place as Makoshika.

While the paleontology and cultural history of Makoshika is fascinating, the geological formations here are absolutely stunning. To fully experience these sights, you must explore deep into the heart of Makoshika. The best place to tent is at Pine-On-Rocks Vista. This out-of-the-way campsite overlooks a large sprawling badlands canyon that stretches for miles into the horizon. The views here are breathtaking and the hiking is phenomenal. The aforementioned Diane Gabriel Trail is an excellent, easy-to-moderate hike, beginning at the campground at Cains Coulee. Park staff like to recommend the Cap Rock Trail. This is a moderately difficult hike down into a rugged badlands valley. You can get up close and personal with numerous hoodoos and wild rock formations.

The best hike, however, is the spectacular Kinney Coulee Trail. This trail is located past Cap Rock Trail, approximately 6.5 kilometres from the visitor centre. Watch for signs because you have to leave the main road. The hike is a slow and arduous 90-metre descent into a so-called coulee, or what I would call a ravine. Walking along the dry riverbed, you will encounter hundreds of breathtaking hoodoos, spires, and strange sandstone formations. You could spend hours here just exploring. Unfortunately, I hiked this trail in near 40-degree-Celsius heat and was light-headed climbing back to the trailhead. The vultures were circling—I'm not kidding! It was an experience not to be forgotten.

RIGHT: THE UNVANQUISHED
These hoodoos look uniquely victorious while everything around them looks to be conquered by the elements.

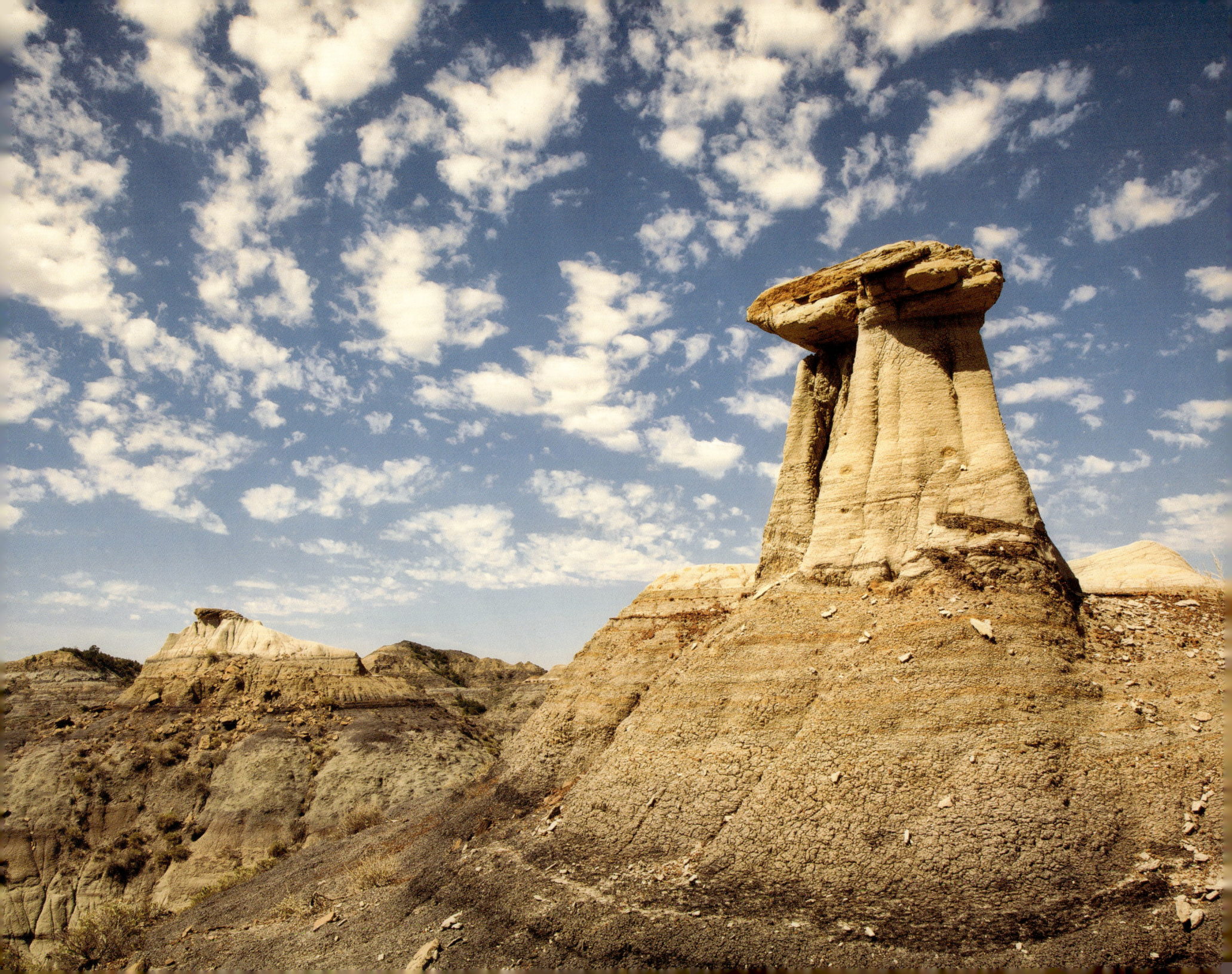

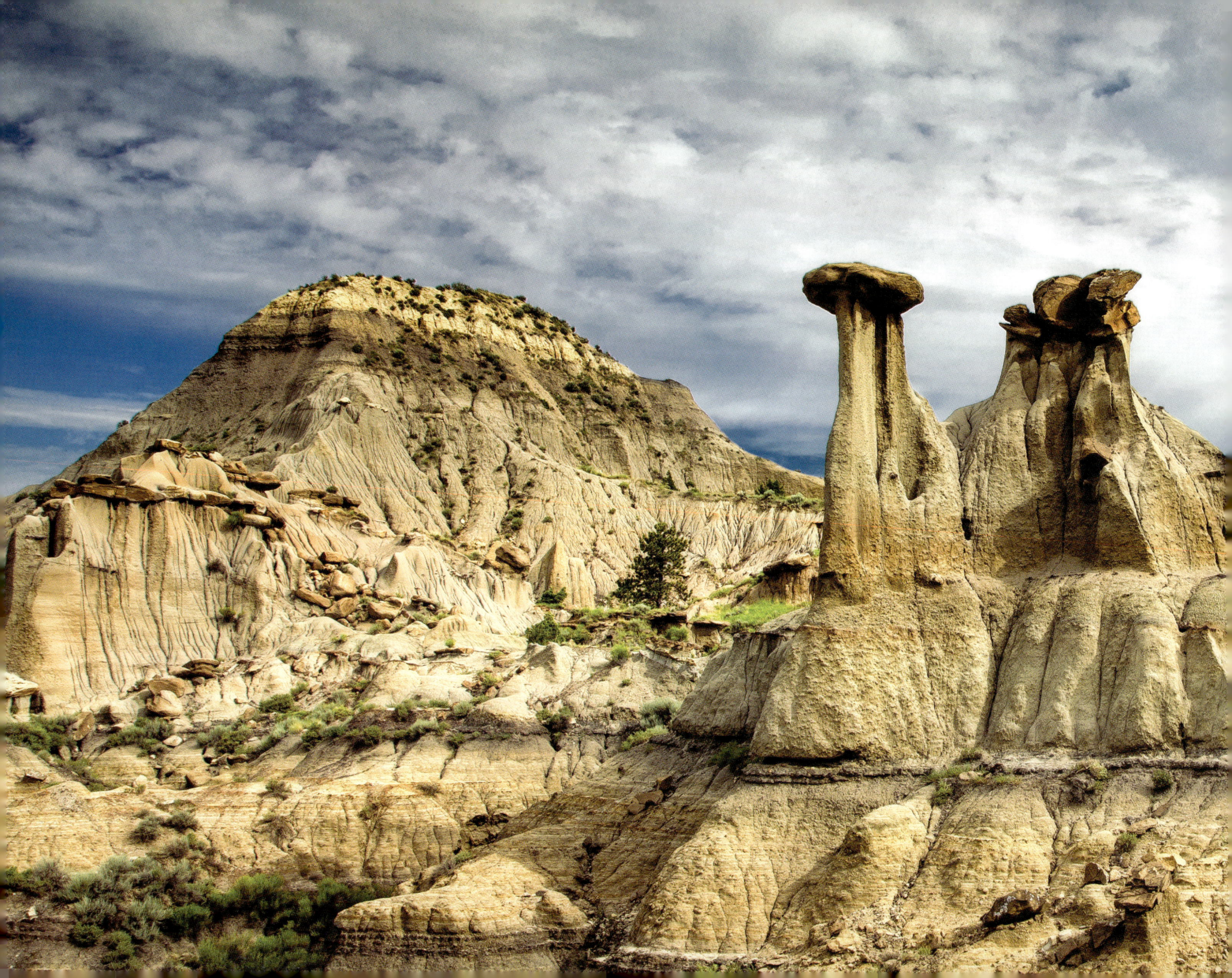

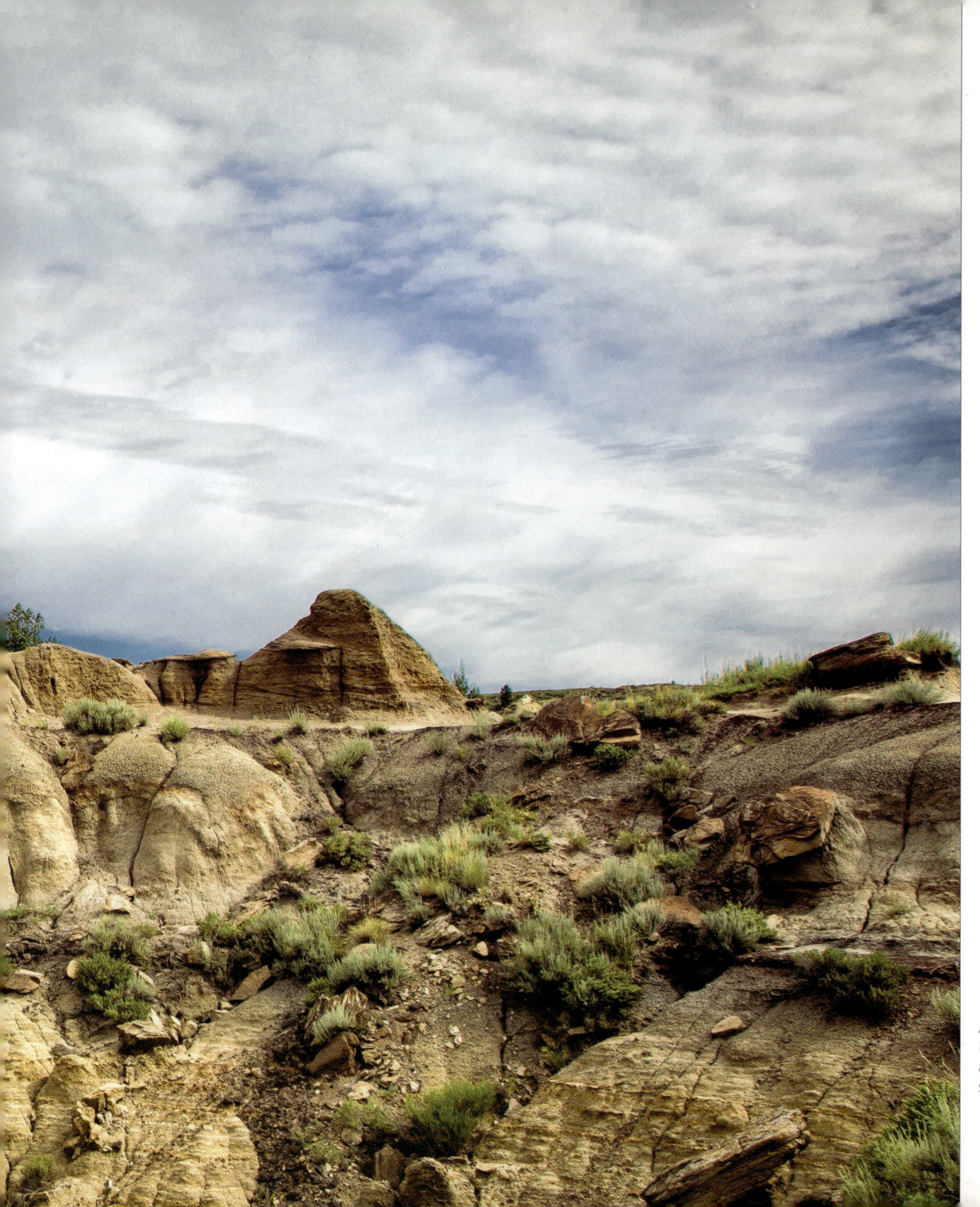

A SECRET HISTORY
Everything in Makoshika seems more grandiose. These hoodoos are the tallest in the Northern Great Plains badlands, and the signature landmark in the park.

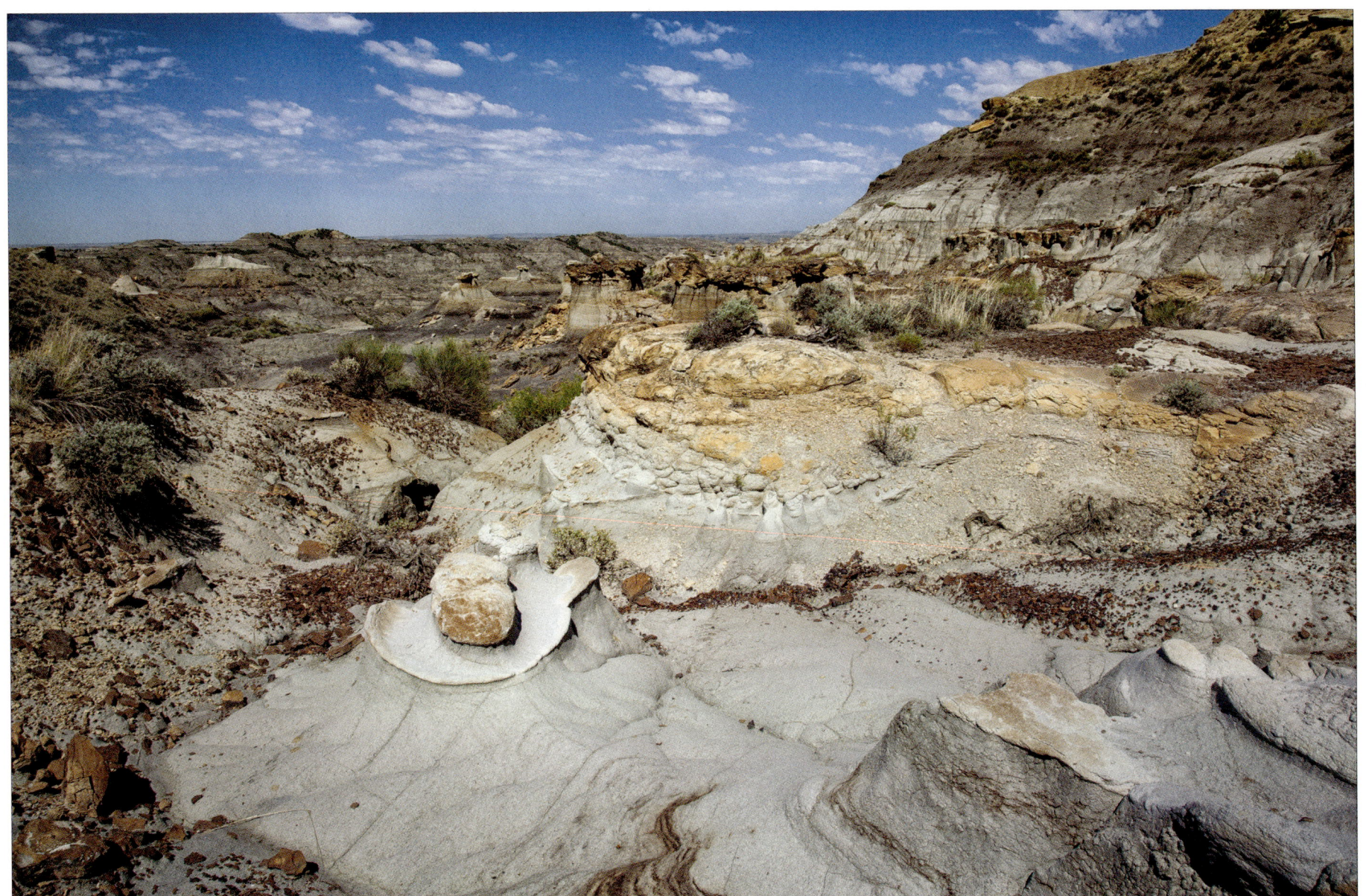

INTRUDER IN THE DUST

This is a busy photograph but it showcases the undulating lines and rhythms found along the Caprock Coulee Trail. Note the sinkhole on the left and the partially exposed stone, as if on an artist's pedestal waiting to be sculpted.

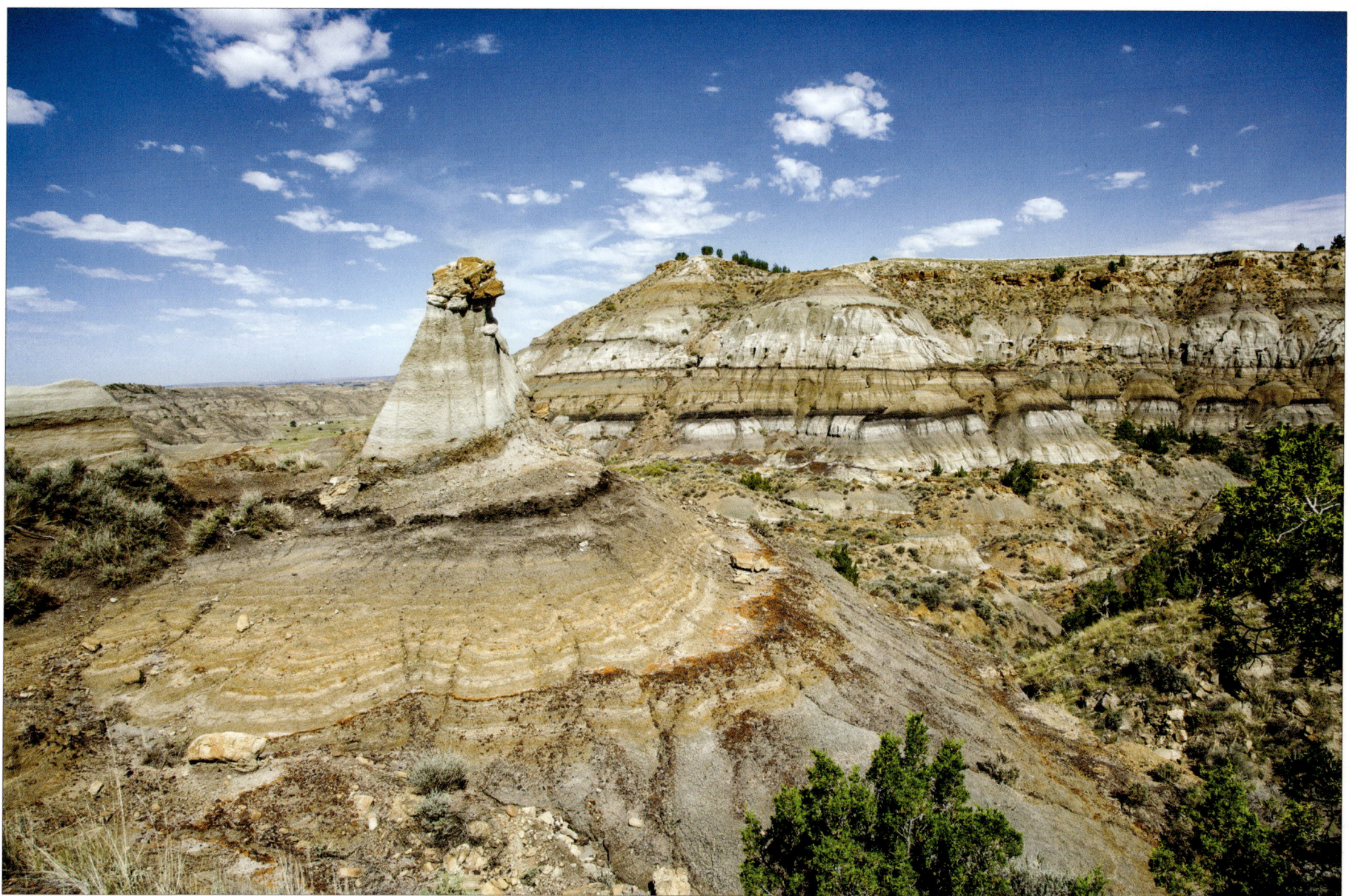

SANCTUARY
I found this solitary hoodoo from an unmarked trail. I was immediately struck by the coloured layers jutting out from the base, as well as the coal striations in the background.

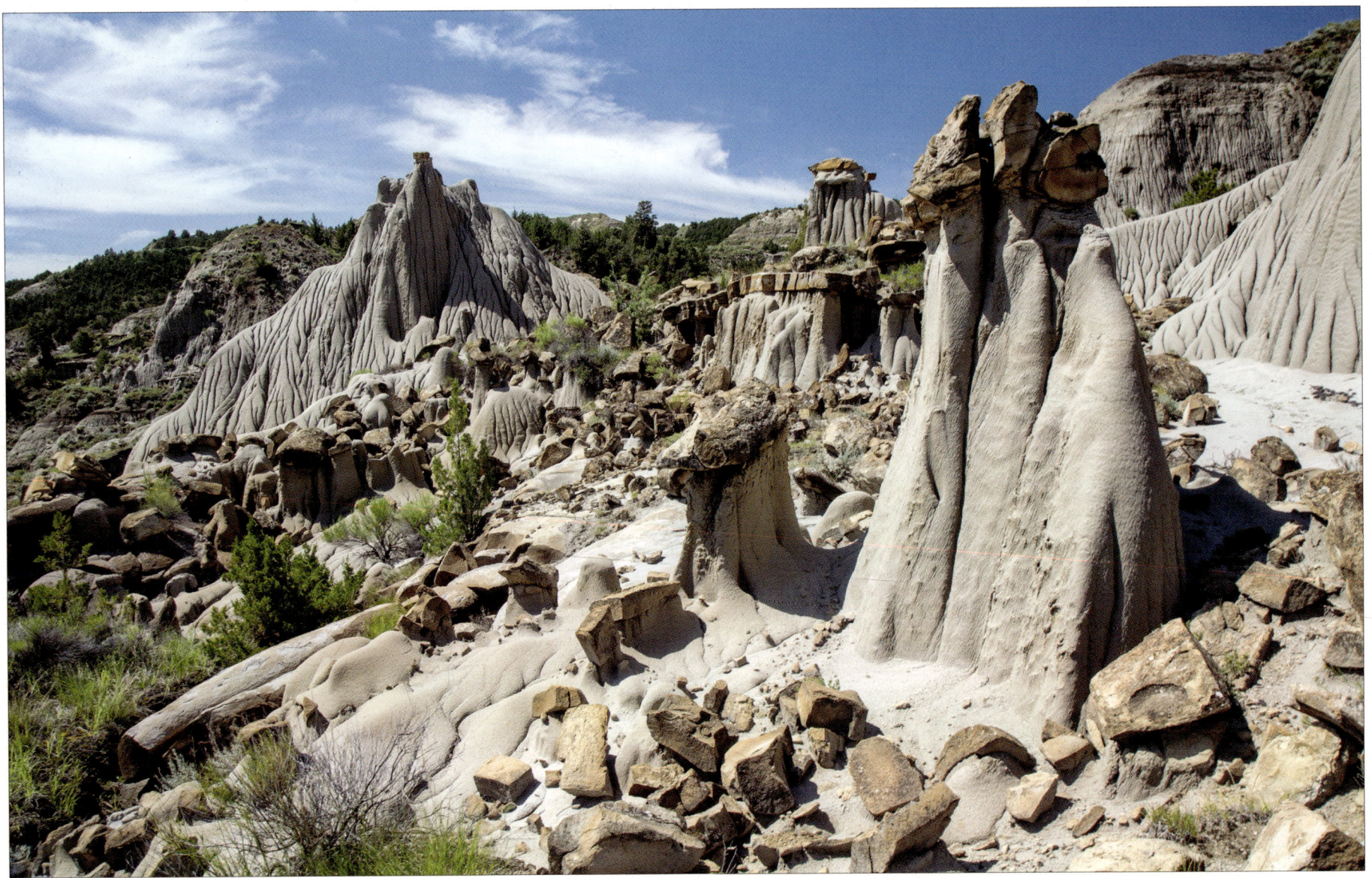

ABOVE: LITERARY LAPSES

RIGHT: PILLARS OF THE EARTH

Erosion is caused by a number of factors, including wind and water. In the Northern Great Plains, the key factor is often frost. With numerous thaw cycles throughout the year, water seeps into cracks, freezes and expands. Repetition turns cracks into gaps and gaps into crevices. So a rock structure like this will likely separate into distinct columns or pillars and, depending on the hardness of the rock, erosion might create a hoodoo or spire.

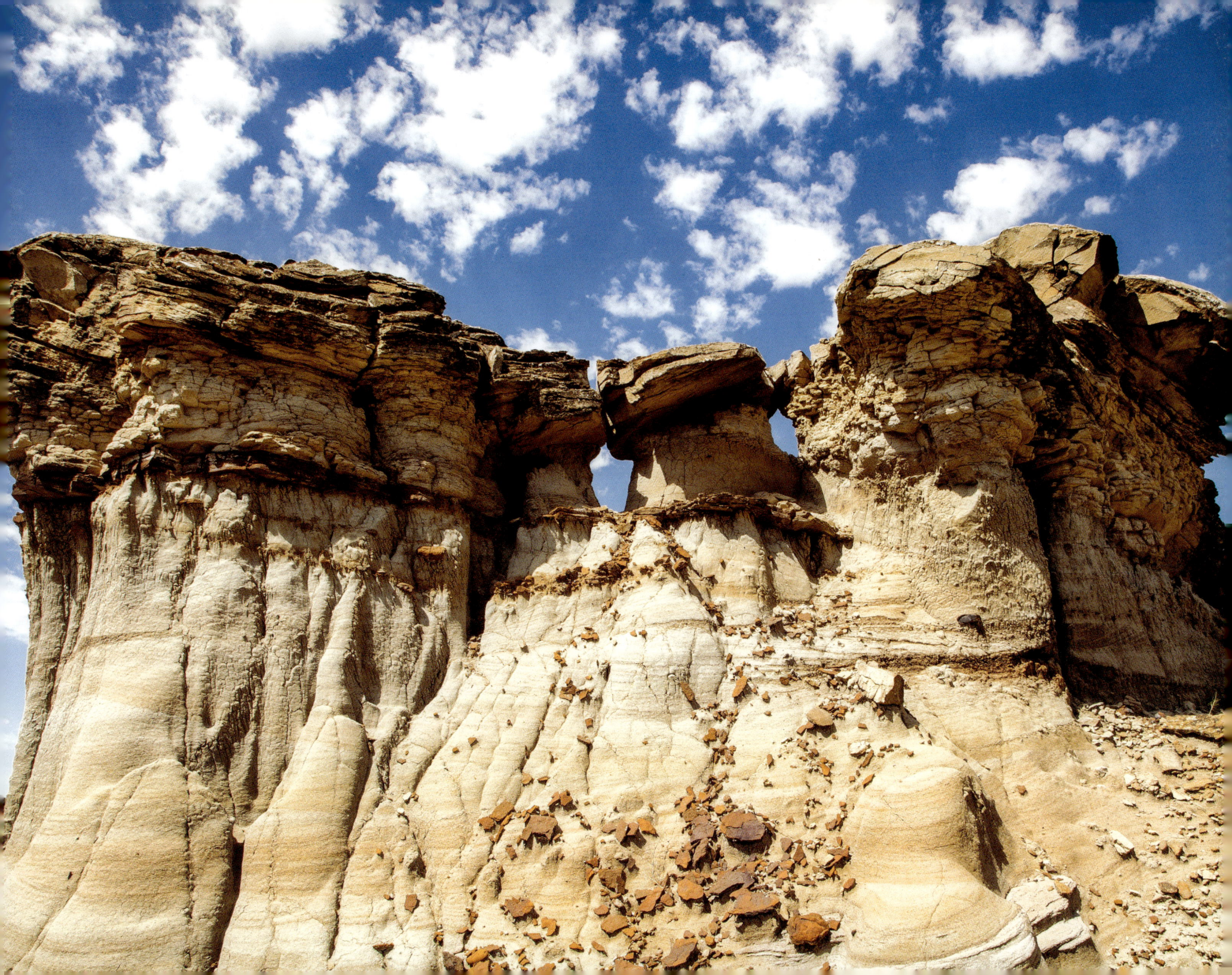

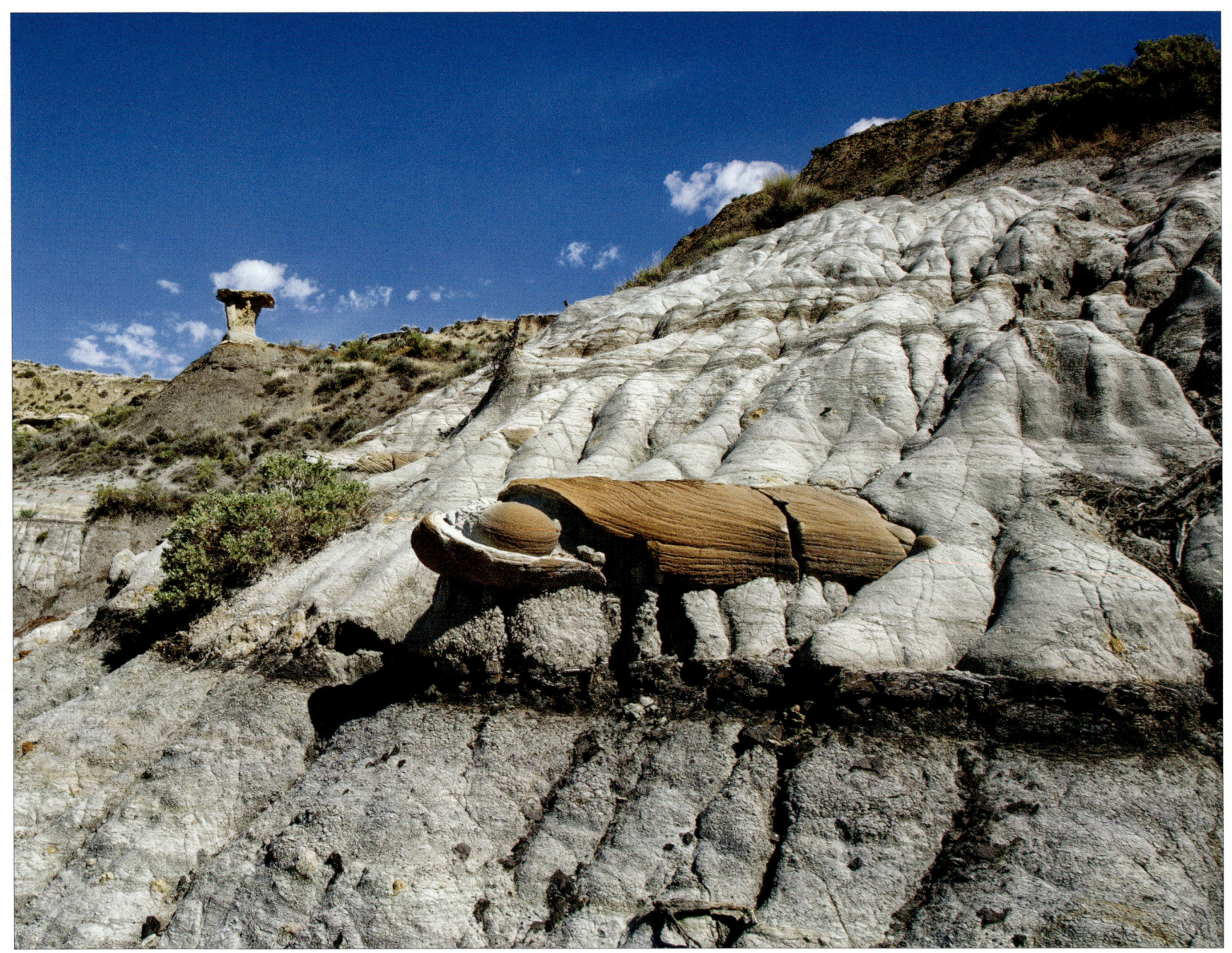

EAST OF EDEN
A grim open casket-like rock formation embedded in the fissured hillside with a majestic hoodoo perched on the hilltop. Once again, Nature seems to masquerade as both creator and destroyer.

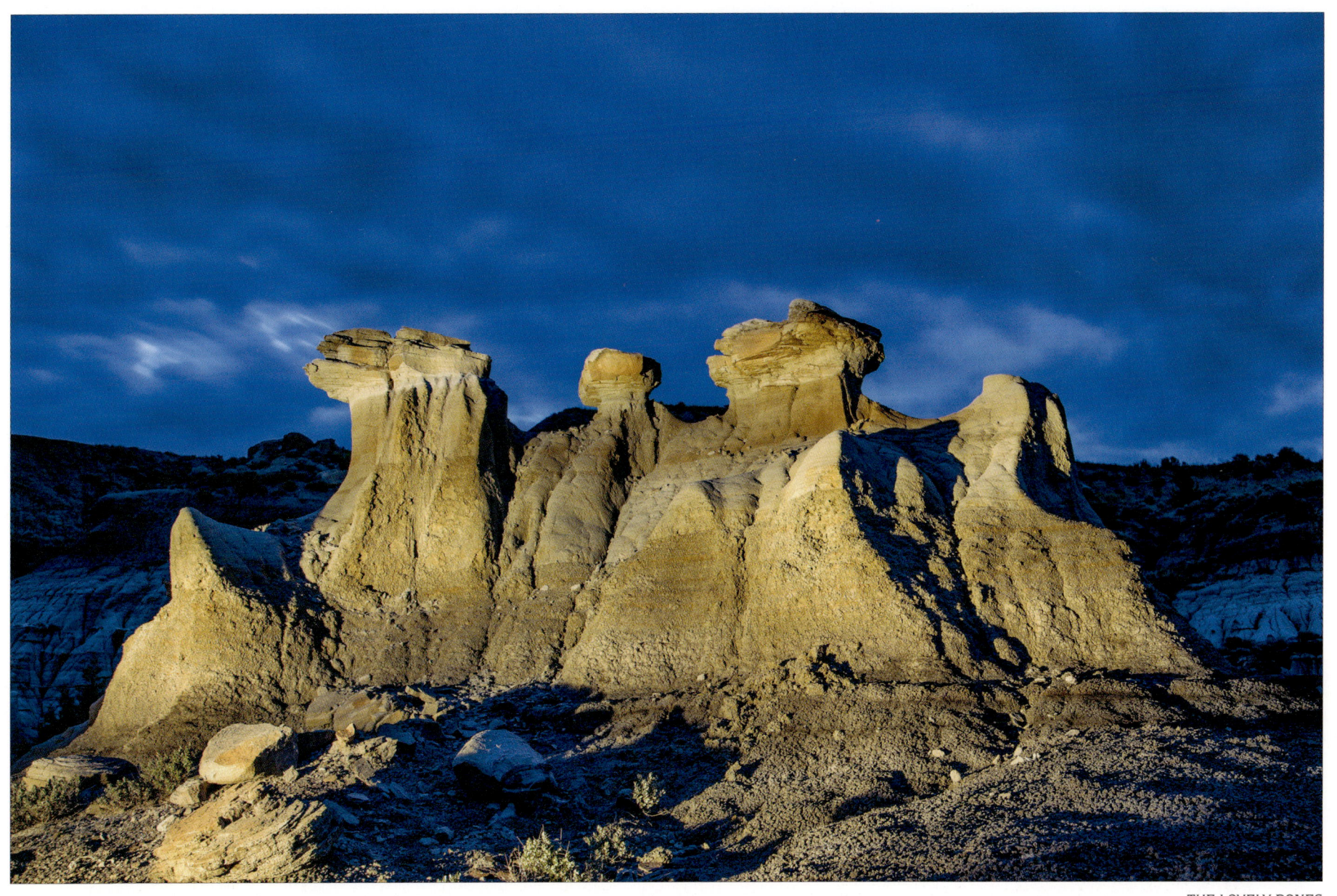

THE LOVELY BONES

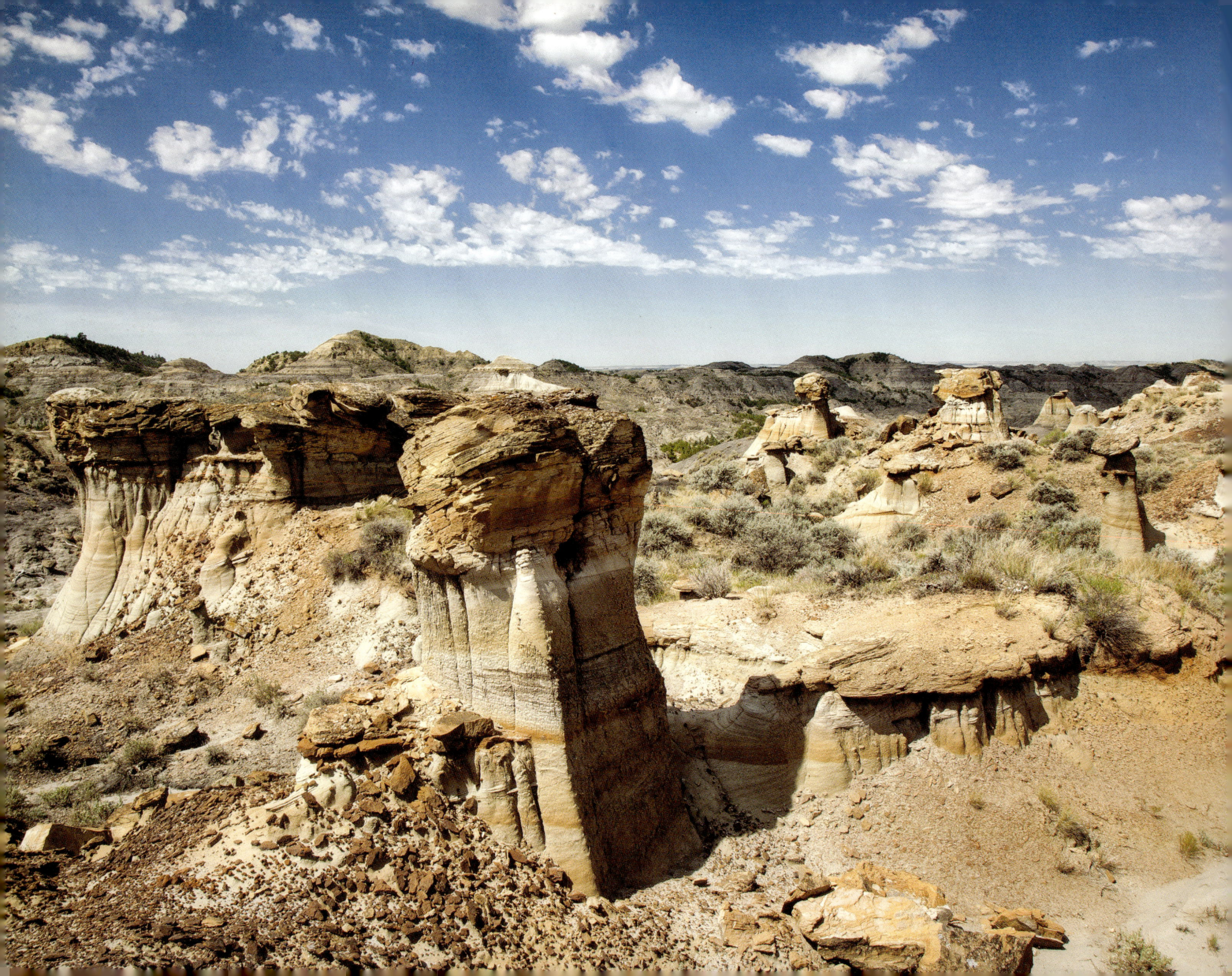

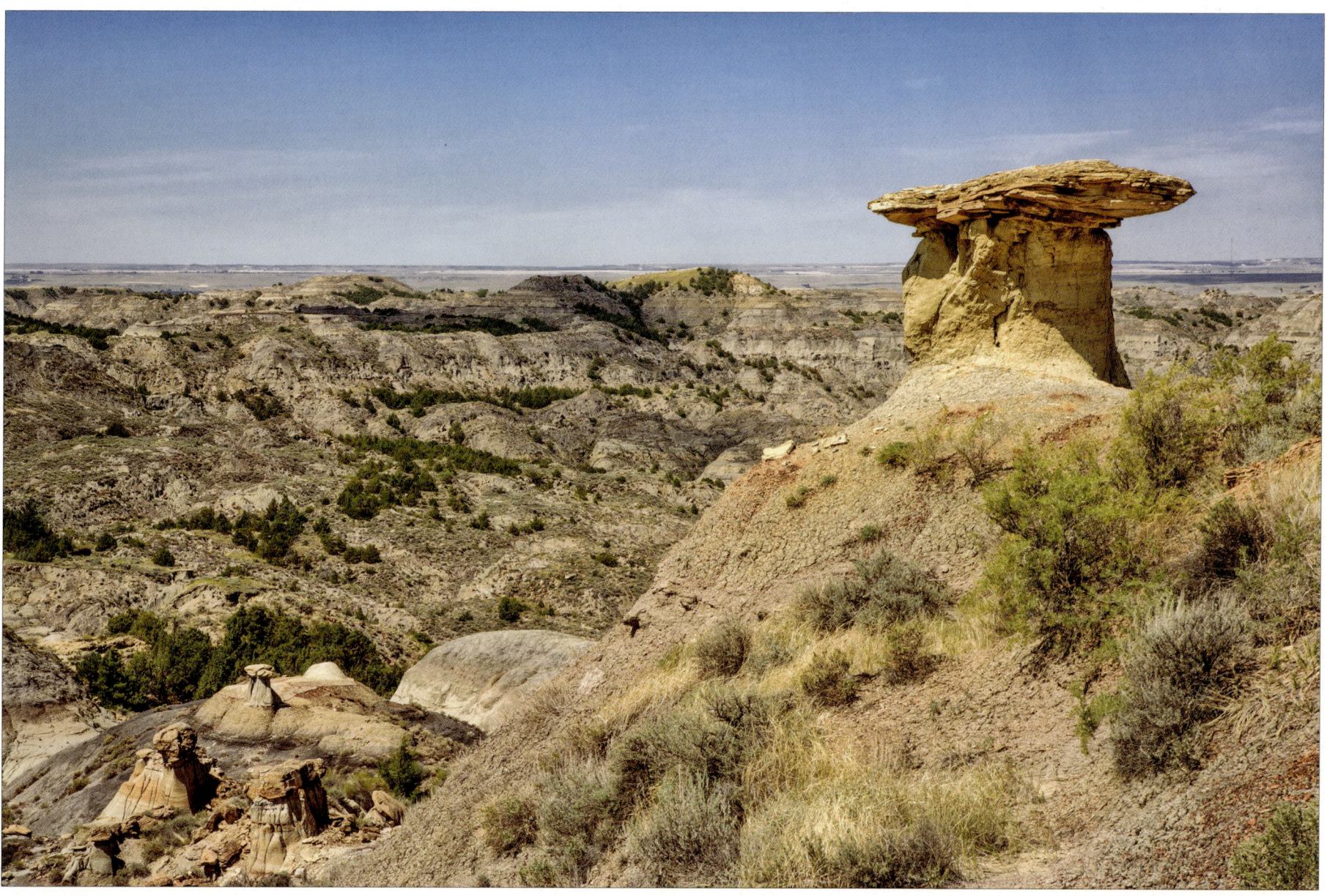

LEFT: A HOUSE DIVIDED

ABOVE: THIS SIDE OF PARADISE
Badlands can be enticingly desolate and despairing, and at other
times, like this, they can be simply majestic.

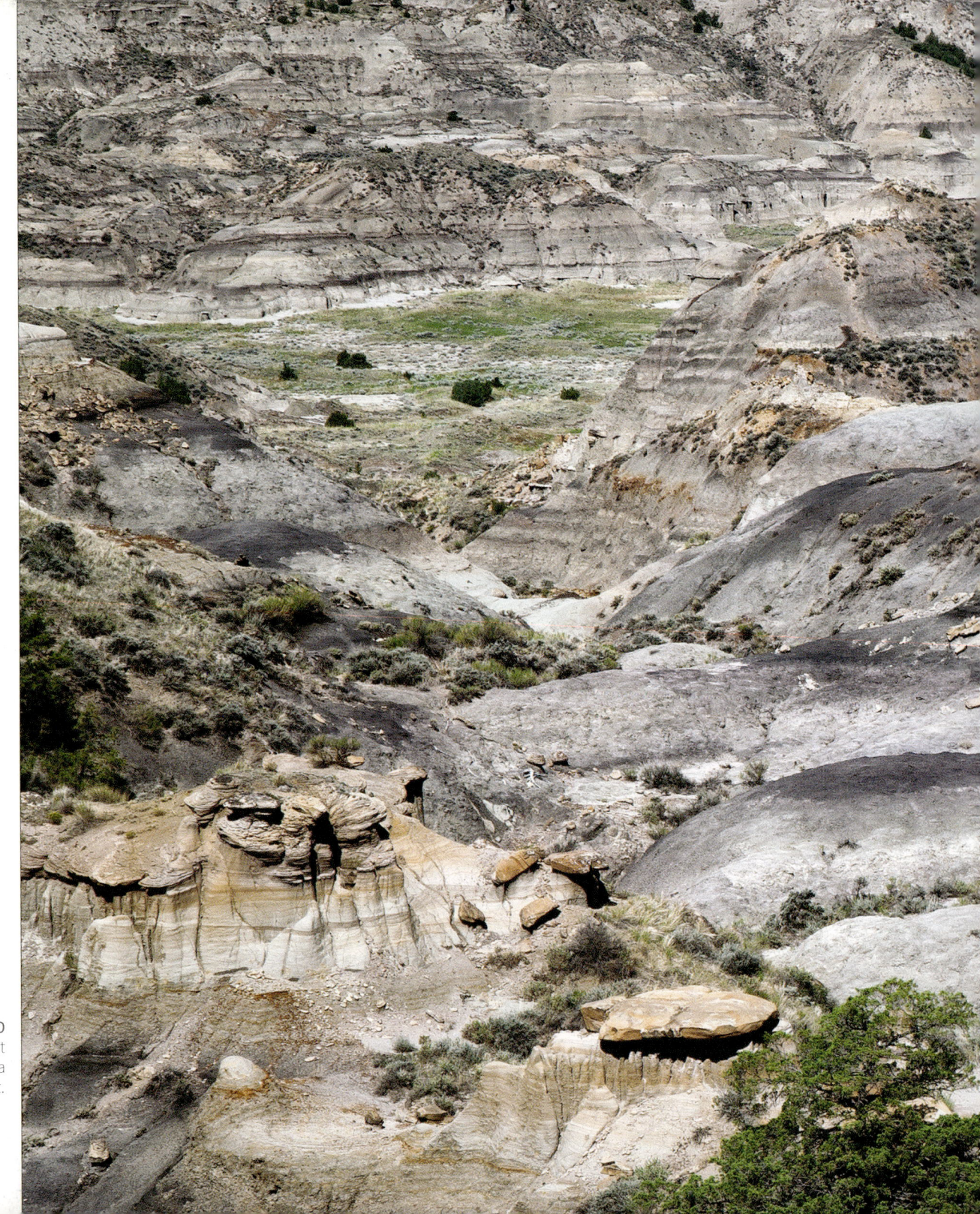

THE WASTELAND
European explorers described Makoshika as "hell without the fire." This landscape with its red shale and coal strata lines stands testament to that.

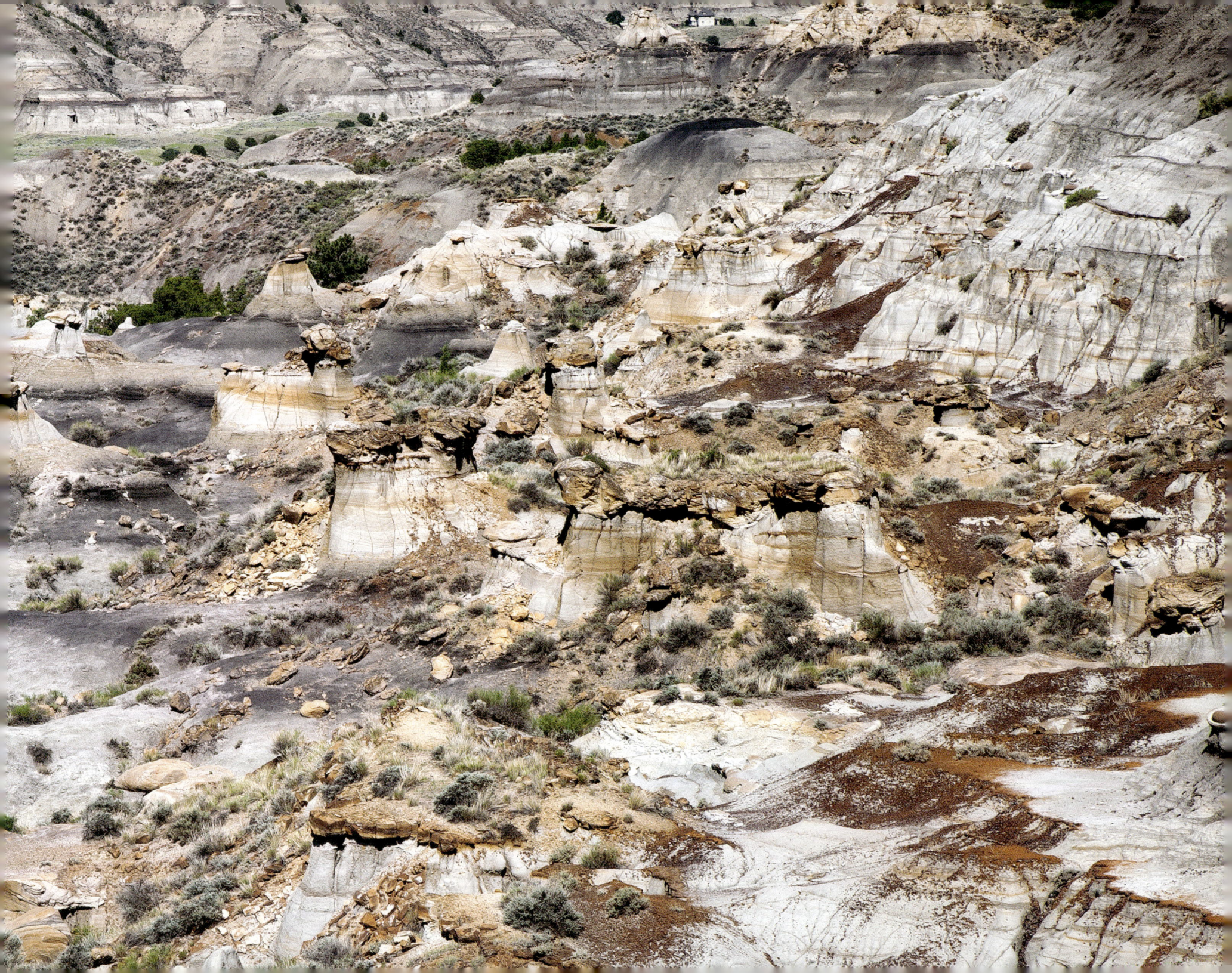

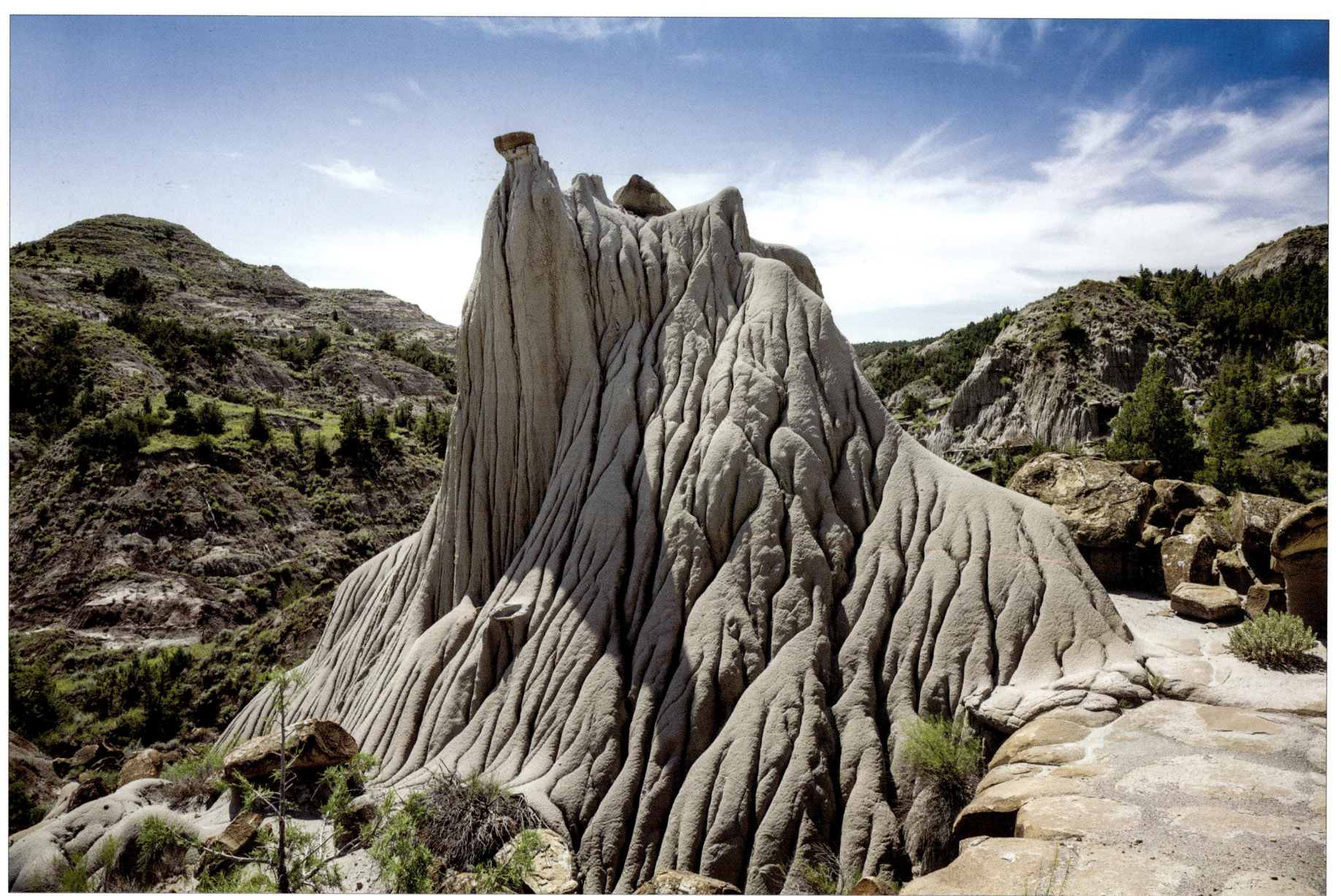

THIS HOUSE OF SKY
Erosion has been tearing away at this haunting sandcastle mountain for eons, but today it seems to be withering in the heat and melting into the ground.

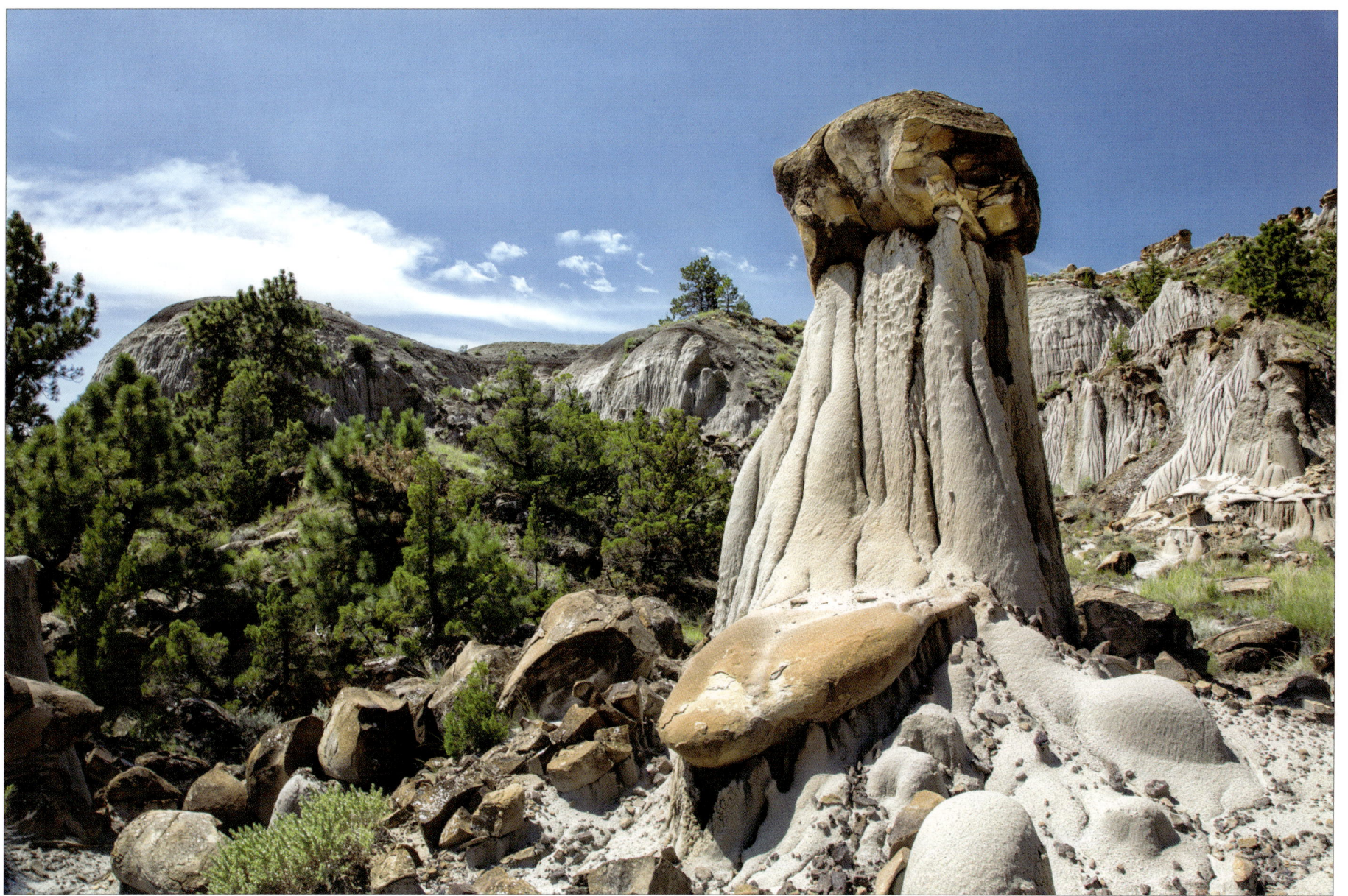

THE TALISMAN

This exotic hoodoo with its tongue-like protuberance is backdropped with lush green pines and barren etched cliffs. Add to the mix peculiar knob-shaped mounds and broken boulders and you have a classic badlands landscape.

BADLANDS BIBLIOGRAPHY

Avonlea Historical Committee. *Arrowheads to Wheatfields; Avonlea, Hearne & Districts.* Avonlea, Saskatchewan: Avonlea Historical Committee, 1983. Print.

Big Muddy Nature Centre & Museum Committee. *Big Muddy Badlands: Just North of the 49th.* Big Beaver, SK: Big Muddy Nature Centre & Museum Committee, 1994. Print.

Brink, Jack. *Archaeology in Southern Alberta: Archaeological Investigations At Writing-on-Stone, Alberta.* Archaeological Survey of Alberta Occasional Paper No. 12 December, 1978

Brune, Nick, Dave Calverley, and Alastair Sweeny. "History of Canada Online." Northern Blue Publishing. Web. <http://canadachannel.ca/HCO/>.

Campbell, Alice A. *Milk River Country.* Lethbridge, AB: Lethbridge Herald Job Printing Department, 1959. Print.

Davis, Leslie B. (ed.). *Avonlea Yesterday and Today: Archaeology and Prehistory.* Saskatoon, SK: Saskatchewan Archaeological Society, 1988. Print.

Di Silvestro, Roger L. *Theodore Roosevelt in the Badlands: A Young Politician's Quest for Recovery in the American West.* New York, NY: Walker & Company, 2011. Print.

Drumheller Valley History Association. *The Hills of Home: Drumheller Valley.* Drumheller, AB: Drumheller Valley History Association, 1973. Print.

The Encyclopedia of Saskatchewan: A Living Legacy. Regina, SK: Canadian Plains Research Center, University of Regina, 2005. Print.

Graspointner, Andreas. Archaeology and Ethno-History of the Milk River in Southern Alberta. Calgary, AB: Western Publishers, 1980. Print.

Gross, Renie. *Dinosaur Country: unearthing the Alberta Badlands.* Wardlow, AB: Alberta Badlands Books, 1998. Print.

Harmsen, Debbie, and Michael Nalepa (eds.). *The Complete Guide to the National Parks of the West.* Toronto, ON: Fodor's Travel, 2007. Print.

Morell, Virginia. "Beyond Nessie." *National Geographic.* 208.6 (2005): 58. Print.

Mussieux, Ron, and Marilyn Nelson. *A Traveller's Guide to Geological wonders in Alberta.* Edmonton, AB: Provincial Museum of Alberta, 1998. Print.

"The U.S. Army and the Sioux - Part 4 Prelude to Little Bighorn: Terry and Custer in the Badlands." *National Park Service U.S. Department of the Interior.* Web. <http://www.nps.gov/thro/history-culture/the-us-army-and-the-sioux-part-4.htm>.

Rubenstein, Dan. "Sounds of Silence." *Canadian Geographic.* Apr 2012: 40. Print.

Thraves, Bernard D., M. L. Lewry, Janis E. Dale, and Hansgeorg Schlichtmann. *Saskatchewan: Geographic Perspectives.* Regina: Canadian Plains Research Center, Department of Geography, University of Regina, 2007. Print.

Tyrrell, J. B. (ed.). *David Thompson's Narrative of His Exploration in Western America, 1784-1812.* Toronto, ON: The Champlain Society, 1916. 321. Print.

White, Mel, Robert Earl Howells, et al. *Secrets of the National Parks: The Expert's Guide to the Best Experiences Beyond the Tourist Trail.* Washington D.C.: National Geographic Society, 2013. Print.

White, Mel. *National Geographic Complete National Parks of the United States.* Washington D.C.: National Geographic Society, 2009. Print.